Means Residential Repair & Remodeling Costs

Contractor's Pricing Guide 2005

Senior Editor
Robert W. Mewis, CCC

Contributing Editors
Barbara Balboni
Robert A. Bastoni
John H. Chiang, PE
Robert J. Kuchta
Melville J. Mossman, PE
John J. Moylan
Jeannene D. Murphy
Stephen C. Plotner
Michael J. Regan
Eugene J. Spencer
Marshall J. Stetson
Phillip R. Waier, PE

Senior Engineering Operations Manager
John H. Ferguson, PE

Product Manager
Jane Crudden

Senior Vice President & General Manager
John Ware

Vice President of Sales
John M. Shea

Production Manager
Michael Kokernak

Production Coordinator
Marion E. Schofield

Technical Support
Thomas J. Dion
Jonathan Forgit
Mary Lou Geary
Gary L. Hoitt
Alice McSharry
Paula Reale-Camelio
Robin Richardson
Kathryn S. Rodriguez
Sheryl A. Rose

Book & Cover Design
Norman R. Forgit

RSMeans

Means Residential Repair & Remodeling Costs

Contractor's Pricing Guide 2005

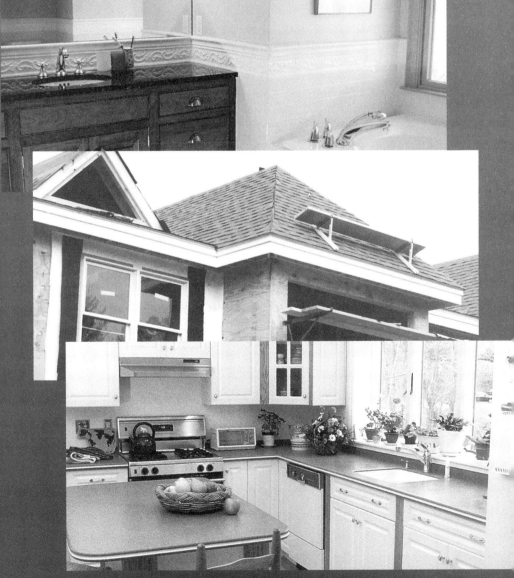

- Quick Costs to Help You Estimate
- New, Easy-to-Use Format . . . Organized the Way You Build
- Hundreds of Work Items with Material, Labor & Equipment

$39.95 per copy. (In United States).
Price subject to change without prior notice.

 Reed Construction Data

Copyright © 2004

Construction Publishers & Consultants
63 Smiths Lane
Kingston, MA 02364-0800
(781) 422-5000

Printed in the United States of America

10 9 8 7 6 5 4 3 2 1

ISSN 1542-9156

ISBN 0-87629-761-0

Foreword

RSMeans is a product line of Reed Construction Data, Inc., a leading provider of construction information, products, and services in North America and globally. Reed Construction Data's project information products include more than 100 regional editions, national construction data, sales leads, and local plan rooms in major business centers. Reed Construction Data's PlansDirect provides surveys, plans and specifications. The First Source suite of products consists of *First Source for Products,* SPEC-DATA™, MANU-SPEC™, CADBlocks, Manufacturer Catalogs and First Source Exchange (www.firstsourceexchange.com) for the selection of nationally available building products. Reed Construction Data also publishes ProFile, a database of more than 20,000 U.S. architectural firms. RSMeans provides construction cost data, training, and consulting services in print, CD-ROM and online. Reed Construction Data, headquartered in Atlanta, is owned by Reed Business Information (www.reedconstructiondata.com), a leading provider of critical information and marketing solutions to business professionals in the media, manufacturing, electronics, construction and retail industries. Its market-leading properties include more than 135 business-to-business publications, over 125 Webzines and Web portals, as well as online services, custom publishing, directories, research and direct-marketing lists. Reed Business Information is a member of the Reed Elsevier plc group (NYSE: RUK and ENL)—a world-leading publisher and information provider operating in the science and medical, legal, education and business-to-business industry sectors.

Our Mission

Since 1942, RSMeans has been actively engaged in construction cost publishing and consulting throughout North America. Today, over 60 years after RSMeans began, our primary objective remains the same: to provide you, the construction and facilities professional, with the most current and comprehensive construction cost data possible.

Whether you are a contractor, an owner, an architect, an engineer, a facilities manager, or anyone else who needs a fast and reliable construction cost estimate, you'll find this publication to be a highly useful and necessary tool.

Today, with the constant flow of new construction methods and materials, it's difficult to find the time to look at and evaluate all the different construction cost possibilities. In addition, because labor and material costs keep changing, last year's cost information is not a reliable basis for today's estimate or budget.

That's why so many construction professionals turn to RSMeans. We keep track of the costs for you, along with a wide range of other key information, from city cost indexes . . . to productivity rates . . . to crew composition . . . to contractor's overhead and profit rates.

RSMeans performs these functions by collecting data from all facets of the industry, and organizing it in a format that is instantly accessible to you. From the preliminary budget to the detailed unit price estimate, you'll find the data in this book useful for all phases of construction cost determination.

The Staff, the Organization, and Our Services

When you purchase one of RSMeans' publications, you are in effect hiring the services of a full-time staff of construction and engineering professionals.

Our thoroughly experienced and highly qualified staff works daily at collecting, analyzing, and disseminating comprehensive cost information for your needs. These staff members have years of practical construction experience and engineering training prior to joining the firm. As a result, you can count on them not only for the cost figures, but also for additional background reference information that will help you create a realistic estimate.

The RSMeans organization is always prepared to help you solve construction problems through its five major divisions: Construction and Cost Data Publishing, Electronic Products and Services, Consulting Services, Insurance Services, and Educational Services. Besides a full array of construction cost

estimating books, Means also publishes a number of other reference works for the construction industry. Subjects include construction estimating and project and business management; special topics such as HVAC, roofing, plumbing, and hazardous waste remediation; and a library of facility management references.

In addition, you can access all of our construction cost data through your computer with RSMeans *CostWorks 2005* CD-ROM, an electronic tool that offers over 50,000 lines of Means detailed construction cost data, along with assembly and whole building cost data. You can also access Means cost information from our Web site at www.rsmeans.com

What's more, you can increase your knowledge and improve your construction estimating and management performance with a Means Construction Seminar or In-House Training Program. These two-day seminar programs offer unparalleled opportunities for everyone in your organization to get updated on a wide variety of construction-related issues.

RSMeans also is a worldwide provider of construction cost management and analysis services for commercial and government owners and of claims and valuation services for insurers.

In short, RSMeans can provide you with the tools and expertise for constructing accurate and dependable construction estimates and budgets in a variety of ways.

Robert Snow Means Established a Tradition of Quality That Continues Today

Robert Snow Means spent years building RSMeans, making certain he always delivered a quality product.

Today, at RSMeans, we do more than talk about the quality of our data and the usefulness of our books. We stand behind all of our data, from historical cost indexes to construction materials and techniques to current costs.

If you have any questions about our products or services, please call us toll-free at 1-800-334-3509. Our customer service representatives will be happy to assist you or visit our Web site at www.rsmeans.com

Table of Contents

How the Book is Built: An Overview

A Powerful Construction Tool

You have in your hands one of the most powerful construction tools available today. A successful project is built on the foundation of an accurate and dependable estimate. This book will enable you to construct just such an estimate.

For the casual user the book is designed to be:

- quickly and easily understood so you can get right to your estimate
- filled with valuable information so you can understand the necessary factors that go into the cost estimate

For the regular user, the book is designed to be:

- a handy desk reference that can be quickly referred to for key costs
- a comprehensive, fully reliable source of current construction costs, so you'll be prepared to estimate any project
- a source book for project cost, product selections, and alternate materials and methods

To meet all of these requirements we have organized the book into the following clearly defined sections.

Unit Price Section

The cost data has been divided into 23 sections representing the order of construction. Within each section is a listing of components and variations to those components applicable to that section. Costs are shown for the various actions that must be done to that component.

Reference Section

This section includes information on Location Factors, and a listing of Abbreviations.

Location Factors: Costs vary depending upon regional economy. You can adjust the "national average" costs in this book to over 930 major cities throughout the U.S. and Canada by using the data in this section.

Abbreviations: A listing of the abbreviations used throughout this book, along with the terms they represent, is included.

Index

A comprehensive listing of all terms and subjects in this book to help you find what you need quickly when you are not sure where it falls in the order of construction.

The Scope of This Book

This book is designed to be as comprehensive and as easy to use as possible. To that end we have made certain assumptions and limited its scope in three key ways:

1. We have established material prices based on a "national average."
2. We have computed labor costs based on a 7 major region average of open shop wage rates.
3. We have targeted the data for projects of a certain size range.

Project Size

This book is intended for use by those involved primarily in Residential Repair and Remodeling construction costing between $10,000-$100,000.

With reasonable exercise of judgment the figures can be used for any building work. For other types of projects, such as new home construction or commercial buildings, consult the appropriate Means publication for more information.

How to Use the Book: The Details

What's Behind the Numbers? The Development of Cost Data

The staff at RSMeans continuously monitors developments in the construction industry in order to ensure reliable, thorough and up-to-date cost information.

While *overall* construction costs may vary relative to general economic conditions, price fluctuations within the industry are dependent upon many factors. Individual price variations may, in fact, be opposite to overall economic trends. Therefore, costs are continually monitored and complete updates are published yearly. Also, new items are frequently added in response to changes in materials and methods.

Costs—$ (U.S.)

All costs represent U.S. national averages and are given in U.S. dollars. The Means Location Factors can be used to adjust costs to a particular location. The Location Factors for Canada can be used to adjust U.S. national averages to local costs in Canadian dollars.

Material Costs

The RSMeans staff contacts manufacturers, dealers, distributors, and contractors all across the U.S. and Canada to determine national average material costs. If you have access to current material costs for your specific location, you may wish to make adjustments to reflect differences from the national average.

Included within material costs are fasteners for a normal installation. RSMeans engineers use manufacturers' recommendations, written specifications and/or standard construction practice for size and spacing of fasteners. Adjustments to material costs may be required for your specific application or location. Material costs do not include sales tax.

Labor Costs

Labor costs are based on the average of open shop wages from across the U.S. for the current year. Rates along with overhead and profit markups are listed on the inside back cover of this book.

- If wage rates in your area vary from those used in this book, or if rate increases are expected within a given year, labor costs should be adjusted accordingly.

Labor costs reflect productivity based on actual working conditions. These figures include time spent during a normal workday on tasks other than actual installation, such as material receiving and handling, mobilization at site, site movement, breaks, and cleanup.

Productivity data is developed over an extended period so as not to be influenced by abnormal variations and reflects a typical average.

Equipment Costs

Equipment costs include not only rental costs, but also operating costs such as fuel, oil, and routine maintenance. Equipment and rental rates are obtained from industry sources throughout North America—contractors, suppliers, dealers, manufacturers, and distributors.

Factors Affecting Costs

Costs can vary depending upon a number of variables. Here's how we have handled the main factors affecting costs.

Quality—The prices for materials and the workmanship upon which productivity is based represent sound construction work. They are also in line with U.S. government specifications.

Overtime—We have made no allowance for overtime. If you anticipate premium time or work beyond normal working hours, be sure to make an appropriate adjustment to your labor costs.

Productivity—The labor costs for each line item are based on working an eight-hour day in daylight hours in moderate temperatures. For work that extends beyond normal work hours or is performed under adverse conditions, productivity may decrease.

Size of Project—The size, scope of work, and type of construction project will have a significant impact on cost. Economies of scale can reduce costs for large projects. Unit costs can often run higher for small projects. Costs in this book are intended for the size and type of project as previously described in "How the Book Is Built: An Overview." Costs for projects of a significantly different size or type should be adjusted accordingly.

Location—Material prices in this book are for metropolitan areas. However, in dense urban areas, traffic and site storage limitations may increase costs. Beyond a 20-mile radius of large cities, extra trucking or transportation charges may also increase the material costs slightly. On the other hand, lower wage rates may be in effect. Be sure to consider both these factors when preparing an estimate, particularly if the job site is located in a central city or remote rural location.

In addition, highly specialized subcontract items may require travel and per diem expenses for mechanics.

Other factors—

- season of year
- contractor management
- weather conditions
- building code requirements
- availability of:
  - adequate energy
  - skilled labor
  - building materials
- owner's special requirements/ restrictions
- safety requirements
- environmental considerations

General Conditions—The extreme right-hand column of each chart gives the "Totals." These figures contain the installing contractor's O&P (in other words, the contractor doing the work). Therefore, it is necessary for a general contractor to add a percentage of all subcontracted items. For a detailed breakdown of O&P see the inside back cover of this book.

Unpredictable Factors—General business conditions influence "in-place" costs of all items. Substitute materials and construction methods may have to be employed. These may affect the installed cost and/or life cycle costs. Such factors may be difficult to evaluate and cannot necessarily be predicted on the basis of the job's location in a particular section of the country. Thus, where these factors apply, you may find significant, but unavoidable cost variations for which you will have to apply a measure of judgment to your estimate.

Final Checklist

Estimating can be a straightforward process provided you remember the basics. Here's a checklist of some of the items you should remember to do before completing your estimate.

Did you remember to . . .

- factor in the Location Factor for your locale
- take into consideration which items have been marked up and by how much
- mark up the entire estimate sufficiently for your purposes
- include all components of your project in the final estimate
- double check your figures to be sure of your accuracy
- call RSMeans if you have any questions about your estimate or the data you've found in our publications

Remember, RSMeans stands behind its publications. If you have any questions about your estimate . . . about the costs you've used from our books . . . or even about the technical aspects of the job that may affect your estimate, feel free to call the RSMeans editors at 1-800-334-3509.

Unit Price Section

Table of Contents

Unit Price Section

Table of Contents

How to Use the Unit Price Pages

The prices in this publication include all overhead and profit mark-ups for the installing contractor. If a general contractor is involved, an additional 10% should be added to the cost shown in the "Total" column.

The costs for material, labor and any equipment are shown here. Material costs include delivery, and a 10% mark-up. No sales taxes are included. Labor costs include all mark-ups as shown on the inside back cover of this book. Equipment costs also include a 10% mark-up.

The cost in the total column should be multiplied by the quantity of each item to determine the total cost for that item. The sum of all item's total costs becomes the project's total cost.

The book is arranged into 23 categories that follow the order of construction for residential projects. The category is shown as white letters on a black background at the top of each page.

A brief specification of the cost item is shown here.

Graphics representing some of the cost items are found at the beginning of each category.

The cost data is arranged in a tree structure format. Major cost sections are left justified and are shown as bold headings within a grey background. Cost items and sub-cost items are found under the bold headings. The complete description should read as "bold heading, cost item, sub-cost item." For example, the identified item's description is "Beam/Girder, Solid Wood Beam, 2″ x 8″."

Each action performed to the cost item is conveniently listed to the right of the description. See the explanation below for the scope of work included as part of the action.

The unit of measure upon which all the costs are based is shown here. An abbreviations list is included within the reference section found at the back of this book.

Rough Frame / Structure

Rough Framing

Beam / Girder		Unit	Material	Labor	Equip.	Total	Specification
Solid Wood Beam							
2" x 8"							
	Demolish	L.F.		.63		.63	Cost includes material and labor to install 2" x 8" beam.
	Install	L.F.	.91	1		1.91	
	Demolish and Install	L.F.	.91	1.63		2.54	
	Clean	L.F.	.10	.21		.31	
	Paint	L.F.	.12	1.55		1.67	
	Minimum Charge	Job		163		163	
2" x 10"							
	Demolish	L.F.		.79		.79	Cost includes material and labor to install 2" x 10" beam.
	Install	L.F.	1.30	1.09		2.39	
	Demolish and Install	L.F.	1.30	1.88		3.18	
	Clean	L.F.	.12	.25		.37	
	Paint	L.F.	.15	.95		1.10	
	Minimum Charge	Job		163		163	
2" x 12"							
	Demolish	L.F.		.95		.95	Cost includes material and labor to install 2" x 12" beam.
	Install	L.F.	1.76	1.19		2.95	
	Demolish and Install	L.F.	1.76	2.14		3.90	
	Clean	L.F.	.13	.29		.42	
	Paint	L.F.	.18	2.29		2.47	
	Minimum Charge	Job		163		163	
4" x 8"							
	Demolish	L.F.		1.97		1.97	Cost includes material and labor to install 4" x 8" beam.
	Install	L.F.	2.96	1.45	.66	5.07	
	Demolish and Install	L.F.	2.96	3.42	.66	7.04	
	Clean	L.F.	.12	.25		.37	
	Paint	L.F.	.15	.95		1.10	
	Minimum Charge	Job		163		163	
4" x 10"							
	Demolish	L.F.		2.47		2.47	Cost includes material and labor to install 4" x 10" beam.
	Install	L.F.	3.71	1.53	.70	5.94	
	Demolish and Install	L.F.	3.71	4	.70	8.41	
	Clean	L.F.	.13	.29		.42	
	Paint	L.F.	.18	1.11		1.29	
	Minimum Charge	Job		163		163	
4" x 12"							
	Demolish	L.F.		2.97		2.97	Cost includes material and labor to install 4" x 12" beam.
	Install	L.F.	4.84	1.62	.74	7.20	
	Demolish and Install	L.F.	4.84	4.59	.74	10.17	
	Clean	L.F.	.15	.34		.49	
	Paint	L.F.	.18	2.29		2.47	
	Minimum Charge	Job		163		163	

Action	Scope
Demolish	Includes removal of the item and hauling the debris to a truck, dumpster, or storage area.
Install	Includes the installation of the item with normal placement and spacing of fasteners.
Demolish and Install	Includes the demolish action and the install action.
Reinstall	Includes the reinstallation of a removed item.
Clean	Includes cleaning of the item.
Paint	Includes up to two coats of finish on the item (unless otherwise indicated).
Minimum Charge	Includes the minimum labor or the minimum labor and equipment cost. Use this charge for small quantities of material.

Job Costs

Scaffolding

Temporary Electrical Service

Air Compressor

Construction Fees		Unit	Material	Labor	Equip.	Total	Specification
Asbestos Test	Minimum Charge	Job		187		187	Minimum charge for testing.
Architectural	Minimum Charge	Day	480			480	Minimum labor charge for architect.
Engineering	Minimum Charge	Day	250			250	Minimum labor charge for field engineer.

Construction Permits		Unit	Material	Labor	Equip.	Total	Specification
Job (less than 5K)	Minimum Charge	Job	110			110	Permit fees, job less than 5K.
Job (5K to 10K)	Minimum Charge	Job	220			220	Permit fees, job more than 5K, but less than 10K.
Job (10K to 25K)	Minimum Charge	Job	550			550	Permit fees, job more than 10K, but less than 25K.
Job (25K to 50K)	Minimum Charge	Job	1100			1100	Permit fees, job more than 25K, but less than 50K.
Job (50K to 100K)	Minimum Charge	Job	2200			2200	Permit fees, job more than 50K, but less than 100K.
Job (100K to 250K)	Minimum Charge	Job	5500			5500	Permit fees, job more than 100K, but less than 250K.
Job (250K to 500K)	Minimum Charge	Job	11000			11000	Permit fees, job more than 250K, but less than 500K.
Job (500K to 1M)	Minimum Charge	Job	22000			22000	Permit fees, job more than 500K, but less than 1 million.

Job Costs

Trade Labor		Unit	Material	Labor	Equip.	Total	Specification
Carpenter							
Daily (8 hours)							
	Install	Day		325		325	Daily labor rate for one carpenter.
	Minimum Charge	Job		163		163	
Weekly (40 hours)							
	Install	Week		1625		1625	Weekly labor rate for one carpenter.
	Minimum Charge	Job		163		163	
Drywaller							
Daily (8 hours)							
	Install	Day		325		325	Daily labor rate for one carpenter.
	Minimum Charge	Job		163		163	
Weekly (40 hours)							
	Install	Week		1625		1625	Weekly labor rate for one carpenter.
	Minimum Charge	Job		163		163	
Roofer							
Daily (8 hours)							
	Install	Day		300		300	Daily labor rate for one skilled roofer.
	Minimum Charge	Job		150		150	
Weekly (40 hours)							
	Install	Week		1500		1500	Weekly labor rate for one skilled roofer.
	Minimum Charge	Job		150		150	
Painter							
Daily (8 hours)							
	Install	Day		286		286	Daily labor rate for one painter.
	Minimum Charge	Job		143		143	
Weekly (40 hours)							
	Install	Week		1425		1425	Weekly labor rate for one painter.
	Minimum Charge	Job		143		143	
Mason							
Daily (8 hours)							
	Install	Day		330		330	Daily labor rate for one skilled mason.
	Minimum Charge	Job		164		164	
Weekly (40 hours)							
	Install	Week		1650		1650	Weekly labor rate for one skilled mason.
	Minimum Charge	Job		164		164	
Electrician							
Daily (8 hours)							
	Install	Day		355		355	Daily labor rate for an electrician.
	Minimum Charge	Job		178		178	
Weekly (40 hours)							
	Install	Week		1775		1775	Weekly labor rate for an electrician.
	Minimum Charge	Job		178		178	
Plumber							
Daily (8 hours)							
	Install	Day		355		355	Daily labor rate for a plumber.
	Minimum Charge	Job		177		177	
Weekly (40 hours)							
	Install	Week		1775		1775	Weekly labor rate for a plumber.
	Minimum Charge	Job		177		177	
Common Laborer							
Daily (8 hours)							
	Install	Day		236		236	Daily labor rate for a common laborer.
	Minimum Charge	Job		118		118	

Job Costs

Trade Labor

Trade Labor		Unit	Material	Labor	Equip.	Total	Specification
Weekly (40 hours)							
	Install	Week		1175		1175	Weekly labor rate for a common
	Minimum Charge	Job		118		118	laborer.

Temporary Utilities

Temporary Utilities		Unit	Material	Labor	Equip.	Total	Specification
Power Pole							
Job Site Electricity							
	Install	Month	33			33	Includes average cost of electricity per
	Minimum Charge	Job		178		178	month.
Water Service							
Job Site Water							
	Install	Month	62.50			62.50	Includes average costs for water for
	Minimum Charge	Month	62.50			62.50	one month.
Telephone							
	Install	Month	224			224	Includes average cost for local and
	Minimum Charge	Month	224			224	long distance telephone service for
							one month.
Chemical Toilet							
	Install	Month			251	251	Monthly rental of portable toilet.
	Minimum Charge	Day			12.55	12.55	
Office Trailer							
Place and Remove							
	Install	Job	239			239	Includes cost to a contractor to deliver,
	Minimum Charge	Job		130		130	place and level an office trailer and
							remove at end of job.
Storage Container							
	Install	Job	155			155	Includes cost to a contractor to deliver
	Minimum Charge	Job		130		130	and place a storage container and
							remove at end of job.
Fencing							
	Install	L.F.	2.34			2.34	Includes material costs for temporary
	Minimum Charge	L.F.	275			275	fencing.
Railing							
	Install	L.F.	8.25	9.15	.51	17.91	Includes material and labor to install
	Minimum Charge	L.F.	275			275	primed steel pipe.

Equipment Rental

Equipment Rental		Unit	Material	Labor	Equip.	Total	Specification
Flatbed Truck							
Daily Rental							
	Install	Day			142	142	Daily rental of flatbed truck.
	Minimum Charge	Job			142	142	
Weekly Rental							
	Install	Week			710	710	Weekly rental of flatbed truck.
	Minimum Charge	Job			142	142	
Monthly Rental							
	Install	Month			2825	2825	Monthly rental of flatbed truck.
	Minimum Charge	Job			142	142	

Job Costs

Equipment Rental		Unit	Material	Labor	Equip.	Total	Specification
Dump Truck							
Daily Rental							
	Install	Day			360	360	Daily rental of dump truck.
	Minimum Charge	Day			360	360	
Weekly Rental							
	Install	Week			1800	1800	Weekly rental of dump truck.
	Minimum Charge	Day			360	360	
Monthly Rental							
	Install	Month			7175	7175	Monthly rental of dump truck.
	Minimum Charge	Day			360	360	
Forklift							
Daily Rental							
	Install	Day			237	237	Daily rental of forklift.
	Minimum Charge	Day			237	237	
Weekly Rental							
	Install	Week			1175	1175	Weekly rental of forklift.
	Minimum Charge	Day			237	237	
Monthly Rental							
	Install	Month			4750	4750	Monthly rental of forklift.
	Minimum Charge	Day			237	237	
Bobcat							
Daily Rental							
	Install	Day			169	169	Daily rental of wheeled, skid steer
	Minimum Charge	Day			169	169	loader.
Weekly Rental							
	Install	Week			845	845	Weekly rental of wheeled, skid steer
	Minimum Charge	Day			169	169	loader.
Monthly Rental							
	Install	Month			3375	3375	Monthly rental of wheeled, skid steer
	Minimum Charge	Day			169	169	loader.
Air Compressor							
Daily Rental							
	Install	Day			61	61	Daily rental of air compressor.
	Minimum Charge	Day			61	61	
Weekly Rental							
	Install	Week			305	305	Weekly rental of air compressor.
	Minimum Charge	Day			61	61	
Monthly Rental							
	Install	Month			1225	1225	Monthly rental of air compressor.
	Minimum Charge	Day			61	61	
Jackhammer							
Daily Rental							
	Install	Day			10.80	10.80	Daily rental of air tool, jackhammer.
	Minimum Charge	Day			10.80	10.80	
Weekly Rental							
	Install	Week			54	54	Weekly rental of air tool, jackhammer.
	Minimum Charge	Day			10.80	10.80	
Monthly Rental							
	Install	Month			216	216	Monthly rental of air tool,
	Minimum Charge	Day			10.80	10.80	jackhammer.
Concrete Mixer							
Daily Rental							
	Install	Day			113	113	Daily rental of concrete mixer.
	Minimum Charge	Day			113	113	

Job Costs

Equipment Rental		Unit	Material	Labor	Equip.	Total	Specification
Weekly Rental							
	Install	Week			565	565	Weekly rental of concrete mixer.
	Minimum Charge	Day			113	113	
Monthly Rental							
	Install	Month			2275	2275	Monthly rental of concrete mixer.
	Minimum Charge	Day			113	113	
Concrete Bucket							
Daily Rental							
	Install	Day			14.95	14.95	Daily rental of concrete bucket.
	Minimum Charge	Day			14.95	14.95	
Weekly Rental							
	Install	Week			75	75	Weekly rental of concrete bucket.
	Minimum Charge	Day			14.95	14.95	
Monthly Rental							
	Install	Month			299	299	Monthly rental of concrete bucket.
	Minimum Charge	Day			14.95	14.95	
Concrete Pump							
Daily Rental							
	Install	Day			775	775	Daily rental of concrete pump.
	Minimum Charge	Day			775	775	
Weekly Rental							
	Install	Week			3875	3875	Weekly rental of concrete pump.
	Minimum Charge	Day			775	775	
Monthly Rental							
	Install	Month			15500	15500	Monthly rental of concrete pump.
	Minimum Charge	Day			775	775	
Generator							
Daily Rental							
	Install	Day			25.50	25.50	Daily rental of generator.
	Minimum Charge	Day			25.50	25.50	
Weekly Rental							
	Install	Week			126	126	Weekly rental of generator.
	Minimum Charge	Day			25.50	25.50	
Monthly Rental							
	Install	Month			505	505	Monthly rental of generator.
	Minimum Charge	Day			25.50	25.50	
Welder							
Daily Rental							
	Install	Day			73	73	Daily rental of welder.
	Minimum Charge	Day			73	73	
Weekly Rental							
	Install	Week			365	365	Weekly rental of welder.
	Minimum Charge	Day			73	73	
Monthly Rental							
	Install	Month			1450	1450	Monthly rental of welder.
	Minimum Charge	Day			73	73	
Sandblaster							
Daily Rental							
	Install	Day			16.70	16.70	Daily rental of sandblaster.
	Minimum Charge	Day			16.70	16.70	
Weekly Rental							
	Install	Week			83.50	83.50	Weekly rental of sandblaster.
	Minimum Charge	Day			16.70	16.70	

Job Costs

Equipment Rental

		Unit	Material	Labor	Equip.	Total	Specification
Monthly Rental							
	Install	Month			335	335	Monthly rental of sandblaster.
	Minimum Charge	Day			16.70	16.70	
Space Heater							
Daily Rental							
	Install	Day			14.55	14.55	Daily rental of space heater.
	Minimum Charge	Day			17.70	17.70	
Weekly Rental							
	Install	Day			73	73	Weekly rental of space heater.
	Minimum Charge	Day			17.70	17.70	
Monthly Rental							
	Install	Day			291	291	Monthly rental of space heater.
	Minimum Charge	Day			17.70	17.70	

Move / Reset Contents

		Unit	Material	Labor	Equip.	Total	Specification
Small Room							
	Install	Room		28.50		28.50	Includes cost to remove and reset the
	Minimum Charge	Job		118		118	contents of a small room.
Average Room							
	Install	Room		34.50		34.50	Includes cost to remove and reset the
	Minimum Charge	Job		118		118	contents of an average room.
Large Room							
	Install	Room		47.50		47.50	Includes cost to remove and reset the
	Minimum Charge	Job		118		118	contents of a large room.
Extra Large Room							
	Install	Room		75.50		75.50	Includes cost to remove and reset the
	Minimum Charge	Job		118		118	contents of a extra large room.
Store Contents							
Weekly							
	Install	Week	152			152	Storage fee, per week.
Monthly							
	Install	Month	455			455	Storage fee, per month.

Cover and Protect

		Unit	Material	Labor	Equip.	Total	Specification
Cover / Protect Walls							
	Install	SF Wall		.09		.09	Includes labor costs to cover and
	Minimum Charge	Job		118		118	protect walls.
Cover / Protect Floors							
	Install	SF Flr.		.09		.09	Includes labor costs to cover and
	Minimum Charge	Job		118		118	protect floors.

Temporary Bracing

		Unit	Material	Labor	Equip.	Total	Specification
Shim Support Piers							
	Install	Ea.	3.41	94		97.41	Includes labor and material to install
	Minimum Charge	Job		282		282	shims support for piers.

Job Costs

Jack and Relevel

		Unit	Material	Labor	Equip.	Total	Specification
Single Story							
	Install	S.F.	1.22	1.13		2.35	Includes labor and equipment to jack
	Minimum Charge	Job		282		282	and relevel single-story building.
Two-story							
	Install	S.F.	1.22	1.88		3.10	Includes labor and equipment to jack
	Minimum Charge	Job		282		282	and relevel two-story building.

Dumpster Rental

		Unit	Material	Labor	Equip.	Total	Specification
Small Load							
	Install	Week			465	465	Dumpster, weekly rental, 2 dumps, 5
	Minimum Charge	Week			465	465	yard container.
Medium Load							
	Install	Week			560	560	Dumpster, weekly rental, 2 dumps, 10
	Minimum Charge	Week			560	560	yard container.
Large Load							
	Install	Week			985	985	Dumpster, weekly rental, 2 dumps, 30
	Minimum Charge	Week			985	985	yard container.

Debris Hauling

		Unit	Material	Labor	Equip.	Total	Specification
Per Ton							
	Install	Ton		73.50	39	112.50	Includes hauling of debris by trailer or
	Minimum Charge	Ton		555	291	846	dump truck.
Per Cubic Yard							
	Install	C.Y.		24.50	12.95	37.45	Includes hauling of debris by trailer or
	Minimum Charge	Job		555	291	846	dump truck.
Per Pick-up Truck Load							
	Install	Ea.		131		131	Includes hauling of debris by pick-up
	Minimum Charge	Job		555	291	846	truck.

Dump Fees

		Unit	Material	Labor	Equip.	Total	Specification
Per Cubic Yard							
	Install	C.Y.	6.95			6.95	Dump charges, building materials.
	Minimum Charge	C.Y.	6.95			6.95	
Per Ton							
	Install	Ton	55.50			55.50	Dump charges, building materials.
	Minimum Charge	Ton	55.50			55.50	

Scaffolding

		Unit	Material	Labor	Equip.	Total	Specification
4' to 6' High							
	Install	Month	51.50			51.50	Includes monthly rental of scaffolding,
	Minimum Charge	Month	51.50			51.50	steel tubular, 30" wide, 7' long, 5' high.
7' to 11' High							
	Install	Month	61.50			61.50	Includes monthly rental of scaffolding,
	Minimum Charge	Month	61.50			61.50	steel tubular, 30" wide, 7' long, 10' high.

Job Costs

Scaffolding

Scaffolding		Unit	Material	Labor	Equip.	Total	Specification
12' to 16' High							
	Install	Month	71			71	Includes monthly rental of scaffolding,
	Minimum Charge	Month	71			71	steel tubular, 30" wide, 7' long, 15' high.
17' to 21' High							
	Install	Month	95			95	Includes monthly rental of scaffolding,
	Minimum Charge	Month	95			95	steel tubular, 30" wide, 7' long, 20' high.
22' to 26' High							
	Install	Month	105			105	Includes monthly rental of scaffolding,
	Minimum Charge	Month	105			105	steel tubular, 30" wide, 7' long, 25' high.
27' to 30' High							
	Install	Month	114			114	Includes monthly rental of scaffolding,
	Minimum Charge	Month	114			114	steel tubular, 30" wide, 7' long, 30' high.

Construction Clean-up

Construction Clean-up		Unit	Material	Labor	Equip.	Total	Specification
Initial							
	Clean	S.F.		.29		.29	Includes labor for general clean-up.
	Minimum Charge	Job		118		118	
Progressive							
	Clean	S.F.		.15		.15	Includes labor for progressive site/job
	Minimum Charge	Job		118		118	clean-up on a weekly basis.

Building Demolition		Unit	Material	Labor	Equip.	Total	Specification
Wood-framed Building							
1 Story Demolition							
	Demolish	S.F.		1.10	1.34	2.44	Includes demolition of 1 floor only light wood-framed building, including hauling of debris. Dump fees not included.
	Minimum Charge	Job		1750	2125	3875	
2nd Story Demolition							
	Demolish	S.F.		1.12	1.37	2.49	Includes demolition of 2nd floor only light wood-framed building, including hauling of debris. Dump fees not included.
	Minimum Charge	Job		1750	2125	3875	
3rd Story Demolition							
	Demolish	S.F.		1.46	1.78	3.24	Includes demolition of 3rd floor only light wood-framed building, including hauling of debris. Dump fees not included.
	Minimum Charge	Job		1750	2125	3875	
Masonry Building							
	Demolish	S.F.		1.57	1.92	3.49	Includes demolition of one story masonry building, including hauling of debris. Dump fees not included.
	Minimum Charge	Job		1750	2125	3875	
Concrete Building							
Unreinforced							
	Demolish	S.F.		3.21	.96	4.17	Includes demolition of a one story non-reinforced concrete building, including hauling of debris. Dump fees not included.
	Minimum Charge	Job		1750	2125	3875	
Reinforced							
	Demolish	S.F.		3.68	1.10	4.78	Includes demolition of a one story reinforced concrete building, including hauling of debris. Dump fees not included.
	Minimum Charge	Job		1750	2125	3875	
Steel Building							
	Demolish	S.F.		2.09	2.55	4.64	Includes demolition of a metal building, including hauling of debris. Dump fees not included. No salvage value assumed for scrap metal.
	Minimum Charge	Job		1750	2125	3875	

Job Preparation

General Clean-up

General Clean-up		Unit	Material	Labor	Equip.	Total	Specification
Remove Debris							
	Clean	S.F.		.72	.14	.86	Includes removal of debris caused by the loss including contents items and loose debris.
	Minimum Charge	Job		360	71	431	
Muck-out							
	Clean	S.F.		.46		.46	Includes labor to remove 2″ of muck and mud after a flood loss.
	Minimum Charge	Job		118		118	
Deodorize / Disinfect							
	Clean	SF Flr.		.16		.16	Includes deodorizing building with application of mildicide and disinfectant to floors and walls - will vary depending on the extent of deodorization required.
	Minimum Charge	Job		118		118	
Clean Walls							
Heavy							
	Clean	S.F.		.11		.11	Includes labor and material for heavy cleaning (multiple applications) with detergent and solvent.
	Minimum Charge	Job		118		118	
Clean Ceiling							
Light							
	Clean	S.F.		.19		.19	Includes labor and material to clean light to moderate smoke and soot.
	Minimum Charge	Job		118		118	
Heavy							
	Clean	S.F.		.29		.29	Includes labor and material to clean heavy smoke and soot.
	Minimum Charge	Job		118		118	

Flood Clean-up

Flood Clean-up		Unit	Material	Labor	Equip.	Total	Specification
Emergency Service Call							
After Hours / Weekend							
	Minimum Charge	Job		157		157	Includes minimum charges for deflooding clean-up after hours or on weekends.
Grey Water / Sewage							
	Minimum Charge	Job		118		118	Includes minimum charges for deflooding clean-up.
Water Extraction							
Clean (non-grey)							
	Clean	SF Flr.	.02	.47		.49	Includes labor to remove clean uncontaminated water from the loss.
	Minimum Charge	Job		118		118	
Grey (non-solids)							
	Clean	SF Flr.	.02	.94		.96	Includes labor to remove non-solid grey water from the loss.
	Minimum Charge	Job		118		118	
Sewage / Solids							
	Clean	SF Flr.	.04	1.18		1.22	Includes labor to remove water from the loss that may contain sewage or solids.
	Minimum Charge	Job		118		118	

Job Preparation

Flood Clean-up

Flood Clean-up		Unit	Material	Labor	Equip.	Total	Specification
Mildicide Walls							
Topographical							
	Clean	SF Wall		.21		.21	Clean-up of deflooding including
	Minimum Charge	Job		118		118	topographical mildicide application.
Injection Treatments							
	Clean	SF Flr.	25	3.62		28.62	Clean-up of deflooding including
	Minimum Charge	Job		118		118	mildicide injection.
Carpet Cleaning							
Uncontaminated							
	Clean	SF Flr.		.15		.15	Includes labor to clean carpet.
	Minimum Charge	Job		118		118	
Contaminated							
	Clean	SF Flr.		.24		.24	Includes labor to clean contaminated
	Minimum Charge	Job		118		118	carpet.
Stairway							
	Clean	Ea.	.10	2.77		2.87	Includes labor to clean contaminated
	Minimum Charge	Job		118		118	carpet on a stairway.
Carpet Treatment							
Disinfect / Deodorize							
	Clean	SF Flr.		.15		.15	Includes disinfecting and deodorizing
	Minimum Charge	Job		118		118	of carpet.
Disinfect For Sewage							
	Clean	SF Flr.		.38		.38	Includes labor to disinfect and
	Minimum Charge	Job		118		118	deodorize carpet exposed to sewage.
Mildicide							
	Clean	SF Flr.	.01	.15		.16	Includes labor and materials to apply
	Minimum Charge	Job		118		118	mildicide to carpet.
Thermal Fog Area							
	Clean	SF Flr.	.03	.36		.39	Includes labor and materials to
	Minimum Charge	Job		118		118	perform thermal fogging during
							deflooding clean-up.
Wet Fog / ULV Area							
	Clean	SF Flr.	.02	.21		.23	Includes labor and materials to
	Minimum Charge	Job		118		118	perform wet fogging / ULV during
							deflooding clean-up.
Steam Clean Fixtures							
	Clean	Ea.	.01	23.50		23.51	Includes labor and materials to steam
	Minimum Charge	Job		118		118	clean fixtures during deflooding
							clean-up.
Pressure Wash Sewage							
	Clean	SF Flr.		2.14		2.14	Includes labor to pressure wash
	Minimum Charge	Job		118		118	sewage contamination.
Airmover							
	Minimum Charge	Day			17.70	17.70	Daily rental of air mover.

Job Preparation

Flood Clean-up

Flood Clean-up		Unit	Material	Labor	Equip.	Total	Specification
Dehumidifier							
Medium Size							
	Minimum Charge	Day			33.50	33.50	Daily rental of dehumidifier, medium.
Large Size							
	Minimum Charge	Day			66.50	66.50	Daily rental of dehumidifier, large.
Air Purifier							
HEPA Filter-Bacteria							
	Minimum Charge	Ea.	195			195	HEPA filter for air purifier.

Asbestos Removal

Asbestos Removal		Unit	Material	Labor	Equip.	Total	Specification
Asbestos Analysis							
	Minimum Charge	Job		187		187	Minimum charge for testing.
Asbestos Pipe Insulation							
1/2" to 3/4" Diameter							
	Demolish	L.F.	.32	4.84		5.16	Minimum charge to remove asbestos. Includes full Tyvek suits for workers changed four times per eight-hour shift with respirators, air monitoring and supervision by qualified professionals.
	Minimum Charge	Job		3000		3000	
1" to 3" Diameter							
	Demolish	L.F.	1	15		16	Minimum charge to remove asbestos. Includes full Tyvek suits for workers changed four times per eight-hour shift with respirators, air monitoring and supervision by qualified professionals.
	Minimum Charge	Job		3000		3000	
Asbestos-based Siding							
	Demolish	S.F.	2.22	33.50		35.72	Minimum charge to remove asbestos. Includes full Tyvek suits for workers changed four times per eight-hour shift with respirators, air monitoring and supervision by qualified professionals.
	Minimum Charge	Job		3000		3000	
Asbestos-based Plaster							
	Demolish	S.F.	1.71	25.50		27.21	Minimum charge to remove asbestos. Includes full Tyvek suits for workers changed four times per eight-hour shift with respirators, air monitoring and supervision by qualified professionals.
	Minimum Charge	Job		3000		3000	
Encapsulate Asbestos Ceiling							
	Install	S.F.	.30	.14		.44	Includes encapsulation with penetrating sealant sprayed onto ceiling and walls with airless sprayer.
	Minimum Charge	Job		3000		3000	
Scrape Acoustical Ceiling							
	Demolish	S.F.	.07	.84		.91	Minimum charge to remove asbestos. Includes full Tyvek suits for workers changed four times per eight-hour shift with respirators, air monitoring and supervision by qualified professionals.
	Minimum Charge	Job		3000		3000	

Job Preparation

Asbestos Removal

		Unit	Material	Labor	Equip.	Total	Specification
Seal Ceiling Asbestos	Install	S.F.	1.12	2.83		3.95	Includes attaching plastic cover to ceiling area and sealing where asbestos exists, taping all seams, caulk or foam insulation in cracks and negative air system.
	Minimum Charge	Job		3000		3000	
Seal Wall Asbestos	Install	S.F.	1.12	2.51		3.63	Includes attaching plastic cover to wall area and sealing where asbestos exists, taping all seams, caulk or foam insulation in cracks and negative air system.
	Minimum Charge	Job		3000		3000	
Seal Floor Asbestos	Install	S.F.	1.79	5		6.79	Includes attaching plastic cover to floor area and sealing where asbestos exists, taping all seams, caulk or foam insulation in cracks and negative air system.
	Minimum Charge	Job		3000		3000	
Negative Air Vent System	Install	Day			43.50	43.50	Daily rental of negative air vent system.
	Minimum Charge	Week			218	218	
HEPA Vacuum Cleaner	Install	Day			18.25	18.25	Daily rental of HEPA vacuum, 16 gallon.
	Minimum Charge	Week			91	91	
Airless Sprayer	Install	Day			40	40	Daily rental of airless sprayer.
	Minimum Charge	Week			201	201	
Decontamination Unit	Install	Day			133	133	Daily rental of decontamination unit.
	Minimum Charge	Week			665	665	
Light Stand	Install	Day			10.55	10.55	Daily rental of floodlight w/tripod.
	Minimum Charge	Week			53	53	

Room Demolition

		Unit	Material	Labor	Equip.	Total	Specification
Strip Typical Room	Demolish	S.F.		2.36		2.36	Includes 4 labor hours for general demolition.
	Minimum Charge	Job		131		131	
Strip Typical Bathroom	Demolish	S.F.		3.14		3.14	Includes 4 labor hours for general demolition.
	Minimum Charge	Job		131		131	
Strip Typical Kitchen	Demolish	S.F.		3.02		3.02	Includes 4 labor hours for general demolition.
	Minimum Charge	Job		131		131	
Strip Typical Laundry	Demolish	S.F.		2.69		2.69	Includes 4 labor hours for general demolition.
	Minimum Charge	Job		131		131	

Selective Demolition		Unit	Material	Labor	Equip.	Total	Specification
Concrete Slab							
4" Unreinforced							
	Demolish	S.F.		1.58	.20	1.78	Includes labor and equipment to
	Minimum Charge	Job		670	86.50	756.50	remove unreinforced concrete slab and haul debris to truck or dumpster.
6" Unreinforced							
	Demolish	S.F.		2.23	.29	2.52	Includes labor and equipment to
	Minimum Charge	Job		670	86.50	756.50	remove unreinforced concrete slab and haul debris to truck or dumpster.
8" Unreinforced							
	Demolish	S.F.		3.67	.47	4.14	Includes labor and equipment to
	Minimum Charge	Job		670	86.50	756.50	remove unreinforced concrete slab and haul debris to truck or dumpster.
4" Reinforced							
	Demolish	S.F.		1.98	.26	2.24	Includes labor and equipment to
	Minimum Charge	Job		670	86.50	756.50	remove reinforced concrete slab and haul debris to truck or dumpster.
6" Reinforced							
	Demolish	S.F.		2.98	.38	3.36	Includes labor and equipment to
	Minimum Charge	Job		670	86.50	756.50	remove reinforced concrete slab and haul debris to truck or dumpster.
8" Reinforced							
	Demolish	S.F.		4.87	.63	5.50	Includes labor and equipment to
	Minimum Charge	Job		670	86.50	756.50	remove reinforced concrete slab and haul debris to truck or dumpster.
Brick Wall							
4" Thick							
	Demolish	S.F.		2.19	.28	2.47	Includes labor and equipment to
	Minimum Charge	Job		670	86.50	756.50	remove brick wall and haul debris to a central location for disposal.
8" Thick							
	Demolish	S.F.		3.01	.39	3.40	Includes labor and equipment to
	Minimum Charge	Job		670	86.50	756.50	remove brick wall and haul debris to a central location for disposal.
12" Thick							
	Demolish	S.F.		4.02	.52	4.54	Includes labor and equipment to
	Minimum Charge	Job		670	86.50	756.50	remove brick wall and haul debris to a central location for disposal.
Concrete Masonry Wall							
4" Thick							
	Demolish	S.F.		2.43	.31	2.74	Includes minimum labor and
	Minimum Charge	Job		670	86.50	756.50	equipment to remove 4" block wall and haul debris to a central location for disposal.
6" Thick							
	Demolish	S.F.		3.01	.39	3.40	Includes minimum labor and
	Minimum Charge	Job		670	86.50	756.50	equipment to remove 6" block wall and haul debris to a central location for disposal.

Job Preparation

Selective Demolition		Unit	Material	Labor	Equip.	Total	Specification
8″ Thick							
	Demolish	S.F.		3.71	.48	4.19	Includes minimum labor and
	Minimum Charge	Job		670	86.50	756.50	equipment to remove 8″ block wall
							and haul debris to a central location
							for disposal.
12″ Thick							
	Demolish	S.F.		5.35	.69	6.04	Includes minimum labor and
	Minimum Charge	Job		670	86.50	756.50	equipment to remove 12″ block wall
							and haul debris to a central location
							for disposal.
Tree Removal							
Small, 4″ - 8″							
	Demolish	Ea.		84.50	62.50	147	Minimum charge, tree removal.
	Minimum Charge	Job		118		118	
Medium, 9″ - 12″							
	Demolish	Ea.		127	93.50	220.50	Minimum charge, tree removal.
	Minimum Charge	Job		118		118	
Large, 13″ - 18″							
	Demolish	Ea.		152	112	264	Minimum charge, tree removal.
	Minimum Charge	Job		385	126	511	
Very Large, 19″ - 24″							
	Demolish	Ea.		190	140	330	Minimum charge, tree removal.
	Minimum Charge	Job		385	126	511	
Stump Removal							
	Demolish	Ea.		52	115	167	Includes minimum charges for a site
	Minimum Charge	Job		214	66	280	demolition crew and backhoe /
							loader.

Backhoe/Loader - Wheel Type	Tractor Loader - Wheel Type	Backhoe - Crawler Type

Excavation	Unit	Material	Labor	Equip.	Total	Specification
Footings						
Hand Trenching						
Install	L.F.		1.47		1.47	Includes hand excavation for footing
Minimum Charge	Job		59		59	or trench up to 10" deep and 12" wide.
Machine Trenching						
Install	L.F.		.55	.24	.79	Includes machine excavation for
Minimum Charge	Job		275	119	394	footing or trench up to 10" deep and 12" wide.
Trenching w / Backfill						
Hand						
Install	L.F.		2.36		2.36	Includes hand excavation, backfill and
Minimum Charge	Job		59		59	compaction.
Machine						
Install	L.F.		2.75	7.80	10.55	Includes machine excavation, backfill
Minimum Charge	Job		118	26.50	144.50	and compaction.

Backfill	Unit	Material	Labor	Equip.	Total	Specification
Hand (no compaction)						
Install	L.C.Y.		19.65		19.65	Includes backfill by hand in average
Minimum Charge	Job		118		118	soil without compaction.
Hand (w / compaction)						
Install	E.C.Y.		27.50		27.50	Includes backfill by hand in average
Minimum Charge	Job		118		118	soil with compaction.
Machine (no compaction)						
Install	L.C.Y.		2.75	1.19	3.94	Includes backfill, by 55 HP wheel
Minimum Charge	Job		275	119	394	loader, of trenches from loose material piled adjacent without compaction.
Machine (w / compaction)						
Install	L.C.Y.		7.35	3.17	10.52	Includes backfill, by 55 HP wheel
Minimum Charge	Job		275	119	394	loader, of trenches from loose material piled adjacent with compaction by vibrating tampers.

Concrete Footing and Formwork

Wood Pile

Piles	Unit	Material	Labor	Equip.	Total	Specification
Treated Wood, 12″ Butt						
To 30′						
Demolish	L.F.		3.73	1.15	4.88	Includes material, labor and
Install	V.L.F.	10.25	3.61	2.69	16.55	equipment to install up to 30′ long
Demolish and Install	V.L.F.	10.25	7.34	3.84	21.43	treated wood piling.
Paint	L.F.	.39	1.63		2.02	
Minimum Charge	Job		2250	1675	3925	
30′ to 40′						
Demolish	L.F.		3.73	1.15	4.88	Includes material, labor and
Install	V.L.F.	10	3.22	2.40	15.62	equipment to install 30′ to 40′ long
Demolish and Install	V.L.F.	10	6.95	3.55	20.50	treated wood piling.
Paint	L.F.	.39	1.63		2.02	
Minimum Charge	Job		2250	1675	3925	
40′ to 50′						
Demolish	L.F.		3.73	1.15	4.88	Includes material, labor and
Install	V.L.F.	10.75	3.13	2.34	16.22	equipment to install 40′ to 50′ long
Demolish and Install	V.L.F.	10.75	6.86	3.49	21.10	treated wood piling.
Paint	L.F.	.39	1.63		2.02	
Minimum Charge	Job		2250	1675	3925	
50′ to 60′						
Demolish	L.F.		3.73	1.15	4.88	Includes material, labor and
Install	V.L.F.	11.85	2.82	2.10	16.77	equipment to install 50′ to 60′ long
Demolish and Install	V.L.F.	11.85	6.55	3.25	21.65	treated wood piling.
Paint	L.F.	.39	1.63		2.02	
Minimum Charge	Job		2250	1675	3925	
60′ to 80′						
Demolish	L.F.		3.73	1.15	4.88	Includes material, labor and
Install	V.L.F.	20.50	3.12	2.56	26.18	equipment to install 60′ to 80′ long
Demolish and Install	V.L.F.	20.50	6.85	3.71	31.06	treated wood piling.
Paint	L.F.	.39	1.63		2.02	
Minimum Charge	Job		2250	1675	3925	
Untreated Wood, 12″ Butt						
To 30′						
Demolish	L.F.		3.73	1.15	4.88	Includes material, labor and
Install	V.L.F.	6.95	3.61	2.69	13.25	equipment to install up to 30′ long
Demolish and Install	V.L.F.	6.95	7.34	3.84	18.13	untreated wood piling.
Paint	L.F.	.39	1.63		2.02	
Minimum Charge	Job		2250	1675	3925	

Foundation

Piles	Unit	Material	Labor	Equip.	Total	Specification
30' to 40'						
Demolish	L.F.		3.73	1.15	4.88	Includes material, labor and
Install	V.L.F.	6.95	3.22	2.40	12.57	equipment to install 30' to 40' long
Demolish and Install	V.L.F.	6.95	6.95	3.55	17.45	untreated wood piling.
Paint	L.F.	.39	1.63		2.02	
Minimum Charge	Job		2250	1675	3925	
40' to 50'						
Demolish	L.F.		3.73	1.15	4.88	Includes material, labor and
Install	V.L.F.	6.95	3.13	2.34	12.42	equipment to install 40' to 50' long
Demolish and Install	V.L.F.	6.95	6.86	3.49	17.30	untreated wood piling.
Paint	L.F.	.39	1.63		2.02	
Minimum Charge	Job		2250	1675	3925	
50' to 60'						
Demolish	L.F.		3.73	1.15	4.88	Includes material, labor and
Install	V.L.F.	7.05	2.82	2.10	11.97	equipment to install 50' to 60' long
Demolish and Install	V.L.F.	7.05	6.55	3.25	16.85	untreated wood piling.
Paint	L.F.	.39	1.63		2.02	
Minimum Charge	Job		2250	1675	3925	
60' to 80'						
Demolish	L.F.		3.73	1.15	4.88	Includes material, labor and
Install	V.L.F.	8.80	2.69	2	13.49	equipment to install 60' to 80' long
Demolish and Install	V.L.F.	8.80	6.42	3.15	18.37	untreated wood piling.
Paint	L.F.	.39	1.63		2.02	
Minimum Charge	Job		2250	1675	3925	
Precast Concrete						
10" Square						
Demolish	L.F.		3.81	1.17	4.98	Includes material, labor and
Install	V.L.F.	9.50	3.22	2.40	15.12	equipment to install a precast,
Demolish and Install	V.L.F.	9.50	7.03	3.57	20.10	prestressed, 40' long, 10" square
Paint	L.F.	.39	1.63		2.02	concrete pile.
Minimum Charge	Job		2250	1675	3925	
12" Square						
Demolish	L.F.		4.08	1.26	5.34	Includes material, labor and
Install	V.L.F.	12	3.32	2.48	17.80	equipment to install a precast,
Demolish and Install	V.L.F.	12	7.40	3.74	23.14	prestressed, 40' long, 12" square
Paint	L.F.	.39	1.63		2.02	concrete pile.
Minimum Charge	Job		2250	1675	3925	
14" Square						
Demolish	L.F.		4.18	1.29	5.47	Includes material, labor and
Install	V.L.F.	14.20	3.76	2.81	20.77	equipment to install a precast,
Demolish and Install	V.L.F.	14.20	7.94	4.10	26.24	prestressed, 40' long, 14" square
Paint	L.F.	.39	1.63		2.02	concrete pile.
Minimum Charge	Job		2250	1675	3925	
16" Square						
Demolish	L.F.		4.40	1.35	5.75	Includes material, labor and
Install	V.L.F.	22.50	4.03	3.01	29.54	equipment to install a precast,
Demolish and Install	V.L.F.	22.50	8.43	4.36	35.29	prestressed, 40' long, 16" square
Paint	L.F.	.39	1.63		2.02	concrete pile.
Minimum Charge	Job		2250	1675	3925	
18" Square						
Demolish	L.F.		4.76	1.47	6.23	Includes material, labor and
Install	V.L.F.	27	5.05	4.13	36.18	equipment to install a precast,
Demolish and Install	V.L.F.	27	9.81	5.60	42.41	prestressed, 40' long, 18" square
Paint	L.F.	.39	1.63		2.02	concrete pile.
Minimum Charge	Job		2250	1675	3925	

Foundation

Piles

	Unit	Material	Labor	Equip.	Total	Specification
Cross Bracing						
Treated 2″ x 6″						
Demolish	L.F.		.77		.77	Cost includes material and labor to
Install	L.F.	.85	4.55		5.40	install 2″ x 6″ lumber for cross-bracing
Demolish and Install	L.F.	.85	5.32		6.17	of foundation piling or posts.
Reinstall	L.F.		3.64		3.64	
Clean	S.F.		.29		.29	
Paint	L.F.	.17	1.10		1.27	
Minimum Charge	Job		565		565	
Treated 2″ x 8″						
Demolish	L.F.		.77		.77	Cost includes material and labor to
Install	L.F.	1.10	4.70		5.80	install 2″ x 8″ lumber for cross-bracing
Demolish and Install	L.F.	1.10	5.47		6.57	of foundation piling or posts.
Reinstall	L.F.		3.76		3.76	
Clean	S.F.		.29		.29	
Paint	L.F.	.22	1.14		1.36	
Minimum Charge	Job		565		565	
Treated 2″ x 10″						
Demolish	L.F.		.79		.79	Cost includes material and labor to
Install	L.F.	1.64	4.86		6.50	install 2″ x 10″ lumber for
Demolish and Install	L.F.	1.64	5.65		7.29	cross-bracing of foundation piling or
Reinstall	L.F.		3.89		3.89	posts.
Clean	S.F.		.29		.29	
Paint	L.F.	.26	1.17		1.43	
Minimum Charge	Job		565		565	
Treated 2″ x 12″						
Demolish	L.F.		.95		.95	Cost includes material and labor to
Install	L.F.	2.10	5.05		7.15	install 2″ x 12″ lumber for
Demolish and Install	L.F.	2.10	6		8.10	cross-bracing of foundation piling or
Reinstall	L.F.		4.03		4.03	posts.
Clean	S.F.		.29		.29	
Paint	L.F.	.31	1.19		1.50	
Minimum Charge	Job		565		565	
Metal Pipe						
6″ Concrete Filled						
Demolish	L.F.		5.30	1.63	6.93	Includes material, labor and
Install	V.L.F.	16.45	4.70	3.51	24.66	equipment to install concrete filled
Demolish and Install	V.L.F.	16.45	10	5.14	31.59	steel pipe pile.
Paint	L.F.	.30	1.63		1.93	
Minimum Charge	Job		2250	1675	3925	
6″ Unfilled						
Demolish	L.F.		5.30	1.63	6.93	Includes material, labor and
Install	V.L.F.	13.95	4.30	3.21	21.46	equipment to install unfilled steel pipe
Demolish and Install	V.L.F.	13.95	9.60	4.84	28.39	pile.
Paint	L.F.	.30	1.63		1.93	
Minimum Charge	Job		2250	1675	3925	
8″ Concrete Filled						
Demolish	L.F.		5.30	1.63	6.93	Includes material, labor and
Install	V.L.F.	14.95	4.90	3.66	23.51	equipment to install concrete filled
Demolish and Install	V.L.F.	14.95	10.20	5.29	30.44	steel pipe pile.
Paint	L.F.	.30	1.63		1.93	
Minimum Charge	Job		2250	1675	3925	
8″ Unfilled						
Demolish	L.F.		5.30	1.63	6.93	Includes material, labor and
Install	V.L.F.	14.10	4.51	3.37	21.98	equipment to install unfilled steel pipe
Demolish and Install	V.L.F.	14.10	9.81	5	28.91	pile.
Paint	L.F.	.30	1.63		1.93	
Minimum Charge	Job		2250	1675	3925	

Foundation

Piles	Unit	Material	Labor	Equip.	Total	Specification
10" Concrete Filled						
Demolish	L.F.		5.45	1.68	7.13	Includes material, labor and
Install	V.L.F.	19.35	5	3.74	28.09	equipment to install concrete filled
Demolish and Install	V.L.F.	19.35	10.45	5.42	35.22	steel pipe pile.
Paint	L.F.	.30	1.63		1.93	
Minimum Charge	Job		2250	1675	3925	
10" Unfilled						
Demolish	L.F.		5.45	1.68	7.13	Includes material, labor and
Install	V.L.F.	17.55	4.51	3.37	25.43	equipment to install unfilled steel pipe
Demolish and Install	V.L.F.	17.55	9.96	5.05	32.56	pile.
Paint	L.F.	.30	1.63		1.93	
Minimum Charge	Job		2250	1675	3925	
12" Concrete Filled						
Demolish	L.F.		5.45	1.68	7.13	Includes material, labor and
Install	V.L.F.	22.50	5.45	4.06	32.01	equipment to install concrete filled
Demolish and Install	V.L.F.	22.50	10.90	5.74	39.14	steel pipe pile.
Paint	L.F.	.30	1.63		1.93	
Minimum Charge	Job		2250	1675	3925	
12" Unfilled						
Demolish	L.F.		5.45	1.68	7.13	Includes material, labor and
Install	V.L.F.	21.50	4.75	3.54	29.79	equipment to install unfilled steel pipe
Demolish and Install	V.L.F.	21.50	10.20	5.22	36.92	pile.
Paint	L.F.	.30	1.63		1.93	
Minimum Charge	Job		2250	1675	3925	
Steel H Section						
H Section 8 x 8 x 36#						
Demolish	L.F.		3.70	1.14	4.84	Includes material, labor and
Install	V.L.F.	13.05	3.53	2.63	19.21	equipment to install steel H section
Demolish and Install	V.L.F.	13.05	7.23	3.77	24.05	pile.
Paint	L.F.	.30	1.63		1.93	
Minimum Charge	Job		2250	1675	3925	
H Section 10 x 10 x 57#						
Demolish	L.F.		4.04	1.24	5.28	Includes material, labor and
Install	V.L.F.	20.50	3.70	2.76	26.96	equipment to install steel H section
Demolish and Install	V.L.F.	20.50	7.74	4	32.24	pile.
Paint	L.F.	.30	1.63		1.93	
Minimum Charge	Job		2250	1675	3925	
H Section 12 x 12 x 74#						
Demolish	L.F.		4.18	1.29	5.47	Includes material, labor and
Install	V.L.F.	27	4.44	3.64	35.08	equipment to install steel H section
Demolish and Install	V.L.F.	27	8.62	4.93	40.55	pile.
Paint	L.F.	.30	1.63		1.93	
Minimum Charge	Job		2250	1675	3925	
H Section 14 x 14 x 89#						
Demolish	L.F.		4.40	1.35	5.75	Includes material, labor and
Install	V.L.F.	33	4.85	3.98	41.83	equipment to install steel H section
Demolish and Install	V.L.F.	33	9.25	5.33	47.58	pile.
Paint	L.F.	.30	1.63		1.93	
Minimum Charge	Job		2250	1675	3925	

Foundation

Foundation Post		Unit	Material	Labor	Equip.	Total	Specification
Wood Foundation Post							
4" x 4"							
	Demolish	L.F.		.60	.18	.78	Cost includes material and labor to
	Install	L.F.	.97	3.93		4.90	install 4" x 4" treated wood post by
	Demolish and Install	L.F.	.97	4.53	.18	5.68	hand in normal soil conditions.
	Reinstall	L.F.		3.14		3.14	
	Minimum Charge	Job		565		565	
4" x 6"							
	Demolish	L.F.		.60	.18	.78	Cost includes material and labor to
	Install	L.F.	1.45	3.93		5.38	install 4" x 6" treated wood post by
	Demolish and Install	L.F.	1.45	4.53	.18	6.16	hand in normal soil conditions.
	Reinstall	L.F.		3.14		3.14	
	Minimum Charge	Job		565		565	
6" x 6"							
	Demolish	L.F.		.60	.18	.78	Cost includes material and labor to
	Install	L.F.	2.17	3.93		6.10	install 6" x 6" treated wood post by
	Demolish and Install	L.F.	2.17	4.53	.18	6.88	hand in normal soil conditions.
	Reinstall	L.F.		3.14		3.14	
	Minimum Charge	Job		565		565	
Water Jet							
	Demolish	Ea.		70		70	Includes material, labor and
	Install	Ea.	57	540		597	equipment to water jet a 6" x 6" wood
	Demolish and Install	Ea.	57	610		667	post under an existing building. Cost
	Reinstall	Ea.		538.51		538.51	may vary depending upon location
	Minimum Charge	Job		470		470	and access.
Shim							
	Install	Ea.	3.18	56.50		59.68	Includes labor and material to trim or
	Minimum Charge	Job		565		565	readjust existing foundation posts
							under an existing building. Access of
							2 - 4 foot assumed.
Cross-bracing							
Treated 2" x 6"							
	Demolish	L.F.		.77		.77	Cost includes material and labor to
	Install	L.F.	.85	4.55		5.40	install 2" x 6" lumber for cross-bracing
	Demolish and Install	L.F.	.85	5.32		6.17	of foundation piling or posts.
	Reinstall	L.F.		3.64		3.64	
	Clean	S.F.		.29		.29	
	Paint	L.F.	.17	1.10		1.27	
	Minimum Charge	Job		565		565	
Treated 2" x 8"							
	Demolish	L.F.		.77		.77	Cost includes material and labor to
	Install	L.F.	1.10	4.70		5.80	install 2" x 8" lumber for cross-bracing
	Demolish and Install	L.F.	1.10	5.47		6.57	of foundation piling or posts.
	Reinstall	L.F.		3.76		3.76	
	Clean	S.F.		.29		.29	
	Paint	L.F.	.22	1.14		1.36	
	Minimum Charge	Job		565		565	
Treated 2" x 10"							
	Demolish	L.F.		.79		.79	Cost includes material and labor to
	Install	L.F.	1.64	4.86		6.50	install 2" x 10" lumber for
	Demolish and Install	L.F.	1.64	5.65		7.29	cross-bracing of foundation piling or
	Reinstall	L.F.		3.89		3.89	posts.
	Clean	S.F.		.29		.29	
	Paint	L.F.	.26	1.17		1.43	
	Minimum Charge	Job		565		565	

Foundation

Foundation Post		Unit	Material	Labor	Equip.	Total	Specification
Treated 2" x 12"							Cost includes material and labor to
	Demolish	L.F.		.95		.95	install 2" x 12" lumber for
	Install	L.F.	2.10	5.05		7.15	cross-bracing of foundation piling or
	Demolish and Install	L.F.	2.10	6		8.10	posts.
	Reinstall	L.F.		4.03		4.03	
	Clean	S.F.		.29		.29	
	Paint	L.F.	.31	1.19		1.50	
	Minimum Charge	Job		565		565	

Concrete Footing		Unit	Material	Labor	Equip.	Total	Specification
Continuous							
12" w x 6" d							
	Demolish	L.F.		1.14	.91	2.05	Includes material and labor to install
	Install	L.F.	3.56	2.67	.02	6.25	continuous reinforced concrete footing
	Demolish and Install	L.F.	3.56	3.81	.93	8.30	poured by chute cast against earth
	Minimum Charge	Job		650		650	including reinforcing, forming and
							finishing.
12" w x 12" d							
	Demolish	L.F.		2.20	1.75	3.95	Includes material and labor to install
	Install	L.F.	8.30	5.35	.03	13.68	continuous reinforced concrete footing
	Demolish and Install	L.F.	8.30	7.55	1.78	17.63	poured by chute cast against earth
	Minimum Charge	Job		650		650	including reinforcing, forming and
							finishing.
18" w x 10" d							
	Demolish	L.F.		2.85	2.27	5.12	Includes material and labor to install
	Install	L.F.	7.65	6.60	.04	14.29	continuous reinforced concrete footing
	Demolish and Install	L.F.	7.65	9.45	2.31	19.41	poured by chute cast against earth
	Minimum Charge	Job		650		650	including reinforcing, forming and
							finishing.
24" w x 24" d							
	Demolish	L.F.		4.76	3.79	8.55	Includes material and labor to install
	Install	L.F.	22.50	21.50	.13	44.13	continuous reinforced concrete footing
	Demolish and Install	L.F.	22.50	26.26	3.92	52.68	poured by chute cast against earth
	Minimum Charge	Job		650		650	including reinforcing, forming and
							finishing.
Stem Wall							
Single Story							
	Demolish	L.F.		10.55	8.40	18.95	Includes material and labor to install
	Install	L.F.	11.45	56	44.50	111.95	stem wall 24" above and 18" below
	Demolish and Install	L.F.	11.45	66.55	52.90	130.90	grade for a single story structure.
	Minimum Charge	Job		650		650	Includes 12" wide and 8" deep
							footing.
Two Story							
	Demolish	L.F.		10.55	8.40	18.95	Includes material and labor to install
	Install	L.F.	14.10	80.50	64	158.60	stem wall 24" above and 18" below
	Demolish and Install	L.F.	14.10	91.05	72.40	177.55	grade typical for a two story structure,
	Minimum Charge	Job		650		650	includes 18" wide and 10" deep
							footing.

Foundation

Concrete Slab

	Unit	Material	Labor	Equip.	Total	Specification
Reinforced						
4"						
Demolish	S.F.		1.98	.26	2.24	Cost includes material and labor to
Install	S.F.	1.98	.42		2.40	install 4" reinforced concrete
Demolish and Install	S.F.	1.98	2.40	.26	4.64	slab-on-grade poured by chute
Clean	S.F.	.04	.26		.30	including forms, vapor barrier, wire
Paint	S.F.	.12	.62		.74	mesh, 3000 PSI concrete, float finish,
Minimum Charge	Job		148		148	and curing.
6"						
Demolish	S.F.		2.98	.38	3.36	Cost includes material and labor to
Install	S.F.	2.67	.63		3.30	install 6" reinforced concrete
Demolish and Install	S.F.	2.67	3.61	.38	6.66	slab-on-grade poured by chute
Clean	S.F.	.04	.26		.30	including forms, vapor barrier, wire
Paint	S.F.	.12	.62		.74	mesh, 3000 PSI concrete, float finish,
Minimum Charge	Job		148		148	and curing.
8"						
Demolish	S.F.		4.87	.63	5.50	Cost includes material and labor to
Install	S.F.	3.39	.83	.01	4.23	install 8" reinforced concrete
Demolish and Install	S.F.	3.39	5.70	.64	9.73	slab-on-grade poured by chute
Clean	S.F.	.04	.26		.30	including forms, vapor barrier, wire
Paint	S.F.	.12	.62		.74	mesh, 3000 PSI concrete, float finish,
Minimum Charge	Job		148		148	and curing.
Unreinforced						
4"						
Demolish	S.F.		1.58	.20	1.78	Cost includes material and labor to
Install	S.F.	1.18	.73	.01	1.92	install 4" concrete slab-on-grade
Demolish and Install	S.F.	1.18	2.31	.21	3.70	poured by chute, including forms,
Clean	S.F.	.04	.26		.30	vapor barrier, 3000 PSI concrete, float
Paint	S.F.	.12	.62		.74	finish, and curing.
Minimum Charge	Job		148		148	
6"						
Demolish	S.F.		2.23	.29	2.52	Cost includes material and labor to
Install	S.F.	1.72	.75	.01	2.48	install 6" concrete slab-on-grade
Demolish and Install	S.F.	1.72	2.98	.30	5	poured by chute, including forms,
Clean	S.F.	.04	.26		.30	vapor barrier, 3000 PSI concrete, float
Paint	S.F.	.12	.62		.74	finish, and curing.
Minimum Charge	Job		148		148	
8"						
Demolish	S.F.		3.67	.47	4.14	Cost includes material and labor to
Install	S.F.	2.34	.79	.01	3.14	install 8" concrete slab-on-grade
Demolish and Install	S.F.	2.34	4.46	.48	7.28	poured by chute, including forms,
Clean	S.F.	.04	.26		.30	vapor barrier, 3000 PSI concrete, float
Paint	S.F.	.12	.62		.74	finish, and curing.
Minimum Charge	Job		148		148	
Lightweight						
Demolish	S.F.		.70	.16	.86	Includes material and labor to install
Install	S.F.	1.43	2.21		3.64	lightweight concrete, 3" - 4" thick.
Demolish and Install	S.F.	1.43	2.91	.16	4.50	
Clean	S.F.	.04	.26		.30	
Paint	S.F.	.12	.62		.74	
Minimum Charge	Job		148		148	

Foundation

Concrete Slab

Concrete Slab	Unit	Material	Labor	Equip.	Total	Specification
Gunite						
Demolish	S.F.		.86	.26	1.12	Includes material and labor to install
Install	S.F.	.28	.94	.39	1.61	gunite on flat plane per inch of
Demolish and Install	S.F.	.28	1.80	.65	2.73	thickness using 1" as base price. No
Clean	S.F.	.04	.26		.30	forms or reinforcing are included.
Paint	S.F.	.12	.62		.74	
Minimum Charge	Job		830		830	
Epoxy Inject						
Install	L.F.	2.17	2.97		5.14	Includes labor and material to inject
Minimum Charge	Job		148		148	epoxy into cracks.
Pressure Grout						
Install	S.F.	2.44	2.20		4.64	Includes labor, material and
Minimum Charge	Job		148		148	equipment to install pressure grout under an existing building.
Scrape and Paint						
Paint	S.F.	.17	.72		.89	Includes labor and materials to scrape
Minimum Charge	Job		143		143	and paint a concrete floor.
Vapor Barrier						
Install	S.F.	.02	.09		.11	Cost includes material and labor to
Minimum Charge	Job		163		163	install 4 mil polyethylene vapor barrier 10' wide sheets with 6" overlaps.

Rough Framing

Beam / Girder		Unit	Material	Labor	Equip.	Total	Specification
Solid Wood Beam							
2″ x 8″							
	Demolish	L.F.		.63		.63	Cost includes material and labor to
	Install	L.F.	.91	1		1.91	install 2″ x 8″ beam.
	Demolish and Install	L.F.	.91	1.63		2.54	
	Clean	L.F.	.10	.21		.31	
	Paint	L.F.	.12	1.55		1.67	
	Minimum Charge	Job		163		163	
2″ x 10″							
	Demolish	L.F.		.79		.79	Cost includes material and labor to
	Install	L.F.	1.30	1.09		2.39	install 2″ x 10″ beam.
	Demolish and Install	L.F.	1.30	1.88		3.18	
	Clean	L.F.	.12	.25		.37	
	Paint	L.F.	.15	.95		1.10	
	Minimum Charge	Job		163		163	
2″ x 12″							
	Demolish	L.F.		.95		.95	Cost includes material and labor to
	Install	L.F.	1.76	1.19		2.95	install 2″ x 12″ beam.
	Demolish and Install	L.F.	1.76	2.14		3.90	
	Clean	L.F.	.13	.29		.42	
	Paint	L.F.	.18	2.29		2.47	
	Minimum Charge	Job		163		163	
4″ x 8″							
	Demolish	L.F.		1.97		1.97	Cost includes material and labor to
	Install	L.F.	2.96	1.45	.66	5.07	install 4″ x 8″ beam.
	Demolish and Install	L.F.	2.96	3.42	.66	7.04	
	Clean	L.F.	.12	.25		.37	
	Paint	L.F.	.15	.95		1.10	
	Minimum Charge	Job		163		163	
4″ x 10″							
	Demolish	L.F.		2.47		2.47	Cost includes material and labor to
	Install	L.F.	3.71	1.53	.70	5.94	install 4″ x 10″ beam.
	Demolish and Install	L.F.	3.71	4	.70	8.41	
	Clean	L.F.	.13	.29		.42	
	Paint	L.F.	.18	1.11		1.29	
	Minimum Charge	Job		163		163	
4″ x 12″							
	Demolish	L.F.		2.97		2.97	Cost includes material and labor to
	Install	L.F.	4.84	1.62	.74	7.20	install 4″ x 12″ beam.
	Demolish and Install	L.F.	4.84	4.59	.74	10.17	
	Clean	L.F.	.15	.34		.49	
	Paint	L.F.	.18	2.29		2.47	
	Minimum Charge	Job		163		163	

Rough Frame / Structure

Beam / Girder		Unit	Material	Labor	Equip.	Total	Specification
6" x 8"							
	Demolish	L.F.		2.97		2.97	Cost includes material and labor to
	Install	L.F.	6.45	2.77	1.26	10.48	install 6" x 8" beam.
	Demolish and Install	L.F.	6.45	5.74	1.26	13.45	
	Clean	L.F.	.13	.29		.42	
	Paint	L.F.	.18	1.11		1.29	
	Minimum Charge	Job		163		163	
6" x 10"							
	Demolish	L.F.		3.71		3.71	Cost includes material and labor to
	Install	L.F.	3.88	1.30		5.18	install 6" x 10" triple beam.
	Demolish and Install	L.F.	3.88	5.01		8.89	
	Clean	L.F.	.15	.34		.49	
	Paint	L.F.	.21	1.27		1.48	
	Minimum Charge	Job		163		163	
6" x 12"							
	Demolish	L.F.		4.41		4.41	Cost includes material and labor to
	Install	L.F.	5.30	1.37		6.67	install 6" x 12" triple beam.
	Demolish and Install	L.F.	5.30	5.78		11.08	
	Clean	L.F.	.18	.38		.56	
	Paint	L.F.	.23	1.43		1.66	
	Minimum Charge	Job		163		163	
Steel I-type							
W8 x 31							
	Demolish	L.F.		6.80	3.02	9.82	Includes material and labor to install
	Install	L.F.	33	2.16	1.35	36.51	W shaped steel beam / girder.
	Demolish and Install	L.F.	33	8.96	4.37	46.33	
	Minimum Charge	Job		395		395	
W8 x 48							
	Demolish	L.F.		6.80	3.02	9.82	Includes material and labor to install
	Install	L.F.	50.50	2.26	1.41	54.17	W shaped steel beam / girder.
	Demolish and Install	L.F.	50.50	9.06	4.43	63.99	
	Minimum Charge	Job		395		395	
W8 x 67							
	Demolish	L.F.		6.80	3.02	9.82	Includes material and labor to install
	Install	L.F.	71	2.37	1.48	74.85	W shaped steel beam / girder.
	Demolish and Install	L.F.	71	9.17	4.50	84.67	
	Minimum Charge	Job		395		395	
W10 x 45							
	Demolish	L.F.		6.80	3.02	9.82	Includes material and labor to install
	Install	L.F.	48	2.26	1.41	51.67	W shaped steel beam / girder.
	Demolish and Install	L.F.	48	9.06	4.43	61.49	
	Minimum Charge	Job		395		395	
W10 x 68							
	Demolish	L.F.		6.80	3.02	9.82	Includes material and labor to install
	Install	L.F.	72	2.37	1.48	75.85	W shaped steel beam / girder.
	Demolish and Install	L.F.	72	9.17	4.50	85.67	
	Minimum Charge	Job		395		395	
W12 x 50							
	Demolish	L.F.		6.80	3.02	9.82	Includes material and labor to install
	Install	L.F.	53	2.26	1.41	56.67	W shaped steel beam / girder.
	Demolish and Install	L.F.	53	9.06	4.43	66.49	
	Minimum Charge	Job		395		395	
W12 x 87							
	Demolish	L.F.		10	4.44	14.44	Includes material and labor to install
	Install	L.F.	92	2.37	1.48	95.85	W shaped steel beam / girder.
	Demolish and Install	L.F.	92	12.37	5.92	110.29	
	Minimum Charge	Job		395		395	

Rough Frame / Structure

Beam / Girder

		Unit	Material	Labor	Equip.	Total	Specification
W12 x 120							
	Demolish	L.F.		10	4.44	14.44	Includes material and labor to install
	Install	L.F.	128	2.43	1.52	131.95	W shaped steel beam / girder.
	Demolish and Install	L.F.	128	12.43	5.96	146.39	
	Minimum Charge	Job		395		395	
W14 x 74							
	Demolish	L.F.		10	4.44	14.44	Includes material and labor to install
	Install	L.F.	78	2.37	1.48	81.85	W shaped steel beam / girder.
	Demolish and Install	L.F.	78	12.37	5.92	96.29	
	Minimum Charge	Job		395		395	
W14 x 120							
	Demolish	L.F.		10	4.44	14.44	Includes material and labor to install
	Install	L.F.	128	2.43	1.52	131.95	W shaped steel beam / girder.
	Demolish and Install	L.F.	128	12.43	5.96	146.39	
	Minimum Charge	Job		395		395	

Column

		Unit	Material	Labor	Equip.	Total	Specification
Concrete							
Small Diameter							
	Demolish	L.F.		11.15	1.44	12.59	Includes material and labor to install
	Install	L.F.	56.50	10.05	2.31	68.86	small diameter concrete column.
	Demolish and Install	L.F.	56.50	21.20	3.75	81.45	
	Minimum Charge	Job		505	115	620	
Large Diameter							
	Demolish	L.F.		33.50	4.32	37.82	Includes material and labor to install
	Install	L.F.	73	11.85	2.71	87.56	large diameter concrete column.
	Demolish and Install	L.F.	73	45.35	7.03	125.38	
	Minimum Charge	Job		505	115	620	

Floor Framing System

		Unit	Material	Labor	Equip.	Total	Specification
12" O.C.							
2" x 6" Joists							
	Demolish	S.F.		.54		.54	Cost includes material and labor to
	Install	S.F.	.69	.26		.95	install 2" x 6" joists, 12" O.C.
	Demolish and Install	S.F.	.69	.80		1.49	including box or band joist. Blocking
	Reinstall	S.F.		.21		.21	or bridging not included.
	Minimum Charge	Job		163		163	
2" x 8" Joists							
	Demolish	S.F.		.56		.56	Cost includes material and labor to
	Install	S.F.	1.07	.30		1.37	install 2" x 8" joists, 12" O.C.
	Demolish and Install	S.F.	1.07	.86		1.93	including box or band joist. Blocking
	Reinstall	S.F.		.24		.24	or bridging not included.
	Minimum Charge	Job		163		163	
2" x 10" Joists							
	Demolish	S.F.		.57		.57	Cost includes material and labor to
	Install	S.F.	1.51	.36		1.87	install 2" x 10" joists, 12" O.C.
	Demolish and Install	S.F.	1.51	.93		2.44	including box or band joist. Blocking
	Reinstall	S.F.		.29		.29	or bridging not included.
	Minimum Charge	Job		163		163	

Floor Framing System		Unit	Material	Labor	Equip.	Total	Specification
2" x 12" Joists							
	Demolish	S.F.		.59		.59	Cost includes material and labor to
	Install	S.F.	2.06	.37		2.43	install 2" x 12" joists, 12" O.C.
	Demolish and Install	S.F.	2.06	.96		3.02	including box or band joist. Blocking
	Reinstall	S.F.		.30		.30	or bridging not included.
	Minimum Charge	Job		163		163	
Block / Bridge							
	Demolish	Ea.		.82		.82	Includes material and labor to install
	Install	Ea.	1.73	1.83		3.56	set of cross bridging or per block of
	Demolish and Install	Ea.	1.73	2.65		4.38	solid bridging for 2" x 10" joists cut to
	Minimum Charge	Job		163		163	size on site.

16" O.C.

		Unit	Material	Labor	Equip.	Total	Specification
2" x 6" Joists							
	Demolish	S.F.		.40		.40	Cost includes material and labor to
	Install	S.F.	.53	.39		.92	install 2" x 6" joists including box or
	Demolish and Install	S.F.	.53	.79		1.32	band joist installed 16" O.C. Does not
	Reinstall	S.F.		.31		.31	include beams, blocking or bridging.
	Minimum Charge	Job		163		163	
2" x 8" Joists							
	Demolish	S.F.		.42		.42	Cost includes material and labor to
	Install	S.F.	.81	.45		1.26	install 2" x 8" joists including box or
	Demolish and Install	S.F.	.81	.87		1.68	band joist installed 16" O.C. Does not
	Reinstall	S.F.		.36		.36	include beams, blocking or bridging.
	Minimum Charge	Job		163		163	
2" x 10" Joists							
	Demolish	S.F.		.43		.43	Cost includes material and labor to
	Install	S.F.	1.17	.54		1.71	install 2" x 10" joists including box or
	Demolish and Install	S.F.	1.17	.97		2.14	band joist 16" O.C. Does not include
	Reinstall	S.F.		.43		.43	beams, blocking or bridging.
	Minimum Charge	Job		163		163	
2" x 12" Joists							
	Demolish	S.F.		.45		.45	Cost includes material and labor to
	Install	S.F.	1.58	.56		2.14	install 2" x 12" joists including box or
	Demolish and Install	S.F.	1.58	1.01		2.59	band joist installed 16" O.C. Does not
	Reinstall	S.F.		.45		.45	include beams, blocking or bridging.
	Minimum Charge	Job		163		163	
Block / Bridge							
	Demolish	Ea.		.82		.82	Includes material and labor to install
	Install	Ea.	1.73	1.83		3.56	set of cross bridging or per block of
	Demolish and Install	Ea.	1.73	2.65		4.38	solid bridging for 2" x 10" joists cut to
	Minimum Charge	Job		163		163	size on site.

24" O.C.

		Unit	Material	Labor	Equip.	Total	Specification
2" x 6" Joists							
	Demolish	S.F.		.27		.27	Cost includes material and labor to
	Install	S.F.	.37	.13		.50	install 2" x 6" joists including box or
	Demolish and Install	S.F.	.37	.40		.77	band joist installed 24" O.C. Does not
	Reinstall	S.F.		.10		.10	include beams, blocking or bridging.
	Minimum Charge	Job		163		163	
2" x 8" Joists							
	Demolish	S.F.		.28		.28	Cost includes material and labor to
	Install	S.F.	.58	.15		.73	install 2" x 8" joists including box or
	Demolish and Install	S.F.	.58	.43		1.01	band joist installed 24" O.C. Does not
	Reinstall	S.F.		.12		.12	include beams, blocking or bridging.
	Minimum Charge	Job		163		163	
2" x 10" Joists							
	Demolish	S.F.		.29		.29	Cost includes material and labor to
	Install	S.F.	.83	.18		1.01	install 2" x 10" joists including box or
	Demolish and Install	S.F.	.83	.47		1.30	band joist installed 24" O.C. Does not
	Reinstall	S.F.		.14		.14	include beams, blocking or bridging.
	Minimum Charge	Job		163		163	

Rough Frame / Structure

Floor Framing System

	Unit	Material	Labor	Equip.	Total	Specification
2" x 12" Joists						
Demolish	S.F.		.30		.30	Cost includes material and labor to
Install	S.F.	1.12	.19		1.31	install 2" x 12" joists including box or
Demolish and Install	S.F.	1.12	.49		1.61	band joist installed 24" O.C. Does not
Reinstall	S.F.		.15		.15	include beams, blocking or bridging.
Minimum Charge	Job		163		163	
Block / Bridge						
Demolish	Ea.		.82		.82	Includes material and labor to install
Install	Ea.	1.73	1.83		3.56	set of cross bridging or per block of
Demolish and Install	Ea.	1.73	2.65		4.38	solid bridging for 2" x 10" joists cut to
Minimum Charge	Job		163		163	size on site.
Engineered Lumber, Joist						
9-1/2"						
Demolish	S.F.		.43		.43	Includes material and labor to install
Install	S.F.	1.45	.33		1.78	engineered lumber truss / joists per
Demolish and Install	S.F.	1.45	.76		2.21	S.F. of floor area based on joists 16"
Reinstall	S.F.		.26		.26	O.C. Beams, supports and bridging
Minimum Charge	Job		163		163	are not included.
11-7/8"						
Demolish	S.F.		.45		.45	Includes material and labor to install
Install	S.F.	1.54	.33		1.87	engineered lumber truss / joists per
Demolish and Install	S.F.	1.54	.78		2.32	S.F. of floor area based on joists 16"
Reinstall	S.F.		.27		.27	O.C. Beams, supports and bridging
Minimum Charge	Job		163		163	are not included.
14"						
Demolish	S.F.		.46		.46	Includes material and labor to install
Install	S.F.	1.77	.36		2.13	engineered lumber truss / joists per
Demolish and Install	S.F.	1.77	.82		2.59	S.F. of floor area based on joists 16"
Reinstall	S.F.		.29		.29	O.C. Beams, supports and bridging
Minimum Charge	Job		163		163	are not included.
16"						
Demolish	S.F.		.48		.48	Includes material and labor to install
Install	S.F.	2.52	.38		2.90	engineered lumber truss / joists per
Demolish and Install	S.F.	2.52	.86		3.38	S.F. of floor area based on joists 16"
Reinstall	S.F.		.30		.30	O.C. Beams, supports and bridging
Minimum Charge	Job		163		163	are not included.
Block / Bridge						
Demolish	Pr.		.82		.82	Includes material and labor to install
Install	Pr.	1.08	2.41		3.49	set of steel, one-nail type cross
Demolish and Install	Pr.	1.08	3.23		4.31	bridging for trusses placed 16" O.C.
Minimum Charge	Job		163		163	
Block / Bridge						
2" x 6"						
Demolish	Ea.		.82		.82	Includes material and labor to install
Install	Ea.	.79	1.47		2.26	set of cross bridging or per block of
Demolish and Install	Ea.	.79	2.29		3.08	solid bridging for 2" x 6" joists cut to
Minimum Charge	Job		163		163	size on site.
2" x 8"						
Demolish	Ea.		.82		.82	Includes material and labor to install
Install	Ea.	1.22	1.63		2.85	set of cross bridging or per block of
Demolish and Install	Ea.	1.22	2.45		3.67	solid bridging for 2" x 8" joists cut to
Minimum Charge	Job		163		163	size on site.

Floor Framing System		Unit	Material	Labor	Equip.	Total	Specification
2″ x 10″							
	Demolish	Ea.		.82		.82	Includes material and labor to install
	Install	Ea.	1.73	1.83		3.56	set of cross bridging or per block of
	Demolish and Install	Ea.	1.73	2.65		4.38	solid bridging for 2″ x 10″ joists cut to
	Minimum Charge	Job		163		163	size on site.
2″ x 12″							
	Demolish	Ea.		.82		.82	Includes material and labor to install
	Install	Ea.	2.35	2.16		4.51	set of cross bridging or per block of
	Demolish and Install	Ea.	2.35	2.98		5.33	solid bridging for 2″ x 12″ joists cut to
	Minimum Charge	Job		163		163	size on site.
Ledger Strips							
1″ x 2″							
	Demolish	L.F.		.22		.22	Cost includes material and labor to
	Install	L.F.	.21	.95		1.16	install 1″ x 2″ ledger strip nailed to the
	Demolish and Install	L.F.	.21	1.17		1.38	face of studs, beams or joist.
	Minimum Charge	Job		163		163	
1″ x 3″							
	Demolish	L.F.		.22		.22	Includes material and labor to install
	Install	L.F.	.66	1.09		1.75	up to 1″ x 4″ ledger strip nailed to the
	Demolish and Install	L.F.	.66	1.31		1.97	face of studs, beams or joist.
	Minimum Charge	Job		163		163	
1″ x 4″							
	Demolish	L.F.		.22		.22	Includes material and labor to install
	Install	L.F.	.66	1.09		1.75	up to 1″ x 4″ ledger strip nailed to the
	Demolish and Install	L.F.	.66	1.31		1.97	face of studs, beams or joist.
	Minimum Charge	Job		163		163	
2″ x 2″							
	Demolish	L.F.		.24		.24	Cost includes material and labor to
	Install	L.F.	.29	.99		1.28	install 2″ x 2″ ledger strip nailed to the
	Demolish and Install	L.F.	.29	1.23		1.52	face of studs, beams or joist.
	Minimum Charge	Job		163		163	
2″ x 4″							
	Demolish	L.F.		.26		.26	Cost includes material and labor to
	Install	L.F.	.37	1.30		1.67	install 2″ x 4″ ledger strip nailed to the
	Demolish and Install	L.F.	.37	1.56		1.93	face of studs, beams or joist.
	Reinstall	L.F.		1.04		1.04	
	Minimum Charge	Job		163		163	
Ledger Boards							
2″ x 6″							
	Demolish	L.F.		.26		.26	Cost includes material and labor to
	Install	L.F.	1.71	2.04		3.75	install 2″ x 6″ ledger board fastened to
	Demolish and Install	L.F.	1.71	2.30		4.01	a wall, joists or studs.
	Reinstall	L.F.		1.63		1.63	
	Minimum Charge	Job		163		163	
4″ x 6″							
	Demolish	L.F.		.44		.44	Cost includes material and labor to
	Install	L.F.	3.58	2.04		5.62	install 4″ x 6″ ledger board fastened to
	Demolish and Install	L.F.	3.58	2.48		6.06	a wall, joists or studs.
	Reinstall	L.F.		1.63		1.63	
	Minimum Charge	Job		163		163	
4″ x 8″							
	Demolish	L.F.		.58		.58	Cost includes material and labor to
	Install	L.F.	4.33	2.04		6.37	install 4″ x 8″ ledger board fastened to
	Demolish and Install	L.F.	4.33	2.62		6.95	a wall, joists or studs.
	Reinstall	L.F.		1.63		1.63	
	Minimum Charge	Job		163		163	

Rough Frame / Structure

Floor Framing System		Unit	Material	Labor	Equip.	Total	Specification
Sill Plate Per L.F.							
2" x 4"							
	Demolish	L.F.		.22		.22	Cost includes material and labor to
	Install	L.F.	.54	1.19		1.73	install 2" x 4" pressure treated lumber,
	Demolish and Install	L.F.	.54	1.41		1.95	drilled and installed on foundation
	Reinstall	L.F.		.95		.95	bolts 48" O.C. Bolts, nuts and washers
	Minimum Charge	Job		163		163	are not included.
2" x 6"							
	Demolish	L.F.		.34		.34	Cost includes material and labor to
	Install	L.F.	.85	1.30		2.15	install 2" x 6" pressure treated lumber,
	Demolish and Install	L.F.	.85	1.64		2.49	drilled and installed on foundation
	Reinstall	L.F.		1.04		1.04	bolts 48" O.C. Bolts, nuts and washers
	Minimum Charge	Job		163		163	not included.
2" x 8"							
	Demolish	L.F.		.45		.45	Cost includes material and labor to
	Install	L.F.	1.10	1.45		2.55	install 2" x 8" pressure treated lumber,
	Demolish and Install	L.F.	1.10	1.90		3	drilled and installed on foundation
	Reinstall	L.F.		1.16		1.16	bolts 48" O.C. Bolts, nuts and washers
	Minimum Charge	Job		163		163	not included.
Earthquake Strapping							
	Install	Ea.	1.80	2.04		3.84	Includes labor and material to replace
	Minimum Charge	Job		163		163	an earthquake strap.

Subflooring		Unit	Material	Labor	Equip.	Total	Specification
Plywood							
1/2"							
	Demolish	S.F.		.68		.68	Cost includes material and labor to
	Install	SF Flr.	.52	.35		.87	install 1/2" plywood subfloor, CD
	Demolish and Install	SF Flr.	.52	1.03		1.55	standard interior grade, plugged and
	Minimum Charge	Job		163		163	touch sanded.
5/8"							
	Demolish	S.F.		.69		.69	Cost includes material and labor to
	Install	SF Flr.	.65	.39		1.04	install 5/8" plywood subfloor, CD
	Demolish and Install	SF Flr.	.65	1.08		1.73	standard interior grade, plugged and
	Reinstall	SF Flr.		.31		.31	touch sanded.
	Minimum Charge	Job		163		163	
3/4"							
	Demolish	S.F.		.70		.70	Cost includes material and labor to
	Install	SF Flr.	.79	.42		1.21	install 3/4" plywood subfloor, CD
	Demolish and Install	SF Flr.	.79	1.12		1.91	standard interior grade, plugged and
	Reinstall	SF Flr.		.34		.34	touch sanded.
	Minimum Charge	Job		163		163	
Particle Board							
1/2"							
	Demolish	S.F.		.34		.34	Cost includes material and labor to
	Install	SF Flr.	.43	.43		.86	install 1/2" particle board subfloor.
	Demolish and Install	SF Flr.	.43	.77		1.20	
	Minimum Charge	Job		163		163	
5/8"							
	Demolish	S.F.		.34		.34	Cost includes material and labor to
	Install	S.F.	.54	.47		1.01	install 5/8" particle board subfloor.
	Demolish and Install	S.F.	.54	.81		1.35	
	Minimum Charge	Job		163		163	

Rough Frame / Structure

Subflooring

Subflooring		Unit	Material	Labor	Equip.	Total	Specification
3/4"							
	Demolish	S.F.		.35		.35	Cost includes material and labor to
	Install	SF Flr.	.61	.52		1.13	install 3/4" particle board subfloor.
	Demolish and Install	SF Flr.	.61	.87		1.48	
	Minimum Charge	Job		163		163	
Plank Board							
	Demolish	S.F.		1.11		1.11	Cost includes material and labor to
	Install	S.F.	1.38	.36		1.74	install 1" x 6" standard grade plank
	Demolish and Install	S.F.	1.38	1.47		2.85	flooring.
	Reinstall	S.F.		.36		.36	
	Minimum Charge	Job		163		163	
Felt Underlay							
	Demolish	S.F.		.06		.06	Includes material and labor to install
	Install	S.F.	.03	.09		.12	#15 felt building paper.
	Demolish and Install	S.F.	.03	.15		.18	
	Minimum Charge	Job		163		163	
Prep (for flooring)							
	Install	S.F.		.28		.28	Preparation of subflooring for
	Minimum Charge	Job		147		147	installation of new finished flooring. Cost reflects average time to prep area.

Wall Framing System

Wall Framing System		Unit	Material	Labor	Equip.	Total	Specification
2" x 4"							
8' High							
	Demolish	L.F.		3.49		3.49	Cost includes material and labor to
	Install	L.F.	3.89	3.84		7.73	install 2" x 4" x 8' high wall system,
	Demolish and Install	L.F.	3.89	7.33		11.22	including studs, treated bottom plate,
	Reinstall	L.F.		3.07		3.07	double top plate and one row of fire
	Minimum Charge	Job		163		163	blocking.
9' High							
	Demolish	L.F.		3.91		3.91	Cost includes material and labor to
	Install	L.F.	4.17	3.84		8.01	install 2" x 4" x 9' wall system
	Demolish and Install	L.F.	4.17	7.75		11.92	including studs, treated bottom plate,
	Reinstall	L.F.		3.07		3.07	double top plate and one row of fire
	Minimum Charge	Job		163		163	blocking.
10' High							
	Demolish	L.F.		4.36		4.36	Cost includes material and labor to
	Install	L.F.	4.44	3.84		8.28	install 2" x 4" x 10' wall system
	Demolish and Install	L.F.	4.44	8.20		12.64	including studs, treated bottom plate,
	Reinstall	L.F.		3.07		3.07	double top plate and one row of fire
	Minimum Charge	Job		163		163	blocking.
12' High							
	Demolish	L.F.		5.25		5.25	Cost includes material and labor to
	Install	L.F.	5	4.66		9.66	install 2" x 4" x 12' wall system
	Demolish and Install	L.F.	5	9.91		14.91	including studs, treated bottom plate,
	Reinstall	L.F.		3.73		3.73	double top plate and one row of fire
	Minimum Charge	Job		163		163	blocking.

Rough Frame / Structure

Wall Framing System	Unit	Material	Labor	Equip.	Total	Specification
2″ x 6″						
8′ High						
Demolish	L.F.		3.27		3.27	Cost includes material and labor to
Install	L.F.	6.15	7.05		13.20	install 2″ x 6″ x 8′ wall system
Demolish and Install	L.F.	6.15	10.32		16.47	including studs, treated bottom plate,
Reinstall	L.F.		5.64		5.64	double top plate and one row of fire
Minimum Charge	Job		163		163	blocking.
9′ High						
Demolish	L.F.		3.69		3.69	Cost includes material and labor to
Install	L.F.	6.60	7.05		13.65	install 2″ x 6″ x 9′ wall system
Demolish and Install	L.F.	6.60	10.74		17.34	including studs, treated bottom plate,
Reinstall	L.F.		5.64		5.64	double top plate and one row of fire
Minimum Charge	Job		163		163	blocking.
10′ High						
Demolish	L.F.		4.03		4.03	Cost includes material and labor to
Install	L.F.	7.05	7.05		14.10	install 2″ x 6″ x 10′ wall system
Demolish and Install	L.F.	7.05	11.08		18.13	including studs, treated bottom plate,
Reinstall	L.F.		5.64		5.64	double top plate and one row of fire
Minimum Charge	Job		163		163	blocking.
12′ High						
Demolish	L.F.		4.76		4.76	Cost includes material and labor to
Install	L.F.	7.90	8.70		16.60	install 2″ x 6″ x 12′ wall system
Demolish and Install	L.F.	7.90	13.46		21.36	including studs, treated bottom plate,
Reinstall	L.F.		6.94		6.94	double top plate and one row of fire
Minimum Charge	Job		163		163	blocking.
Fireblock						
2″ x 4″, 16″ O.C. System						
Demolish	L.F.		.44		.44	Cost includes material and labor to
Install	L.F.	.37	1.09		1.46	install 2″ x 4″ fireblocks in wood frame
Demolish and Install	L.F.	.37	1.53		1.90	walls per L.F. of wall to be blocked.
Reinstall	L.F.		.87		.87	
Minimum Charge	Job		163		163	
2″ x 4″, 24″ O.C. System						
Demolish	L.F.		.44		.44	Cost includes material and labor to
Install	L.F.	.37	1.09		1.46	install 2″ x 4″ fireblocks in wood frame
Demolish and Install	L.F.	.37	1.53		1.90	walls per L.F. of wall to be blocked.
Reinstall	L.F.		.87		.87	
Minimum Charge	Job		163		163	
2″ x 6″, 16″ O.C. System						
Demolish	L.F.		.65		.65	Cost includes material and labor to
Install	L.F.	.59	1.09		1.68	install 2″ x 6″ fireblocks in wood frame
Demolish and Install	L.F.	.59	1.74		2.33	walls per L.F. of wall to be blocked.
Reinstall	L.F.		.87		.87	
Minimum Charge	Job		163		163	
2″ x 6″, 24″ O.C. System						
Demolish	L.F.		.65		.65	Cost includes material and labor to
Install	L.F.	.59	1.09		1.68	install 2″ x 6″ fireblocks in wood frame
Demolish and Install	L.F.	.59	1.74		2.33	walls per L.F. of wall to be blocked.
Reinstall	L.F.		.87		.87	
Minimum Charge	Job		163		163	
Bracing						
1″ x 3″						
Demolish	L.F.		.25		.25	Cost includes material and labor to
Install	L.F.	.31	2.04		2.35	install 1″ x 3″ let-in bracing.
Demolish and Install	L.F.	.31	2.29		2.60	
Minimum Charge	Job		163		163	

Wall Framing System		Unit	Material	Labor	Equip.	Total	Specification
1" x 4"							
	Demolish	L.F.		.25		.25	Cost includes material and labor to
	Install	L.F.	.45	2.04		2.49	install 1" x 4" let-in bracing.
	Demolish and Install	L.F.	.45	2.29		2.74	
	Minimum Charge	Job		163		163	
1" x 6"							
	Demolish	L.F.		.25		.25	Cost includes material and labor to
	Install	L.F.	.64	2.17		2.81	install 1" x 6" let-in bracing.
	Demolish and Install	L.F.	.64	2.42		3.06	
	Minimum Charge	Job		163		163	
2" x 3"							
	Demolish	L.F.		.33		.33	Cost includes material and labor to
	Install	L.F.	.29	2.17		2.46	install 2" x 3" let-in bracing.
	Demolish and Install	L.F.	.29	2.50		2.79	
	Reinstall	L.F.		1.74		1.74	
	Minimum Charge	Job		163		163	
2" x 4"							
	Demolish	L.F.		.33		.33	Cost includes material and labor to
	Install	L.F.	.37	2.17		2.54	install 2" x 4" let-in bracing.
	Demolish and Install	L.F.	.37	2.50		2.87	
	Reinstall	L.F.		1.74		1.74	
	Minimum Charge	Job		163		163	
2" x 6"							
	Demolish	L.F.		.33		.33	Cost includes material and labor to
	Install	L.F.	.59	2.33		2.92	install 2" x 6" let-in bracing.
	Demolish and Install	L.F.	.59	2.66		3.25	
	Reinstall	L.F.		1.86		1.86	
	Minimum Charge	Job		163		163	
2" x 8"							
	Demolish	L.F.		.33		.33	Cost includes material and labor to
	Install	L.F.	.91	2.33		3.24	install 2" x 8" let-in bracing.
	Demolish and Install	L.F.	.91	2.66		3.57	
	Reinstall	L.F.		1.86		1.86	
	Minimum Charge	Job		163		163	
Earthquake Strapping							
	Install	Ea.	1.80	2.04		3.84	Includes labor and material to replace
	Minimum Charge	Job		163		163	an earthquake strap.
Hurricane Clips							
	Install	Ea.	1.58	2.25		3.83	Includes labor and material to install a
	Minimum Charge	Job		163		163	hurricane clip.
Stud							
2" x 4"							
	Demolish	L.F.		.26		.26	Cost includes material and labor to
	Install	L.F.	.37	.59		.96	install 2" x 4" wall stud per L.F. of stud.
	Demolish and Install	L.F.	.37	.85		1.22	
	Reinstall	L.F.		.47		.47	
	Minimum Charge	Job		163		163	
2" x 6"							
	Demolish	L.F.		.33		.33	Cost includes material and labor to
	Install	L.F.	.59	.65		1.24	install 2" x 6" stud per L.F. of stud.
	Demolish and Install	L.F.	.59	.98		1.57	
	Reinstall	L.F.		.52		.52	
	Minimum Charge	Job		163		163	

Rough Frame / Structure

Wall Framing System		Unit	Material	Labor	Equip.	Total	Specification
Plates							
2″ x 4″							
	Demolish	L.F.		.24		.24	Cost includes material and labor to
	Install	L.F.	.37	.82		1.19	install 2″ x 4″ plate per L.F. of plate.
	Demolish and Install	L.F.	.37	1.06		1.43	
	Reinstall	L.F.		.65		.65	
	Minimum Charge	Job		163		163	
2″ x 6″							
	Demolish	L.F.		.25		.25	Cost includes material and labor to
	Install	L.F.	.59	.87		1.46	install 2″ x 6″ plate per L.F. of plate.
	Demolish and Install	L.F.	.59	1.12		1.71	
	Reinstall	L.F.		.70		.70	
	Minimum Charge	Job		163		163	
Headers							
	Demolish	L.F.		.23		.23	Includes material and labor to install
	Install	L.F.	.59	1.81		2.40	header over wall openings and
	Demolish and Install	L.F.	.59	2.04		2.63	around floor, ceiling and roof
	Clean	L.F.	.01	.30		.31	openings or flush beam.
	Paint	L.F.	.09	1.14		1.23	
	Minimum Charge	Job		163		163	
2″ x 8″							
	Demolish	L.F.		.25		.25	Includes material and labor to install
	Install	L.F.	.91	1.92		2.83	header over wall openings and
	Demolish and Install	L.F.	.91	2.17		3.08	around floor, ceiling and roof
	Clean	L.F.	.02	.30		.32	openings or flush beam.
	Paint	L.F.	.12	1.55		1.67	
	Minimum Charge	Job		163		163	
2″ x 10″							
	Demolish	L.F.		.25		.25	Includes material and labor to install
	Install	L.F.	1.30	2.04		3.34	header over wall openings and
	Demolish and Install	L.F.	1.30	2.29		3.59	around floor, ceiling and roof
	Clean	L.F.	.02	.30		.32	openings or flush beam.
	Paint	L.F.	.15	1.91		2.06	
	Minimum Charge	Job		163		163	
2″ x 12″							
	Demolish	L.F.		.26		.26	Includes material and labor to install
	Install	L.F.	1.76	2.17		3.93	header over wall openings and
	Demolish and Install	L.F.	1.76	2.43		4.19	around floor, ceiling and roof
	Clean	L.F.	.03	.30		.33	openings or flush beam.
	Paint	L.F.	.18	2.29		2.47	
	Minimum Charge	Job		163		163	
4″ x 8″							
	Demolish	L.F.		.47		.47	Includes material and labor to install
	Install	L.F.	2.96	2.51		5.47	header over wall openings and
	Demolish and Install	L.F.	2.96	2.98		5.94	around floor, ceiling and roof
	Clean	L.F.	.02	.30		.32	openings or flush beam.
	Paint	L.F.	.12	1.55		1.67	
	Minimum Charge	Job		163		163	
4″ x 10″							
	Demolish	L.F.		.50		.50	Includes material and labor to install
	Install	L.F.	3.71	2.72		6.43	header over wall openings and
	Demolish and Install	L.F.	3.71	3.22		6.93	around floor, ceiling and roof
	Clean	L.F.	.02	.30		.32	openings or flush beam.
	Paint	L.F.	.15	1.91		2.06	
	Minimum Charge	Job		163		163	

Rough Frame / Structure

Wall Framing System

	Unit	Material	Labor	Equip.	Total	Specification
4″ x 12″						
Demolish	L.F.		.52		.52	Includes material and labor to install
Install	L.F.	4.84	3.43		8.27	header over wall openings and
Demolish and Install	L.F.	4.84	3.95		8.79	around floor, ceiling and roof
Clean	L.F.	.03	.30		.33	openings or flush beam.
Paint	L.F.	.18	2.29		2.47	
Minimum Charge	Job		163		163	
6″ x 8″						
Demolish	L.F.		.47		.47	Includes material and labor to install
Install	L.F.	6.45	1.81		8.26	header over wall openings and
Demolish and Install	L.F.	6.45	2.28		8.73	around floor, ceiling and roof
Clean	L.F.	.02	.30		.32	openings or flush beam.
Paint	L.F.	.12	1.55		1.67	
Minimum Charge	Job		163		163	
6″ x 10″						
Demolish	L.F.		.50		.50	Includes material and labor to install
Install	L.F.	10.45	3.95		14.40	header over wall openings and
Demolish and Install	L.F.	10.45	4.45		14.90	around floor, ceiling and roof
Clean	L.F.	.02	.30		.32	openings or flush beam.
Paint	L.F.	.15	1.91		2.06	
Minimum Charge	Job		163		163	
6″ x 12″						
Demolish	L.F.		.52		.52	Includes material and labor to install
Install	L.F.	10	4.66		14.66	header over wall openings and
Demolish and Install	L.F.	10	5.18		15.18	around floor, ceiling and roof
Clean	L.F.	.03	.30		.33	openings or flush beam.
Paint	L.F.	.18	.79		.97	
Minimum Charge	Job		163		163	

Rough-in Opening

	Unit	Material	Labor	Equip.	Total	Specification
Door w / 2″ x 4″ Lumber						
3′ Wide						
Demolish	Ea.		8.20		8.20	Includes material and labor to install
Install	Ea.	15.80	10.20		26	header, double studs each side,
Demolish and Install	Ea.	15.80	18.40		34.20	cripples, blocking and nails, up to 3′
Reinstall	Ea.		8.15		8.15	opening in 2″ x 4″ stud wall 8′ high.
Minimum Charge	Job		163		163	
4′ Wide						
Demolish	Ea.		8.20		8.20	Includes material and labor to install
Install	Ea.	16.95	10.20		27.15	header, double studs each side,
Demolish and Install	Ea.	16.95	18.40		35.35	cripples, blocking and nails, up to 4′
Reinstall	Ea.		8.15		8.15	opening in 2″ x 4″ stud wall 8′ high.
Minimum Charge	Job		163		163	
5′ Wide						
Demolish	Ea.		8.20		8.20	Includes material and labor to install
Install	Ea.	21.50	10.20		31.70	header, double studs each side,
Demolish and Install	Ea.	21.50	18.40		39.90	cripples, blocking and nails, up to 5′
Reinstall	Ea.		8.15		8.15	opening in 2″ x 4″ stud wall 8′ high.
Minimum Charge	Job		163		163	
6′ Wide						
Demolish	Ea.		8.20		8.20	Includes material and labor to install
Install	Ea.	23	10.20		33.20	header, double studs each side,
Demolish and Install	Ea.	23	18.40		41.40	cripples, blocking and nails, up to 6′
Reinstall	Ea.		8.15		8.15	opening in 2″ x 4″ stud wall 8′ high.
Minimum Charge	Job		163		163	

Rough Frame / Structure

Rough-in Opening

	Unit	Material	Labor	Equip.	Total	Specification
8' Wide						
Demolish	Ea.		8.75		8.75	Includes material and labor to install
Install	Ea.	33	10.85		43.85	header, double studs each side,
Demolish and Install	Ea.	33	19.60		52.60	cripples, blocking and nails, up to 8'
Reinstall	Ea.		8.69		8.69	opening in 2" x 4" stud wall 8' high.
Minimum Charge	Job		163		163	
10' Wide						
Demolish	Ea.		8.75		8.75	Includes material and labor to install
Install	Ea.	48	10.85		58.85	header, double studs each side,
Demolish and Install	Ea.	48	19.60		67.60	cripples, blocking and nails, up to 10'
Reinstall	Ea.		8.69		8.69	opening in 2" x 4" stud wall 8' high.
Minimum Charge	Job		163		163	
12' Wide						
Demolish	Ea.		8.75		8.75	Includes material and labor to install
Install	Ea.	70	10.85		80.85	header, double studs each side,
Demolish and Install	Ea.	70	19.60		89.60	cripples, blocking and nails, up to 12'
Reinstall	Ea.		8.69		8.69	opening in 2" x 4" stud wall 8' high.
Minimum Charge	Job		163		163	

Door w / 2" x 6" Lumber

	Unit	Material	Labor	Equip.	Total	Specification
3' Wide						
Demolish	Ea.		8.20		8.20	Includes material and labor to install
Install	Ea.	22.50	10.20		32.70	header, double studs each side,
Demolish and Install	Ea.	22.50	18.40		40.90	cripples, blocking and nails, up to 3'
Reinstall	Ea.		8.15		8.15	opening in 2" x 6" stud wall 8' high.
Minimum Charge	Job		163		163	
4' Wide						
Demolish	Ea.		8.20		8.20	Includes material and labor to install
Install	Ea.	24	10.20		34.20	header, double studs each side,
Demolish and Install	Ea.	24	18.40		42.40	cripples, blocking and nails, up to 4'
Reinstall	Ea.		8.15		8.15	opening in 2" x 6" stud wall 8' high.
Minimum Charge	Job		163		163	
5' Wide						
Demolish	Ea.		8.20		8.20	Includes material and labor to install
Install	Ea.	28.50	10.20		38.70	header, double studs each side,
Demolish and Install	Ea.	28.50	18.40		46.90	cripples, blocking and nails up to 5'
Reinstall	Ea.		8.15		8.15	opening in 2" x 6" stud wall 8' high.
Minimum Charge	Job		163		163	
6' Wide						
Demolish	Ea.		8.20		8.20	Includes material and labor to install
Install	Ea.	30.50	10.20		40.70	header, double studs each side,
Demolish and Install	Ea.	30.50	18.40		48.90	cripples, blocking and nails up to 6'
Reinstall	Ea.		8.15		8.15	opening in 2" x 6" stud wall 8' high.
Minimum Charge	Job		163		163	
8' Wide						
Demolish	Ea.		8.75		8.75	Includes material and labor to install
Install	Ea.	40	10.85		50.85	header, double studs each side,
Demolish and Install	Ea.	40	19.60		59.60	cripples, blocking and nails up to 8'
Reinstall	Ea.		8.69		8.69	opening in 2" x 6" stud wall 8' high.
Minimum Charge	Job		163		163	
10' Wide						
Demolish	Ea.		8.75		8.75	Includes material and labor to install
Install	Ea.	55	10.85		65.85	header, double studs each side,
Demolish and Install	Ea.	55	19.60		74.60	cripples, blocking and nails up to 10'
Reinstall	Ea.		8.69		8.69	opening in 2" x 6" stud wall 8' high.
Minimum Charge	Job		163		163	
12' Wide						
Demolish	Ea.		8.75		8.75	Includes material and labor to install
Install	Ea.	77	10.85		87.85	header, double studs each side,
Demolish and Install	Ea.	77	19.60		96.60	cripples, blocking and nails, up to 12'
Reinstall	Ea.		8.69		8.69	opening in 2" x 6" stud wall 8' high.
Minimum Charge	Job		163		163	

Rough-in Opening

	Unit	Material	Labor	Equip.	Total	Specification
Window w / 2″ x 4″ Lumber						
2′ Wide						
Demolish	Ea.		10.90		10.90	Includes material and labor to install
Install	Ea.	16.55	13.60		30.15	header, double studs each side of
Demolish and Install	Ea.	16.55	24.50		41.05	opening, cripples, blocking, nails and
Reinstall	Ea.		10.87		10.87	sub-sills for opening up to 2′ in a 2″ x
Minimum Charge	Job		163		163	4″ stud wall 8′ high.
3′ Wide						
Demolish	Ea.		10.90		10.90	Includes material and labor to install
Install	Ea.	19.45	13.60		33.05	header, double studs each side of
Demolish and Install	Ea.	19.45	24.50		43.95	opening, cripples, blocking, nails and
Reinstall	Ea.		10.87		10.87	sub-sills for opening up to 3′ in a 2″ x
Minimum Charge	Job		163		163	4″ stud wall 8′ high.
4′ Wide						
Demolish	Ea.		10.90		10.90	Includes material and labor to install
Install	Ea.	21.50	13.60		35.10	header, double studs each side of
Demolish and Install	Ea.	21.50	24.50		46	opening, cripples, blocking, nails and
Reinstall	Ea.		10.87		10.87	sub-sills for opening up to 4′ in a 2″ x
Minimum Charge	Job		163		163	4″ stud wall 8′ high.
5′ Wide						
Demolish	Ea.		10.90		10.90	Includes material and labor to install
Install	Ea.	26	13.60		39.60	header, double studs each side of
Demolish and Install	Ea.	26	24.50		50.50	opening, cripples, blocking, nails and
Reinstall	Ea.		10.87		10.87	sub-sills for opening up to 5′ in a 2″ x
Minimum Charge	Job		163		163	4″ stud wall 8′ high.
6′ Wide						
Demolish	Ea.		10.90		10.90	Includes material and labor to install
Install	Ea.	29	13.60		42.60	header, double studs each side of
Demolish and Install	Ea.	29	24.50		53.50	opening, cripples, blocking, nails and
Reinstall	Ea.		10.87		10.87	sub-sills for opening up to 6′ in a 2″ x
Minimum Charge	Job		163		163	4″ stud wall 8′ high.
7′ Wide						
Demolish	Ea.		10.90		10.90	Includes material and labor to install
Install	Ea.	37.50	13.60		51.10	header, double studs each side of
Demolish and Install	Ea.	37.50	24.50		62	opening, cripples, blocking, nails and
Reinstall	Ea.		10.87		10.87	sub-sills for opening up to 7′ in a 2″ x
Minimum Charge	Job		163		163	4″ stud wall 8′ high.
8′ Wide						
Demolish	Ea.		11.90		11.90	Includes material and labor to install
Install	Ea.	41.50	14.80		56.30	header, double studs each side of
Demolish and Install	Ea.	41.50	26.70		68.20	opening, cripples, blocking, nails and
Reinstall	Ea.		11.85		11.85	sub-sills for opening up to 8′ in a 2″ x
Minimum Charge	Job		163		163	4″ stud wall 8′ high.
10′ Wide						
Demolish	Ea.		11.90		11.90	Includes material and labor to install
Install	Ea.	57	14.80		71.80	header, double studs each side of
Demolish and Install	Ea.	57	26.70		83.70	opening, cripples, blocking, nails and
Reinstall	Ea.		11.85		11.85	sub-sills for opening up to 10′ in a 2″
Minimum Charge	Job		163		163	x 4″ stud wall 8′ high.

Rough-in Opening	Unit	Material	Labor	Equip.	Total	Specification
12' Wide						
Demolish	Ea.		11.90		11.90	Includes material and labor to install
Install	Ea.	81.50	14.80		96.30	header, double studs each side of
Demolish and Install	Ea.	81.50	26.70		108.20	opening, cripples, blocking, nails and
Reinstall	Ea.		11.85		11.85	sub-sills for opening up to 12' in a 2"
Minimum Charge	Job		163		163	x 4" stud wall 8' high.
Window w / 2" x 6" Lumber						
2' Wide						
Demolish	Ea.		10.90		10.90	Includes material and labor to install
Install	Ea.	25.50	13.60		39.10	header, double studs each side of
Demolish and Install	Ea.	25.50	24.50		50	opening, cripples, blocking, nails and
Reinstall	Ea.		10.87		10.87	sub-sills for opening up to 2' in a 2" x
Minimum Charge	Job		163		163	6" stud wall 8' high.
3' Wide						
Demolish	Ea.		10.90		10.90	Includes material and labor to install
Install	Ea.	28.50	13.60		42.10	header, double studs each side of
Demolish and Install	Ea.	28.50	24.50		53	opening, cripples, blocking, nails and
Reinstall	Ea.		10.87		10.87	sub-sills for opening up to 3' in a 2" x
Minimum Charge	Job		163		163	6" stud wall 8' high.
4' Wide						
Demolish	Ea.		10.90		10.90	Includes material and labor to install
Install	Ea.	31	13.60		44.60	header, double studs each side of
Demolish and Install	Ea.	31	24.50		55.50	opening, cripples, blocking, nails and
Reinstall	Ea.		10.87		10.87	sub-sills for opening up to 4' in a 2" x
Minimum Charge	Job		163		163	6" stud wall 8' high.
5' Wide						
Demolish	Ea.		10.90		10.90	Includes material and labor to install
Install	Ea.	36	13.60		49.60	header, double studs each side of
Demolish and Install	Ea.	36	24.50		60.50	opening, cripples, blocking, nails and
Reinstall	Ea.		10.87		10.87	sub-sills for opening up to 5' in a 2" x
Minimum Charge	Job		163		163	6" stud wall 8' high.
6' Wide						
Demolish	Ea.		10.90		10.90	Includes material and labor to install
Install	Ea.	39	13.60		52.60	header, double studs each side of
Demolish and Install	Ea.	39	24.50		63.50	opening, cripples, blocking, nails and
Reinstall	Ea.		10.87		10.87	sub-sills for opening up to 6' in a 2" x
Minimum Charge	Job		163		163	6" stud wall 8' high.
7' Wide						
Demolish	Ea.		10.90		10.90	Includes material and labor to install
Install	Ea.	48.50	13.60		62.10	header, double studs each side of
Demolish and Install	Ea.	48.50	24.50		73	opening, cripples, blocking, nails and
Reinstall	Ea.		10.87		10.87	sub-sills for opening up to 7' in a 2" x
Minimum Charge	Job		163		163	6" stud wall 8' high.
8' Wide						
Demolish	Ea.		11.90		11.90	Includes material and labor to install
Install	Ea.	53	14.80		67.80	header, double studs each side of
Demolish and Install	Ea.	53	26.70		79.70	opening, cripples, blocking, nails and
Reinstall	Ea.		11.85		11.85	sub-sills for opening up to 8' in a 2" x
Minimum Charge	Job		163		163	6" stud wall 8' high.

Rough Frame / Structure

Rough-in Opening		Unit	Material	Labor	Equip.	Total	Specification
10' Wide							
	Demolish	Ea.		11.90		11.90	Includes material and labor to install
	Install	Ea.	69.50	14.80		84.30	header, double studs each side of
	Demolish and Install	Ea.	69.50	26.70		96.20	opening, cripples, blocking, nails and
	Reinstall	Ea.		11.85		11.85	sub-sills for opening up to 10' in a 2"
	Minimum Charge	Job		163		163	x 6" stud wall 8' high.
12' Wide							
	Demolish	Ea.		11.90		11.90	Includes material and labor to install
	Install	Ea.	94.50	14.80		109.30	header, double studs each side of
	Demolish and Install	Ea.	94.50	26.70		121.20	opening, cripples, blocking, nails and
	Reinstall	Ea.		11.85		11.85	sub-sills for opening up to 12' in a 2"
	Minimum Charge	Job		163		163	x 6" stud wall 8' high.

Glue-Laminated Beams		Unit	Material	Labor	Equip.	Total	Specification
3-1/2" x 6"							
	Demolish	L.F.		1.97		1.97	Includes material and labor to install
	Install	L.F.	4.65	2.42	1.10	8.17	glue-laminated wood beam.
	Demolish and Install	L.F.	4.65	4.39	1.10	10.14	
	Reinstall	L.F.		1.94	.88	2.82	
	Clean	L.F.	.10	.29		.39	
	Paint	L.F.	.29	.84		1.13	
	Minimum Charge	Job		163		163	
3-1/2" x 9"							
	Demolish	L.F.		1.97		1.97	Includes material and labor to install
	Install	L.F.	6.20	2.42	1.10	9.72	glue-laminated wood beam.
	Demolish and Install	L.F.	6.20	4.39	1.10	11.69	
	Reinstall	L.F.		1.94	.88	2.82	
	Clean	L.F.	.13	.30		.43	
	Paint	L.F.	.36	1.09		1.45	
	Minimum Charge	Job		163		163	
3-1/2" x 12"							
	Demolish	L.F.		2.97		2.97	Includes material and labor to install
	Install	L.F.	8.30	2.42	1.10	11.82	glue-laminated wood beam.
	Demolish and Install	L.F.	8.30	5.39	1.10	14.79	
	Reinstall	L.F.		1.94	.88	2.82	
	Clean	L.F.	.15	.31		.46	
	Paint	L.F.	.45	1.36		1.81	
	Minimum Charge	Job		163		163	
3-1/2" x 15"							
	Demolish	L.F.		2.97		2.97	Includes material and labor to install
	Install	L.F.	10.40	2.51	1.14	14.05	glue-laminated wood beam.
	Demolish and Install	L.F.	10.40	5.48	1.14	17.02	
	Reinstall	L.F.		2.01	.91	2.92	
	Clean	L.F.	.19	.32		.51	
	Paint	L.F.	.53	1.59		2.12	
	Minimum Charge	Job		163		163	
3-1/2" x 18"							
	Demolish	L.F.		3.71		3.71	Includes material and labor to install
	Install	L.F.	13.95	2.51	1.14	17.60	glue-laminated wood beam.
	Demolish and Install	L.F.	13.95	6.22	1.14	21.31	
	Reinstall	L.F.		2.01	.91	2.92	
	Clean	L.F.	.21	.34		.55	
	Paint	L.F.	.62	1.85		2.47	
	Minimum Charge	Job		163		163	

Rough Frame / Structure

Glue-Laminated Beams		Unit	Material	Labor	Equip.	Total	Specification
5-1/8″ x 6″							
	Demolish	L.F.		2.97		2.97	Includes material and labor to install
	Install	L.F.	6.80	2.42	1.10	10.32	glue-laminated wood beam.
	Demolish and Install	L.F.	6.80	5.39	1.10	13.29	
	Reinstall	L.F.		1.94	.88	2.82	
	Clean	L.F.	.11	.29		.40	
	Paint	L.F.	.30	.92		1.22	
	Minimum Charge	Job		163		163	
5-1/8″ x 9″							
	Demolish	L.F.		3.71		3.71	Includes material and labor to install
	Install	L.F.	10.25	2.42	1.10	13.77	glue-laminated wood beam.
	Demolish and Install	L.F.	10.25	6.13	1.10	17.48	
	Reinstall	L.F.		1.94	.88	2.82	
	Clean	L.F.	.13	.30		.43	
	Paint	L.F.	.40	1.18		1.58	
	Minimum Charge	Job		163		163	
5-1/8″ x 12″							
	Demolish	L.F.		4.41		4.41	Includes material and labor to install
	Install	L.F.	13.65	2.42	1.10	17.17	glue-laminated wood beam.
	Demolish and Install	L.F.	13.65	6.83	1.10	21.58	
	Reinstall	L.F.		1.94	.88	2.82	
	Clean	L.F.	.17	.31		.48	
	Paint	L.F.	.48	1.43		1.91	
	Minimum Charge	Job		163		163	
5-1/8″ x 18″							
	Demolish	L.F.		7.40		7.40	Includes material and labor to install
	Install	L.F.	20.50	2.51	1.14	24.15	glue-laminated wood beam.
	Demolish and Install	L.F.	20.50	9.91	1.14	31.55	
	Reinstall	L.F.		2.01	.91	2.92	
	Clean	L.F.	.22	.34		.56	
	Paint	L.F.	.56	1.93		2.49	
	Minimum Charge	Job		163		163	
6-3/4″ x 12″							
	Demolish	L.F.		4.41		4.41	Includes material and labor to install
	Install	L.F.	17.95	2.51	1.14	21.60	glue-laminated wood beam.
	Demolish and Install	L.F.	17.95	6.92	1.14	26.01	
	Reinstall	L.F.		2.01	.91	2.92	
	Clean	L.F.	.19	.32		.51	
	Paint	L.F.	.53	1.59		2.12	
	Minimum Charge	Job		163		163	
6-3/4″ x 15″							
	Demolish	L.F.		7.40		7.40	Includes material and labor to install
	Install	L.F.	22.50	2.51	1.14	26.15	glue-laminated wood beam.
	Demolish and Install	L.F.	22.50	9.91	1.14	33.55	
	Reinstall	L.F.		2.01	.91	2.92	
	Clean	L.F.	.21	.34		.55	
	Paint	L.F.	.62	1.85		2.47	
	Minimum Charge	Job		163		163	
6-3/4″ x 18″							
	Demolish	L.F.		7.40		7.40	Includes material and labor to install
	Install	L.F.	27	2.60	1.18	30.78	glue-laminated wood beam.
	Demolish and Install	L.F.	27	10	1.18	38.18	
	Reinstall	L.F.		2.08	.95	3.03	
	Clean	L.F.	.24	.35		.59	
	Paint	L.F.	.70	2.10		2.80	
	Minimum Charge	Job		163		163	

Rough Frame / Structure

Glue-Laminated Beams

Glue-Laminated Beams		Unit	Material	Labor	Equip.	Total	Specification
Hardware							
5-1/4" Glue-Lam Seat							
	Install	Ea.	52.50	1.81		54.31	Includes labor and material to install
	Minimum Charge	Job		163		163	beam hangers.
6-3/4" Glue-Lam Seat							
	Install	Ea.	54	1.81		55.81	Includes labor and material to install
	Minimum Charge	Job		163		163	beam hangers.
8-3/4" Glue-Lam Seat							
	Install	Ea.	61.50	1.81		63.31	Includes labor and material to install
	Minimum Charge	Job		163		163	beam hangers.
Earthquake Strapping							
	Install	Ea.	1.80	2.04		3.84	Includes labor and material to replace
	Minimum Charge	Job		163		163	an earthquake strap.
Hurricane Clips							
	Install	Ea.	1.58	2.25		3.83	Includes labor and material to install a
	Minimum Charge	Job		163		163	hurricane clip.

Metal Stud Framing

Metal Stud Framing		Unit	Material	Labor	Equip.	Total	Specification
16" O.C. System							
4", 16 Ga.							
	Demolish	S.F.		.97		.97	Includes material and labor to install
	Install	S.F.	1.45	1		2.45	load bearing cold rolled metal stud
	Demolish and Install	S.F.	1.45	1.97		3.42	walls, to 10' high, including studs, top
	Reinstall	S.F.		.80		.80	and bottom track and screws.
	Minimum Charge	Job		163		163	
6", 16 Ga.							
	Demolish	S.F.		.97		.97	Includes material and labor to install
	Install	S.F.	1.83	1.02		2.85	load bearing cold rolled metal stud
	Demolish and Install	S.F.	1.83	1.99		3.82	walls, to 10' high, including studs, top
	Reinstall	S.F.		.82		.82	and bottom track and screws.
	Minimum Charge	Job		163		163	
4", 25 Ga.							
	Demolish	S.F.		.67		.67	Includes material and labor to install
	Install	S.F.	.57	.69		1.26	cold rolled metal stud walls, to 10'
	Demolish and Install	S.F.	.57	1.36		1.93	high, including studs, top and bottom
	Reinstall	S.F.		.55		.55	track and screws.
	Minimum Charge	Job		163		163	
6", 25 Ga.							
	Demolish	S.F.		.67		.67	Includes material and labor to install
	Install	S.F.	.68	.69		1.37	cold rolled metal stud walls, to 10'
	Demolish and Install	S.F.	.68	1.36		2.04	high, including studs, top and bottom
	Reinstall	S.F.		.55		.55	track and screws.
	Minimum Charge	Job		163		163	
24" O.C. System							
4", 25 Ga.							
	Demolish	S.F.		.57		.57	Includes material and labor to install
	Install	S.F.	.43	.44		.87	cold rolled metal stud walls, to 10'
	Demolish and Install	S.F.	.43	1.01		1.44	high, including studs, top and bottom
	Reinstall	S.F.		.35		.35	track and screws.
	Minimum Charge	Job		163		163	

Rough Frame / Structure

Metal Stud Framing

		Unit	Material	Labor	Equip.	Total	Specification
6″, 25 Ga.							
	Demolish	S.F.		.57		.57	Includes material and labor to install
	Install	S.F.	.51	.45		.96	cold rolled metal stud walls, to 10′
	Demolish and Install	S.F.	.51	1.02		1.53	high, including studs, top and bottom
	Reinstall	S.F.		.36		.36	track and screws.
	Minimum Charge	Job		163		163	
4″, 16 Ga.							
	Demolish	L.F.		6.50		6.50	Includes material and labor to install
	Install	S.F.	1.06	.72		1.78	load bearing cold rolled metal stud
	Demolish and Install	S.F.	1.06	7.22		8.28	walls, to 10′ high, including studs, top
	Reinstall	S.F.		.58		.58	and bottom track and screws.
	Minimum Charge	Job		163		163	
6″, 16 Ga.							
	Demolish	L.F.		6.50		6.50	Includes material and labor to install
	Install	S.F.	1.33	.74		2.07	load bearing cold rolled metal stud
	Demolish and Install	S.F.	1.33	7.24		8.57	walls, to 10′ high, including studs, top
	Reinstall	S.F.		.59		.59	and bottom track and screws.
	Minimum Charge	Job		163		163	

Metal Joist

		Unit	Material	Labor	Equip.	Total	Specification
Bar Joist							
18K9							
	Demolish	L.F.		.77	.34	1.11	Includes material, labor and
	Install	L.F.	6.80	1.76	.82	9.38	equipment to install open web joist.
	Demolish and Install	L.F.	6.80	2.53	1.16	10.49	
	Reinstall	L.F.		1.40	.65	2.05	
	Minimum Charge	Job		1450	665	2115	
16K6							
	Demolish	L.F.		.77	.34	1.11	Includes material, labor and
	Install	L.F.	5.65	1.19	.07	6.91	equipment to install open web joist.
	Demolish and Install	L.F.	5.65	1.96	.41	8.02	
	Reinstall	L.F.		.95	.05	1	
	Minimum Charge	Job		197		197	

Exterior Sheathing

		Unit	Material	Labor	Equip.	Total	Specification
CDX Plywood							
5/16″							
	Demolish	S.F.		.45		.45	Cost includes material and labor to
	Install	S.F.	.69	.41		1.10	install 5/16″ CDX plywood sheathing.
	Demolish and Install	S.F.	.69	.86		1.55	
	Minimum Charge	Job		163		163	
3/8″							
	Demolish	S.F.		.46		.46	Cost includes material and labor to
	Install	S.F.	.72	.54		1.26	install 3/8″ CDX plywood sheathing.
	Demolish and Install	S.F.	.72	1		1.72	
	Minimum Charge	Job		163		163	
1/2″							
	Demolish	S.F.		.47		.47	Cost includes material and labor to
	Install	S.F.	.68	.58		1.26	install 1/2″ CDX plywood sheathing.
	Demolish and Install	S.F.	.68	1.05		1.73	
	Minimum Charge	Job		163		163	

Rough Frame / Structure

Exterior Sheathing		Unit	Material	Labor	Equip.	Total	Specification
5/8"							
	Demolish	S.F.		.48		.48	Cost includes material and labor to
	Install	S.F.	.86	.62		1.48	install 5/8" CDX plywood sheathing.
	Demolish and Install	S.F.	.86	1.10		1.96	
	Minimum Charge	Job		163		163	
3/4"							
	Demolish	S.F.		.49		.49	Cost includes material and labor to
	Install	S.F.	1.03	.67		1.70	install 3/4" CDX plywood sheathing.
	Demolish and Install	S.F.	1.03	1.16		2.19	
	Minimum Charge	Job		163		163	
OSB							
1/2"							
	Demolish	S.F.		.47		.47	Cost includes material and labor to
	Install	S.F.	.68	.25		.93	install 4' x 8' x 1/2" OSB sheathing.
	Demolish and Install	S.F.	.68	.72		1.40	
	Minimum Charge	Job		163		163	
5/8"							
	Demolish	S.F.		.48		.48	Cost includes material and labor to
	Install	S.F.	.86	.26		1.12	install 4' x 8' x 5/8" OSB sheathing.
	Demolish and Install	S.F.	.86	.74		1.60	
	Minimum Charge	Job		163		163	
Vapor Barrier							
Black Paper							
	Install	S.F.	.03	.09		.12	Includes material and labor to install
	Minimum Charge	Job		163		163	#15 felt building paper.

Plywood Sheathing		Unit	Material	Labor	Equip.	Total	Specification
Finish Plywood							
5/16"							
	Demolish	S.F.		.45		.45	Includes material and labor to install
	Install	S.F.	.96	.20		1.16	exterior 5/16" AC plywood on walls.
	Demolish and Install	S.F.	.96	.65		1.61	
	Minimum Charge	Job		163		163	
3/8"							
	Demolish	S.F.		.46		.46	Includes material and labor to install
	Install	S.F.	.96	.27		1.23	exterior 3/8" AC plywood on walls.
	Demolish and Install	S.F.	.96	.73		1.69	
	Minimum Charge	Job		163		163	
1/2"							
	Demolish	S.F.		.47		.47	Includes material and labor to install
	Install	S.F.	1.17	.29		1.46	exterior 1/2" AC plywood on walls.
	Demolish and Install	S.F.	1.17	.76		1.93	
	Minimum Charge	Job		163		163	
5/8"							
	Demolish	S.F.		.48		.48	Includes material and labor to install
	Install	S.F.	1.34	.29		1.63	exterior 5/8" AC plywood on walls.
	Demolish and Install	S.F.	1.34	.77		2.11	
	Reinstall	S.F.		.29		.29	
	Minimum Charge	Job		163		163	
3/4"							
	Demolish	S.F.		.49		.49	Includes material and labor to install
	Install	S.F.	1.54	.29		1.83	exterior 3/4" AC plywood on walls.
	Demolish and Install	S.F.	1.54	.78		2.32	
	Reinstall	S.F.		.29		.29	
	Minimum Charge	Job		163		163	

Rough Frame / Structure

Stairs

	Unit	Material	Labor	Equip.	Total	Specification
Job-built						
Treads and Risers						Includes material and labor to install
Demolish	Ea.		32.50		32.50	three 2" x 12" stringers, treads and
Install	Ea.	455	41		496	risers of 3/4" CDX plywood, installed
Demolish and Install	Ea.	455	73.50		528.50	in a straight or "L" shaped run
Clean	Flight	2.92	20.50		23.42	including carpet.
Paint	Ea.	.09	57		57.09	
Minimum Charge	Job		163		163	
Landing						
Demolish	S.F.		.26		.26	Includes material and labor to install
Install	S.F.	49	2.26		51.26	landing framing, 3/4" CDX plywood
Demolish and Install	S.F.	49	2.52		51.52	surface and carpet.
Clean	S.F.		.29		.29	
Paint	S.F.	.26	.71		.97	
Minimum Charge	Job		163		163	

Ceiling Framing

	Unit	Material	Labor	Equip.	Total	Specification
16" O.C. System						
2" x 6" Joists						
Demolish	S.F.		.40		.40	Cost includes material and labor to
Install	S.F.	.45	.54		.99	install 2" x 6" joists including end and
Demolish and Install	S.F.	.45	.94		1.39	header joist installed 16" O.C. Does
Reinstall	S.F.		.43		.43	not include beams, ledger strips,
Minimum Charge	Job		163		163	blocking or bridging.
2" x 8" Joists						
Demolish	S.F.		.42		.42	Cost includes material and labor to
Install	S.F.	.69	.66		1.35	install 2" x 8" joists including end and
Demolish and Install	S.F.	.69	1.08		1.77	header joist installed 16" O.C. Does
Reinstall	S.F.		.53		.53	not include beams, ledger strips,
Minimum Charge	Job		163		163	blocking or bridging.
2" x 10" Joists						
Demolish	S.F.		.43		.43	Cost includes material and labor to
Install	S.F.	.98	.78		1.76	install 2" x 10" joists including end
Demolish and Install	S.F.	.98	1.21		2.19	and header joist installed 16" O.C.
Reinstall	S.F.		.62		.62	Does not include beams, ledger strips,
Minimum Charge	Job		163		163	blocking or bridging.
2" x 12" Joists						
Demolish	S.F.		.45		.45	Cost includes material and labor to
Install	S.F.	1.34	.95		2.29	install 2" x 12" joists including end
Demolish and Install	S.F.	1.34	1.40		2.74	and header joist installed 16" O.C.
Reinstall	S.F.		.76		.76	Does not include beams, ledger strips,
Minimum Charge	Job		163		163	blocking or bridging.
Block / Bridge						
Demolish	Ea.		.82		.82	Includes material and labor to install
Install	Ea.	1.73	1.83		3.56	set of cross bridging or per block of
Demolish and Install	Ea.	1.73	2.65		4.38	solid bridging for 2" x 10" joists cut to
Minimum Charge	Job		163		163	size on site.

Rough Frame / Structure

Ceiling Framing		Unit	Material	Labor	Equip.	Total	Specification
24″ O.C. System							
2″ x 6″ Joists							
	Demolish	S.F.		.27		.27	Cost includes material and labor to
	Install	S.F.	.30	.36		.66	install 2″ x 6″ joists including end and
	Demolish and Install	S.F.	.30	.63		.93	header joist installed 24″ O.C. Does
	Reinstall	S.F.		.29		.29	not include beams, ledger strips,
	Minimum Charge	Job		163		163	blocking or bridging.
2″ x 8″ Joists							
	Demolish	S.F.		.28		.28	Cost includes material and labor to
	Install	S.F.	.46	.44		.90	install 2″ x 8″ joists including end and
	Demolish and Install	S.F.	.46	.72		1.18	header joist installed 24″ O.C. Does
	Reinstall	S.F.		.35		.35	not include beams, ledger strips,
	Minimum Charge	Job		163		163	blocking or bridging.
2″ x 10″ Joists							
	Demolish	S.F.		.29		.29	Cost includes material and labor to
	Install	S.F.	.66	.52		1.18	install 2″ x 10″ joists including end
	Demolish and Install	S.F.	.66	.81		1.47	and header joist installed 24″ O.C.
	Reinstall	S.F.		.42		.42	Does not include beams, ledger strips,
	Minimum Charge	Job		163		163	blocking or bridging.
2″ x 12″ Joists							
	Demolish	S.F.		.30		.30	Cost includes material and labor to
	Install	S.F.	.90	.64		1.54	install 2″ x 12″ joists including end
	Demolish and Install	S.F.	.90	.94		1.84	and header joist installed 24″ O.C.
	Reinstall	S.F.		.51		.51	Does not include beams, ledger strips,
	Minimum Charge	Job		163		163	blocking or bridging.
Block / Bridge							
	Demolish	Ea.		.82		.82	Includes material and labor to install
	Install	Ea.	1.73	1.83		3.56	set of cross bridging or per block of
	Demolish and Install	Ea.	1.73	2.65		4.38	solid bridging for 2″ x 10″ joists cut to
	Minimum Charge	Job		163		163	size on site.
Ledger Strips							
1″ x 2″							
	Demolish	L.F.		.22		.22	Cost includes material and labor to
	Install	L.F.	.21	.95		1.16	install 1″ x 2″ ledger strip nailed to the
	Demolish and Install	L.F.	.21	1.17		1.38	face of studs, beams or joist.
	Minimum Charge	Job		163		163	
1″ x 3″							
	Demolish	L.F.		.22		.22	Includes material and labor to install
	Install	L.F.	.66	1.09		1.75	up to 1″ x 4″ ledger strip nailed to the
	Demolish and Install	L.F.	.66	1.31		1.97	face of studs, beams or joist.
	Minimum Charge	Job		163		163	
1″ x 4″							
	Demolish	L.F.		.22		.22	Includes material and labor to install
	Install	L.F.	.66	1.09		1.75	up to 1″ x 4″ ledger strip nailed to the
	Demolish and Install	L.F.	.66	1.31		1.97	face of studs, beams or joist.
	Minimum Charge	Job		163		163	
2″ x 2″							
	Demolish	L.F.		.24		.24	Cost includes material and labor to
	Install	L.F.	.29	.99		1.28	install 2″ x 2″ ledger strip nailed to the
	Demolish and Install	L.F.	.29	1.23		1.52	face of studs, beams or joist.
	Minimum Charge	Job		163		163	
2″ x 4″							
	Demolish	L.F.		.26		.26	Cost includes material and labor to
	Install	L.F.	.37	1.30		1.67	install 2″ x 4″ ledger strip nailed to the
	Demolish and Install	L.F.	.37	1.56		1.93	face of studs, beams or joist.
	Reinstall	L.F.		1.04		1.04	
	Minimum Charge	Job		163		163	

Rough Frame / Structure

Ceiling Framing

Ceiling Framing		Unit	Material	Labor	Equip.	Total	Specification
Ledger Boards							
	Demolish	L.F.		.26		.26	Cost includes material and labor to
	Install	L.F.	1.67	1.81		3.48	install 2" x 4" ledger board fastened to
	Demolish and Install	L.F.	1.67	2.07		3.74	a wall, joists or studs.
	Reinstall	L.F.		1.45		1.45	
	Minimum Charge	Job		163		163	
4" x 6"							
	Demolish	L.F.		.44		.44	Cost includes material and labor to
	Install	L.F.	3.58	2.04		5.62	install 4" x 6" ledger board fastened to
	Demolish and Install	L.F.	3.58	2.48		6.06	a wall, joists or studs.
	Reinstall	L.F.		1.63		1.63	
	Minimum Charge	Job		163		163	
4" x 8"							
	Demolish	L.F.		.58		.58	Cost includes material and labor to
	Install	L.F.	4.33	2.04		6.37	install 4" x 8" ledger board fastened to
	Demolish and Install	L.F.	4.33	2.62		6.95	a wall, joists or studs.
	Reinstall	L.F.		1.63		1.63	
	Minimum Charge	Job		163		163	
Earthquake Strapping							
	Install	Ea.	1.80	2.04		3.84	Includes labor and material to replace
	Minimum Charge	Job		163		163	an earthquake strap.
Hurricane Clips							
	Install	Ea.	1.58	2.25		3.83	Includes labor and material to install a
	Minimum Charge	Job		163		163	hurricane clip.

Roof Framing

Roof Framing		Unit	Material	Labor	Equip.	Total	Specification
16" O.C. System							
2" x 4" Rafters							
	Demolish	S.F.		.59		.59	Cost includes material and labor to
	Install	S.F.	.28	.29		.57	install 2" x 4", 16" O.C. rafter framing
	Demolish and Install	S.F.	.28	.88		1.16	for flat, shed or gable roofs, 25' span,
	Reinstall	S.F.		.23		.23	up to 5/12 slope, per S.F. of roof.
	Minimum Charge	Job		163		163	
2" x 6" Rafters							
	Demolish	S.F.		.62		.62	Cost includes material and labor to
	Install	S.F.	.44	.31		.75	install 2" x 6", 16" O.C. rafter framing
	Demolish and Install	S.F.	.44	.93		1.37	for flat, shed or gable roofs, 25' span,
	Reinstall	S.F.		.24		.24	up to 5/12 slope, per S.F. of roof.
	Minimum Charge	Job		163		163	
2" x 8" Rafters							
	Demolish	S.F.		.64		.64	Cost includes material and labor to
	Install	S.F.	.68	.33		1.01	install 2" x 8", 16" O.C. rafter framing
	Demolish and Install	S.F.	.68	.97		1.65	for flat, shed or gable roofs, 25' span,
	Reinstall	S.F.		.26		.26	up to 5/12 slope, per S.F. of roof.
	Minimum Charge	Job		163		163	
2" x 10" Rafters							
	Demolish	S.F.		.64		.64	Cost includes material and labor to
	Install	S.F.	.97	.49		1.46	install 2" x 10", 16" O.C. rafter
	Demolish and Install	S.F.	.97	1.13		2.10	framing for flat, shed or gable roofs,
	Reinstall	S.F.		.40		.40	25' span, up to 5/12 slope, per S.F.
	Minimum Charge	Job		163		163	of roof.

Roof Framing		Unit	Material	Labor	Equip.	Total	Specification
2" x 12" Rafters							
	Demolish	S.F.		.65		.65	Cost includes material and labor to
	Install	S.F.	1.32	.54		1.86	install 2" x 12", 16" O.C. rafter
	Demolish and Install	S.F.	1.32	1.19		2.51	framing for flat, shed or gable roofs,
	Reinstall	S.F.		.43		.43	25' span, up to 5/12 slope, per S.F.
	Minimum Charge	Job		163		163	of roof.
24" O.C. System							
2" x 4" Rafters							
	Demolish	S.F.		.45		.45	Cost includes material and labor to
	Install	S.F.	.19	.22		.41	install 2" x 4", 24" O.C. rafter framing
	Demolish and Install	S.F.	.19	.67		.86	for flat, shed or gable roofs, 25' span,
	Reinstall	S.F.		.18		.18	up to 5/12 slope, per S.F. of roof.
	Minimum Charge	Job		163		163	
2" x 6" Rafters							
	Demolish	S.F.		.47		.47	Cost includes material and labor to
	Install	S.F.	.30	.23		.53	install 2" x 6", 24" O.C. rafter framing
	Demolish and Install	S.F.	.30	.70		1	for flat, shed or gable roofs, 25' span,
	Reinstall	S.F.		.18		.18	up to 5/12 slope, per S.F. of roof.
	Minimum Charge	Job		163		163	
2" x 8" Rafters							
	Demolish	S.F.		.48		.48	Cost includes material and labor to
	Install	S.F.	.45	.25		.70	install 2" x 8", 24" O.C. rafter framing
	Demolish and Install	S.F.	.45	.73		1.18	for flat, shed or gable roofs, 25' span,
	Reinstall	S.F.		.20		.20	up to 5/12 slope, per S.F. of roof.
	Minimum Charge	Job		163		163	
2" x 10" Rafters							
	Demolish	S.F.		.48		.48	Cost includes material and labor to
	Install	S.F.	.65	.37		1.02	install 2" x 10", 24" O.C. rafter
	Demolish and Install	S.F.	.65	.85		1.50	framing for flat, shed or gable roofs,
	Reinstall	S.F.		.30		.30	25' span, up to 5/12 slope, per S.F.
	Minimum Charge	Job		163		163	of roof.
2" x 12" Rafters							
	Demolish	S.F.		.49		.49	Cost includes material and labor to
	Install	S.F.	.88	.40		1.28	install 2" x 12", 24" O.C. rafter
	Demolish and Install	S.F.	.88	.89		1.77	framing for flat, shed or gable roofs,
	Reinstall	S.F.		.32		.32	25' span, up to 5/12 slope, per S.F.
	Minimum Charge	Job		163		163	of roof.
Rafter							
2" x 4"							
	Demolish	L.F.		.61		.61	Cost includes material and labor to
	Install	L.F.	.37	.95		1.32	install 2" x 4" rafters for flat, shed or
	Demolish and Install	L.F.	.37	1.56		1.93	gable roofs, up to 5/12 slope, 25'
	Reinstall	L.F.		.76		.76	span, per L.F.
	Minimum Charge	Job		163		163	
2" x 6"							
	Demolish	L.F.		.62		.62	Cost includes material and labor to
	Install	L.F.	.59	.82		1.41	install 2" x 6" rafters for flat, shed or
	Demolish and Install	L.F.	.59	1.44		2.03	gable roofs, up to 5/12 slope, 25'
	Reinstall	L.F.		.65		.65	span, per L.F.
	Minimum Charge	Job		163		163	
2" x 8"							
	Demolish	L.F.		.63		.63	Cost includes material and labor to
	Install	L.F.	.91	.87		1.78	install 2" x 8" rafters for flat, shed or
	Demolish and Install	L.F.	.91	1.50		2.41	gable roofs, up to 5/12 slope, 25'
	Reinstall	L.F.		.70		.70	span, per L.F.
	Minimum Charge	Job		163		163	

Rough Frame / Structure

Roof Framing		Unit	Material	Labor	Equip.	Total	Specification
2" x 10"							
	Demolish	L.F.		.63		.63	Cost includes material and labor to
	Install	L.F.	1.30	1.32		2.62	install 2" x 10" rafters for flat, shed or
	Demolish and Install	L.F.	1.30	1.95		3.25	gable roofs, up to 5/12 slope, 25'
	Reinstall	L.F.		1.05		1.05	span, per L.F.
	Minimum Charge	Job		163		163	
2" x 12"							
	Demolish	L.F.		.64		.64	Cost includes material and labor to
	Install	L.F.	1.76	1.43		3.19	install 2" x 12" rafters for flat, shed or
	Demolish and Install	L.F.	1.76	2.07		3.83	gable roofs, up to 5/12 slope, 25'
	Reinstall	L.F.		1.15		1.15	span, per L.F.
	Minimum Charge	Job		163		163	
2" x 4" Valley / Jack							
	Demolish	L.F.		.61		.61	Includes material and labor to install
	Install	L.F.	.59	1.37		1.96	up to 2" x 6" valley/jack rafters for
	Demolish and Install	L.F.	.59	1.98		2.57	flat, shed or gable roofs, up to 5/12
	Reinstall	L.F.		1.10		1.10	slope, 25' span, per L.F.
	Minimum Charge	Job		163		163	
2" x 6" Valley / Jack							
	Demolish	L.F.		.62		.62	Includes material and labor to install
	Install	L.F.	.59	1.37		1.96	up to 2" x 6" valley/jack rafters for
	Demolish and Install	L.F.	.59	1.99		2.58	flat, shed or gable roofs, up to 5/12
	Reinstall	L.F.		1.10		1.10	slope, 25' span, per L.F.
	Minimum Charge	Job		163		163	
Ridgeboard							
2" x 4"							
	Demolish	L.F.		.58		.58	Cost includes material and labor to
	Install	L.F.	.37	.59		.96	install 2" x 4" ridgeboard for flat, shed
	Demolish and Install	L.F.	.37	1.17		1.54	or gable roofs, up to 5/12 slope, 25'
	Reinstall	L.F.		.47		.47	span, per L.F.
	Minimum Charge	Job		163		163	
2" x 6"							
	Demolish	L.F.		.60		.60	Cost includes material and labor to
	Install	L.F.	.59	1.30		1.89	install 2" x 6" ridgeboard for flat, shed
	Demolish and Install	L.F.	.59	1.90		2.49	or gable roofs, up to 5/12 slope, 25'
	Reinstall	L.F.		1.04		1.04	span, per L.F.
	Minimum Charge	Job		163		163	
2" x 8"							
	Demolish	L.F.		.62		.62	Cost includes material and labor to
	Install	L.F.	.91	1.45		2.36	install 2" x 8" ridgeboard for flat, shed
	Demolish and Install	L.F.	.91	2.07		2.98	or gable roofs, up to 5/12 slope, 25'
	Reinstall	L.F.		1.16		1.16	span, per L.F.
	Minimum Charge	Job		163		163	
2" x 10"							
	Demolish	L.F.		.63		.63	Cost includes material and labor to
	Install	L.F.	1.30	1.63		2.93	install 2" x 10" ridgeboard for flat,
	Demolish and Install	L.F.	1.30	2.26		3.56	shed or gable roofs, up to 5/12
	Reinstall	L.F.		1.30		1.30	slope, 25' span, per L.F.
	Minimum Charge	Job		163		163	
2" x 12"							
	Demolish	L.F.		.65		.65	Cost includes material and labor to
	Install	L.F.	1.76	.93		2.69	install 2" x 12" ridgeboard for flat,
	Demolish and Install	L.F.	1.76	1.58		3.34	shed or gable roofs, up to 5/12
	Reinstall	L.F.		.75		.75	slope, 25' span, per L.F.
	Minimum Charge	Job		163		163	

Rough Frame / Structure

Roof Framing	Unit	Material	Labor	Equip.	Total	Specification
Collar Beam						
1″ x 6″						
Demolish	L.F.		.46		.46	Cost includes material and labor to
Install	L.F.	.64	.41		1.05	install 1″ x 6″ collar beam or tie for
Demolish and Install	L.F.	.64	.87		1.51	roof framing.
Reinstall	L.F.		.33		.33	
Minimum Charge	Job		163		163	
2″ x 6″						
Demolish	L.F.		.54		.54	Cost includes material and labor to
Install	L.F.	.59	.41		1	install 2″ x 6″ collar beam or tie for
Demolish and Install	L.F.	.59	.95		1.54	roof framing.
Reinstall	L.F.		.33		.33	
Minimum Charge	Job		163		163	
Purlins						
Demolish	L.F.		.52		.52	Cost includes material and labor to
Install	L.F.	.59	.36		.95	install 2″ x 6″ purlins below roof
Demolish and Install	L.F.	.59	.88		1.47	rafters.
Reinstall	L.F.		.29		.29	
Minimum Charge	Job		163		163	
2″ x 8″						
Demolish	L.F.		.53		.53	Cost includes material and labor to
Install	L.F.	.91	.37		1.28	install 2″ x 8″ purlins below roof
Demolish and Install	L.F.	.91	.90		1.81	rafters.
Reinstall	L.F.		.29		.29	
Minimum Charge	Job		163		163	
2″ x 10″						
Demolish	L.F.		.53		.53	Cost includes material and labor to
Install	L.F.	1.30	.37		1.67	install 2″ x 10″ purlins below roof
Demolish and Install	L.F.	1.30	.90		2.20	rafters.
Reinstall	L.F.		.30		.30	
Minimum Charge	Job		163		163	
2″ x 12″						
Demolish	L.F.		.54		.54	Cost includes material and labor to
Install	L.F.	1.76	.37		2.13	install 2″ x 12″ purlins below roof
Demolish and Install	L.F.	1.76	.91		2.67	rafters.
Reinstall	L.F.		.30		.30	
Minimum Charge	Job		163		163	
4″ x 6″						
Demolish	L.F.		.54		.54	Cost includes material and labor to
Install	L.F.	2.20	.37		2.57	install 4″ x 6″ purlins below roof
Demolish and Install	L.F.	2.20	.91		3.11	rafters.
Reinstall	L.F.		.30		.30	
Minimum Charge	Job		163		163	
4″ x 8″						
Demolish	L.F.		.55		.55	Cost includes material and labor to
Install	L.F.	2.96	.38		3.34	install 4″ x 8″ purlins below roof
Demolish and Install	L.F.	2.96	.93		3.89	rafters.
Reinstall	L.F.		.31		.31	
Minimum Charge	Job		163		163	
Ledger Board						
2″ x 6″						
Demolish	L.F.		.26		.26	Cost includes material and labor to
Install	L.F.	.59	1.47		2.06	install 2″ x 6″ ledger board nailed to
Demolish and Install	L.F.	.59	1.73		2.32	the face of a rafter.
Reinstall	L.F.		1.17		1.17	
Minimum Charge	Job		163		163	

Rough Frame / Structure

Roof Framing		Unit	Material	Labor	Equip.	Total	Specification
4″ x 6″							
	Demolish	L.F.		.44		.44	Cost includes material and labor to
	Install	L.F.	2.20	2.16		4.36	install 4″ x 6″ ledger board nailed to
	Demolish and Install	L.F.	2.20	2.60		4.80	the face of a rafter.
	Reinstall	L.F.		1.73		1.73	
	Minimum Charge	Job		163		163	
4″ x 8″							
	Demolish	L.F.		.58		.58	Cost includes material and labor to
	Install	L.F.	2.96	2.49		5.45	install 4″ x 8″ ledger board nailed to
	Demolish and Install	L.F.	2.96	3.07		6.03	the face of a rafter.
	Reinstall	L.F.		1.99		1.99	
	Minimum Charge	Job		163		163	
Outriggers							
2″ x 4″							
	Demolish	L.F.		.18		.18	Cost includes material and labor to
	Install	L.F.	.37	1.30		1.67	install 2″ x 4″ outrigger rafters for flat,
	Demolish and Install	L.F.	.37	1.48		1.85	shed or gabled roofs, up to 5/12
	Reinstall	L.F.		1.04		1.04	slope, per L.F.
	Minimum Charge	Job		163		163	
2″ x 6″							
	Demolish	L.F.		.27		.27	Cost includes material and labor to
	Install	L.F.	.59	1.30		1.89	install 2″ x 6″ outrigger rafters for flat,
	Demolish and Install	L.F.	.59	1.57		2.16	shed or gabled roofs, up to 5/12
	Reinstall	L.F.		1.04		1.04	slope, per L.F.
	Minimum Charge	Job		163		163	
Lookout Rafter							
2″ x 4″							
	Demolish	L.F.		.61		.61	Cost includes material and labor to
	Install	L.F.	.37	.54		.91	install 2″ x 4″ lookout rafters for flat,
	Demolish and Install	L.F.	.37	1.15		1.52	shed or gabled roofs, up to 5/12
	Reinstall	L.F.		.43		.43	slope, per L.F.
	Minimum Charge	Job		163		163	
2″ x 6″							
	Demolish	L.F.		.62		.62	Cost includes material and labor to
	Install	L.F.	.59	.55		1.14	install 2″ x 6″ lookout rafters for flat,
	Demolish and Install	L.F.	.59	1.17		1.76	shed or gabled roofs, up to 5/12
	Reinstall	L.F.		.44		.44	slope, per L.F.
	Minimum Charge	Job		163		163	
Fly Rafter							
2″ x 4″							
	Demolish	L.F.		.61		.61	Cost includes material and labor to
	Install	L.F.	.37	.54		.91	install 2″ x 4″ fly rafters for flat, shed
	Demolish and Install	L.F.	.37	1.15		1.52	or gabled roofs, up to 5/12 slope,
	Reinstall	L.F.		.43		.43	per L.F.
	Minimum Charge	Job		163		163	
2″ x 6″							
	Demolish	L.F.		.62		.62	Cost includes material and labor to
	Install	L.F.	.59	.55		1.14	install 2″ x 6″ fly rafters for flat, shed
	Demolish and Install	L.F.	.59	1.17		1.76	or gabled roofs, up to 5/12 slope,
	Reinstall	L.F.		.44		.44	per L.F.
	Minimum Charge	Job		163		163	

Rough Frame / Structure

Residential Trusses	Unit	Material	Labor	Equip.	Total	Specification
W or Fink Truss						
24' Span						
Demolish	Ea.		24.50	11.15	35.65	Includes material, labor and
Install	Ea.	56	24	11.05	91.05	equipment to install wood gang-nailed
Demolish and Install	Ea.	56	48.50	22.20	126.70	residential truss, up to 24' span.
Reinstall	Ea.		24.15	11.05	35.20	
Minimum Charge	Job		163		163	
28' Span						
Demolish	Ea.		26	11.90	37.90	Includes material, labor and
Install	Ea.	67.50	27.50	12.50	107.50	equipment to install wood gang-nailed
Demolish and Install	Ea.	67.50	53.50	24.40	145.40	residential truss, up to 28' span.
Reinstall	Ea.		27.34	12.51	39.85	
Minimum Charge	Job		163		163	
32' Span						
Demolish	Ea.		29	13.15	42.15	Includes material, labor and
Install	Ea.	95.50	29	13.25	137.75	equipment to install wood gang-nailed
Demolish and Install	Ea.	95.50	58	26.40	179.90	residential truss, up to 32' span.
Reinstall	Ea.		28.98	13.26	42.24	
Minimum Charge	Job		163		163	
36' Span						
Demolish	Ea.		31	14.15	45.15	Includes material, labor and
Install	Ea.	120	31.50	14.40	165.90	equipment to install wood gang-nailed
Demolish and Install	Ea.	120	62.50	28.55	211.05	residential truss, up to 36' span.
Reinstall	Ea.		31.50	14.41	45.91	
Minimum Charge	Job		163		163	
40' Span						
Demolish	Ea.		33.50	15.35	48.85	Includes material, labor and
Install	Ea.	134	33.50	15.40	182.90	equipment to install wood gang-nailed
Demolish and Install	Ea.	134	67	30.75	231.75	residential truss, up to 40' span.
Reinstall	Ea.		33.70	15.41	49.11	
Minimum Charge	Job		163		163	
Gable End						
24' Span						
Demolish	Ea.		24.50	11.15	35.65	Includes material, labor and
Install	Ea.	79	26	11.85	116.85	equipment to install wood gang-nailed
Demolish and Install	Ea.	79	50.50	23	152.50	residential gable end truss, up to 24'
Reinstall	Ea.		25.88	11.84	37.72	span.
Clean	Ea.	1.21	31.50		32.71	
Paint	Ea.	3.52	38		41.52	
Minimum Charge	Job		163		163	
28' Span						
Demolish	Ea.		26	11.90	37.90	Includes material, labor and
Install	Ea.	96.50	31	14.10	141.60	equipment to install wood gang-nailed
Demolish and Install	Ea.	96.50	57	26	179.50	residential gable end truss, up to 30'
Reinstall	Ea.		30.83	14.10	44.93	span.
Clean	Ea.	1.27	43		44.27	
Paint	Ea.	3.67	52		55.67	
Minimum Charge	Job		163		163	
32' Span						
Demolish	Ea.		29	13.15	42.15	Includes material, labor and
Install	Ea.	107	31.50	14.40	152.90	equipment to install wood gang-nailed
Demolish and Install	Ea.	107	60.50	27.55	195.05	residential gable end truss, up to 32'
Reinstall	Ea.		31.50	14.41	45.91	span.
Clean	Ea.	1.32	56		57.32	
Paint	Ea.	3.84	68		71.84	
Minimum Charge	Job		163		163	

Rough Frame / Structure

Residential Trusses

	Unit	Material	Labor	Equip.	Total	Specification
36' Span						
Demolish	Ea.		31	14.15	45.15	Includes material, labor and
Install	Ea.	128	36	16.55	180.55	equipment to install wood gang-nailed
Demolish and Install	Ea.	128	67	30.70	225.70	residential gable end truss, up to 36'
Reinstall	Ea.		36.23	16.57	52.80	span.
Clean	Ea.	1.39	71		72.39	
Paint	Ea.	4.03	86		90.03	
Minimum Charge	Job		163		163	
40' Span						
Demolish	Ea.		33.50	15.35	48.85	Includes material, labor and
Install	Ea.	142	38	17.45	197.45	equipment to install wood gang-nailed
Demolish and Install	Ea.	142	71.50	32.80	246.30	residential gable end truss, up to 40'
Reinstall	Ea.		38.14	17.44	55.58	span.
Clean	Ea.	4.22	87.50		91.72	
Paint	Ea.	4.22	106		110.22	
Minimum Charge	Job		163		163	

Truss Hardware

	Unit	Material	Labor	Equip.	Total	Specification
5-1/4" Glue-Lam Seat						
Install	Ea.	52.50	1.81		54.31	Includes labor and material to install
Minimum Charge	Job		163		163	beam hangers.
6-3/4" Glue-Lam Seat						
Install	Ea.	54	1.81		55.81	Includes labor and material to install
Minimum Charge	Job		163		163	beam hangers.
8-3/4" Glue-Lam Seat						
Install	Ea.	61.50	1.81		63.31	Includes labor and material to install
Minimum Charge	Job		163		163	beam hangers.
2" x 4" Joist Hanger						
Install	Ea.	.64	1.86		2.50	Includes labor and material to install
Minimum Charge	Job		163		163	joist and beam hangers.
2" x 12" Joist Hanger						
Install	Ea.	.74	1.98		2.72	Cost includes labor and material to
Minimum Charge	Job		163		163	install 2" x 12" joist hanger.

Commercial Trusses

	Unit	Material	Labor	Equip.	Total	Specification
Engineered Lumber, Truss / Joist						
9-1/2"						
Demolish	S.F.		.43		.43	Includes material and labor to install
Install	S.F.	1.45	.33		1.78	engineered lumber truss / joists per
Demolish and Install	S.F.	1.45	.76		2.21	S.F. of floor area based on joists 16"
Reinstall	S.F.		.26		.26	O.C. Beams, supports and bridging
Minimum Charge	Job		163		163	are not included.
11-7/8"						
Demolish	S.F.		.45		.45	Includes material and labor to install
Install	S.F.	1.54	.33		1.87	engineered lumber truss / joists per
Demolish and Install	S.F.	1.54	.78		2.32	S.F. of floor area based on joists 16"
Reinstall	S.F.		.27		.27	O.C. Beams, supports and bridging
Minimum Charge	Job		163		163	are not included.
14"						
Demolish	S.F.		.46		.46	Includes material and labor to install
Install	S.F.	1.77	.36		2.13	engineered lumber truss / joists per
Demolish and Install	S.F.	1.77	.82		2.59	S.F. of floor area based on joists 16"
Reinstall	S.F.		.29		.29	O.C. Beams, supports and bridging
Minimum Charge	Job		163		163	are not included.

Rough Frame / Structure

Commercial Trusses

	Unit	Material	Labor	Equip.	Total	Specification
16″						
Demolish	S.F.		.48		.48	Includes material and labor to install
Install	S.F.	2.52	.38		2.90	engineered lumber truss / joists per
Demolish and Install	S.F.	2.52	.86		3.38	S.F. of floor area based on joists 16″
Reinstall	S.F.		.30		.30	O.C. Beams, supports and bridging
Minimum Charge	Job		163		163	are not included.
Bowstring Truss						
100′ Clear Span						
Demolish	SF Flr.		.44	.30	.74	Includes material, labor and
Install	SF Flr.	6.10	.36	.25	6.71	equipment to install bow string truss
Demolish and Install	SF Flr.	6.10	.80	.55	7.45	system up to 100′ clear span per S.F.
Minimum Charge	Job		163		163	of floor area.
120′ Clear Span						
Demolish	SF Flr.		.48	.34	.82	Includes material, labor and
Install	SF Flr.	6.55	.40	.27	7.22	equipment to install bow string truss
Demolish and Install	SF Flr.	6.55	.88	.61	8.04	system up to 120′ clear span per S.F.
Minimum Charge	Job		163		163	of floor area.

Roof Sheathing

	Unit	Material	Labor	Equip.	Total	Specification
Plywood						
3/8″						
Demolish	S.F.		.34		.34	Cost includes material and labor to
Install	S.F.	.72	.43		1.15	install 3/8″ CDX plywood roof
Demolish and Install	S.F.	.72	.77		1.49	sheathing.
Reinstall	S.F.		.43		.43	
Minimum Charge	Job		163		163	
1/2″						
Demolish	S.F.		.34		.34	Cost includes material and labor to
Install	S.F.	.68	.47		1.15	install 1/2″ CDX plywood roof
Demolish and Install	S.F.	.68	.81		1.49	sheathing.
Minimum Charge	Job		163		163	
5/8″						
Demolish	S.F.		.34		.34	Cost includes material and labor to
Install	S.F.	.86	.50		1.36	install 5/8″ CDX plywood roof
Demolish and Install	S.F.	.86	.84		1.70	sheathing.
Minimum Charge	Job		163		163	
3/4″						
Demolish	S.F.		.34		.34	Cost includes material and labor to
Install	S.F.	1.03	.54		1.57	install 3/4″ CDX plywood roof
Demolish and Install	S.F.	1.03	.88		1.91	sheathing.
Minimum Charge	Job		163		163	
OSB						
1/2″						
Demolish	S.F.		.34		.34	Cost includes material and labor to
Install	S.F.	.68	.25		.93	install 4′ x 8′ x 1/2″ OSB sheathing.
Demolish and Install	S.F.	.68	.59		1.27	
Minimum Charge	Job		163		163	
5/8″						
Demolish	S.F.		.34		.34	Cost includes material and labor to
Install	S.F.	.86	.26		1.12	install 4′ x 8′ x 5/8″ OSB sheathing.
Demolish and Install	S.F.	.86	.60		1.46	
Minimum Charge	Job		163		163	

Rough Frame / Structure

Roof Sheathing

Roof Sheathing	Unit	Material	Labor	Equip.	Total	Specification
Plank						
Demolish	S.F.		.34		.34	Cost includes material and labor to
Install	S.F.	1.22	.90		2.12	install 1" x 6" or 1" x 8" utility T&G
Demolish and Install	S.F.	1.22	1.24		2.46	board sheathing laid diagonal.
Reinstall	S.F.		.90		.90	
Minimum Charge	Job		163		163	
Skip-type						
1" x 4"						
Demolish	S.F.		.52		.52	Includes material and labor to install
Install	S.F.	.56	.27		.83	sheathing board material 1" x 4", 7"
Demolish and Install	S.F.	.56	.79		1.35	O.C.
Minimum Charge	Job		163		163	
1" x 6"						
Demolish	S.F.		.52		.52	Includes material and labor to install
Install	S.F.	.64	.22		.86	sheathing board material 1" x 6", 7"
Demolish and Install	S.F.	.64	.74		1.38	or 9" O.C.
Minimum Charge	Job		163		163	

Wood Deck

Wood Deck	Unit	Material	Labor	Equip.	Total	Specification
Treated						
Demolish	S.F.		1.09		1.09	Includes material and labor to install
Install	S.F.	4.52	8.15		12.67	unfinished deck using 2" x 4" treated
Demolish and Install	S.F.	4.52	9.24		13.76	decking, with 2" x 6" double beams
Reinstall	S.F.		6.52		6.52	24" O.C., 4" x 4" posts 5' O.C. set in
Clean	S.F.	.06	.24		.30	concrete.
Paint	S.F.	.06	.22		.28	
Minimum Charge	Job		163		163	
Redwood						
Demolish	S.F.		1.09		1.09	Includes material and labor to install
Install	S.F.	16.45	8.15		24.60	unfinished deck using 2" x 4" select
Demolish and Install	S.F.	16.45	9.24		25.69	redwood decking, with 2" x 6" double
Reinstall	S.F.		6.52		6.52	beams 24" O.C., 4" x 4" posts 5'
Clean	S.F.	.06	.24		.30	O.C. set in concrete.
Paint	S.F.	.06	.22		.28	
Minimum Charge	Job		163		163	
Cedar						
Demolish	S.F.		1.09		1.09	Includes material and labor to install
Install	S.F.	7.60	8.15		15.75	unfinished deck using 2" x 4" select
Demolish and Install	S.F.	7.60	9.24		16.84	cedar decking, with 2" x 6" double
Reinstall	S.F.		6.52		6.52	beams 24" O.C., 4" x 4" posts 5'
Clean	S.F.	.06	.24		.30	O.C. set in concrete.
Paint	S.F.	.06	.22		.28	
Minimum Charge	Job		163		163	
Stairs						
Demolish	Ea.		315		315	Includes material and labor to install
Install	Ea.	165	565		730	wood steps. 5 - 7 risers included.
Demolish and Install	Ea.	165	880		1045	
Reinstall	Ea.		450.30		450.30	
Clean	Flight	2.92	20.50		23.42	
Paint	Flight	10.35	38		48.35	
Minimum Charge	Job		163		163	

Rough Frame / Structure

Wood Deck

Wood Deck	Unit	Material	Labor	Equip.	Total	Specification
Railing						
Demolish	L.F.		4.03		4.03	Includes material and labor to install
Install	L.F.	15.55	14.80		30.35	porch rail with balusters.
Demolish and Install	L.F.	15.55	18.83		34.38	
Clean	L.F.	.58	1.47		2.05	
Paint	L.F.	.50	2.38		2.88	
Minimum Charge	Job		163		163	
Bench Seating						
Demolish	L.F.		4.36		4.36	Includes material and labor to install
Install	L.F.	2.23	16.30		18.53	pressure treated bench seating.
Demolish and Install	L.F.	2.23	20.66		22.89	
Reinstall	L.F.		13.04		13.04	
Clean	L.F.	.12	1.18		1.30	
Paint	L.F.	.51	2.29		2.80	
Minimum Charge	Job		163		163	
Privacy Wall						
Demolish	SF Flr.		1.05		1.05	Includes material, labor and
Install	SF Flr.	.45	.55	.11	1.11	equipment to install privacy wall.
Demolish and Install	SF Flr.	.45	1.60	.11	2.16	
Reinstall	SF Flr.		.44	.09	.53	
Clean	SF Flr.	.06	.24		.30	
Paint	S.F.	.06	.22		.28	
Minimum Charge	Job		163		163	

Porch

Porch	Unit	Material	Labor	Equip.	Total	Specification
Treated						
Demolish	S.F.		2.18		2.18	Includes material and labor to install
Install	S.F.	5.55	16.30		21.85	floor and roof frame and decking
Demolish and Install	S.F.	5.55	18.48		24.03	using 2″ x 4″ treated decking, with 2″
Reinstall	S.F.		13.04		13.04	x 6″ double beams 24″ O.C., 4″ x 4″
Clean	S.F.	.12	.49		.61	posts 5′ O.C. set in concrete.
Paint	S.F.	.17	.57		.74	
Minimum Charge	Job		163		163	
Redwood						
Demolish	S.F.		2.18		2.18	Includes material and labor to install
Install	S.F.	17.75	16.30		34.05	floor and roof frame and decking
Demolish and Install	S.F.	17.75	18.48		36.23	using 2″ x 4″ redwood decking, with
Reinstall	S.F.		13.04		13.04	2″ x 6″ double beams 24″ O.C., 4″ x
Clean	S.F.	.12	.49		.61	4″ posts 5′ O.C. set in concrete.
Paint	S.F.	.17	.57		.74	
Minimum Charge	Job		163		163	
Cedar						
Demolish	S.F.		2.18		2.18	Includes material and labor to install
Install	S.F.	8.85	16.30		25.15	floor and roof frame and cedar
Demolish and Install	S.F.	8.85	18.48		27.33	decking using 2″ x 4″ select cedar,
Reinstall	S.F.		13.04		13.04	with 2″ x 6″ double beams 24″ O.C.,
Clean	S.F.	.12	.49		.61	4″ x 4″ posts 5′ O.C. set in concrete.
Paint	S.F.	.17	.57		.74	
Minimum Charge	Job		163		163	
Stairs						
Demolish	Ea.		315		315	Includes material and labor to install
Install	Ea.	165	565		730	wood steps. 5 - 7 risers included.
Demolish and Install	Ea.	165	880		1045	
Reinstall	Ea.		450.30		450.30	
Clean	Flight	2.92	20.50		23.42	
Paint	Flight	10.35	38		48.35	
Minimum Charge	Job		163		163	

Rough Frame / Structure

Porch	Unit	Material	Labor	Equip.	Total	Specification
Railing						
Demolish	L.F.		4.03		4.03	Includes material and labor to install
Install	L.F.	15.55	14.80		30.35	porch rail with balusters.
Demolish and Install	L.F.	15.55	18.83		34.38	
Clean	L.F.	.58	1.84		2.42	
Paint	L.F.	.50	2.38		2.88	
Minimum Charge	Job		163		163	
Bench Seating						
Demolish	L.F.		4.36		4.36	Includes material and labor to install
Install	L.F.	2.23	16.30		18.53	pressure treated bench seating.
Demolish and Install	L.F.	2.23	20.66		22.89	
Reinstall	L.F.		13.04		13.04	
Clean	L.F.	.12	1.18		1.30	
Paint	L.F.	.51	2.29		2.80	
Minimum Charge	Job		163		163	
Privacy Wall						
Demolish	SF Flr.		1.05		1.05	Includes material, labor and
Install	SF Flr.	.45	.55	.11	1.11	equipment to install privacy wall.
Demolish and Install	SF Flr.	.45	1.60	.11	2.16	
Reinstall	SF Flr.		.44	.09	.53	
Clean	SF Flr.	.06	.24		.30	
Paint	S.F.	.06	.22		.28	
Minimum Charge	Job		163		163	
Wood Board Ceiling						
Demolish	S.F.		.44		.44	Includes material and labor to install
Install	S.F.	1.91	1.45		3.36	wood board on ceiling.
Demolish and Install	S.F.	1.91	1.89		3.80	
Clean	S.F.	.02	.14		.16	
Paint	S.F.	.15	.36		.51	
Minimum Charge	Job		163		163	
Re-screen						
Install	S.F.	.53	1.30		1.83	Includes labor and material to
Minimum Charge	Job		163		163	re-screen wood frame.

Exterior Trim

Exterior Trim		Unit	Material	Labor	Equip.	Total	Specification
Stock Lumber							
1″ x 2″							
	Demolish	L.F.		.22		.22	Cost includes material and labor to
	Install	L.F.	.36	.99		1.35	install 1″ x 2″ pine trim.
	Demolish and Install	L.F.	.36	1.21		1.57	
	Clean	L.F.	.01	.16		.17	
	Paint	L.F.	.02	.45		.47	
	Minimum Charge	Job		163		163	
1″ x 3″							
	Demolish	L.F.		.22		.22	Cost includes material and labor to
	Install	L.F.	.31	1.09		1.40	install 1″ x 3″ pine trim.
	Demolish and Install	L.F.	.31	1.31		1.62	
	Clean	L.F.	.01	.16		.17	
	Paint	L.F.	.02	.45		.47	
	Minimum Charge	Job		163		163	
1″ x 4″							
	Demolish	L.F.		.22		.22	Cost includes material and labor to
	Install	L.F.	.75	1.30		2.05	install 1″ x 4″ pine trim.
	Demolish and Install	L.F.	.75	1.52		2.27	
	Clean	L.F.	.01	.16		.17	
	Paint	L.F.	.02	.45		.47	
	Minimum Charge	Job		163		163	
1″ x 6″							
	Demolish	L.F.		.22		.22	Cost includes material and labor to
	Install	L.F.	1.31	1.30		2.61	install 1″ x 6″ pine trim.
	Demolish and Install	L.F.	1.31	1.52		2.83	
	Clean	L.F.	.01	.16		.17	
	Paint	L.F.	.07	.45		.52	
	Minimum Charge	Job		163		163	
1″ x 8″							
	Demolish	L.F.		.24		.24	Cost includes material and labor to
	Install	L.F.	1.61	1.63		3.24	install 1″ x 8″ pine trim.
	Demolish and Install	L.F.	1.61	1.87		3.48	
	Reinstall	L.F.		1.30		1.30	
	Clean	L.F.	.02	.17		.19	
	Paint	L.F.	.07	.45		.52	
	Minimum Charge	Job		163		163	

Exterior Trim

Exterior Trim		Unit	Material	Labor	Equip.	Total	Specification
1" x 10"							
	Demolish	L.F.		.36		.36	Cost includes material and labor to
	Install	L.F.	1.78	1.81		3.59	install 1" x 10" pine trim.
	Demolish and Install	L.F.	1.78	2.17		3.95	
	Minimum Charge	Job		163		163	
1" x 12"							
	Demolish	L.F.		.36		.36	Cost includes material and labor to
	Install	L.F.	1.95	1.81		3.76	install 1" x 12" pine trim.
	Demolish and Install	L.F.	1.95	2.17		4.12	
	Minimum Charge	Job		163		163	

Fascia		Unit	Material	Labor	Equip.	Total	Specification
Aluminum							
	Demolish	L.F.		.46		.46	Includes material and labor to install
	Install	S.F.	1.11	1.55		2.66	aluminum fascia.
	Demolish and Install	S.F.	1.11	2.01		3.12	
	Reinstall	S.F.		1.24		1.24	
	Clean	L.F.		.29		.29	
	Minimum Charge	Job		163		163	
Vinyl							
	Demolish	L.F.		.46		.46	Cost includes material and labor to
	Install	L.F.	2.34	1.86		4.20	install 6" vinyl fascia with 12" vinyl
	Demolish and Install	L.F.	2.34	2.32		4.66	soffit and J-channel.
	Reinstall	L.F.		1.49		1.49	
	Clean	L.F.		.29		.29	
	Minimum Charge	Job		163		163	
Plywood							
	Demolish	L.F.		.45		.45	Cost includes material and labor to
	Install	L.F.	1.13	.72		1.85	install 12" wide plywood fascia.
	Demolish and Install	L.F.	1.13	1.17		2.30	
	Reinstall	L.F.		.58		.58	
	Clean	L.F.	.03	.20		.23	
	Paint	L.F.	.14	.72		.86	
	Minimum Charge	Job		163		163	
Cedar							
	Demolish	L.F.		.45		.45	Cost includes material and labor to
	Install	L.F.	1.77	.72		2.49	install 12" wide cedar fascia.
	Demolish and Install	L.F.	1.77	1.17		2.94	
	Reinstall	L.F.		.58		.58	
	Clean	L.F.	.03	.20		.23	
	Paint	L.F.	.14	.72		.86	
	Minimum Charge	Job		163		163	
Redwood							
1" x 6"							
	Demolish	L.F.		.35		.35	Cost includes material and labor to
	Install	L.F.	2.41	1.30		3.71	install 1" x 6" redwood fascia board.
	Demolish and Install	L.F.	2.41	1.65		4.06	
	Reinstall	L.F.		1.04		1.04	
	Minimum Charge	Job		163		163	
1" x 8"							
	Demolish	L.F.		.39		.39	Cost includes material and labor to
	Install	L.F.	3.30	1.45		4.75	install 1" x 8" redwood fascia board.
	Demolish and Install	L.F.	3.30	1.84		5.14	
	Reinstall	L.F.		1.16		1.16	
	Minimum Charge	Job		163		163	

Exterior Trim

Fascia

	Unit	Material	Labor	Equip.	Total	Specification
2″ x 6″						
Demolish	L.F.		.35		.35	Cost includes material and labor to
Install	L.F.	2.33	1.45		3.78	install 2″ x 6″ redwood fascia board.
Demolish and Install	L.F.	2.33	1.80		4.13	
Reinstall	L.F.		1.16		1.16	
Minimum Charge	Job		163		163	
2″ x 8″						
Demolish	L.F.		.39		.39	Cost includes material and labor to
Install	L.F.	3.16	1.63		4.79	install 2″ x 8″ redwood fascia board.
Demolish and Install	L.F.	3.16	2.02		5.18	
Reinstall	L.F.		1.30		1.30	
Minimum Charge	Job		163		163	

Hem-fir Std & Better

	Unit	Material	Labor	Equip.	Total	Specification
2″ x 6″						
Demolish	L.F.		.35		.35	Cost includes material and labor to
Install	L.F.	.59	2.61		3.20	install 2″ x 6″ hem-fir exterior trim
Demolish and Install	L.F.	.59	2.96		3.55	board.
Reinstall	L.F.		2.09		2.09	
Clean	L.F.	.01	.20		.21	
Paint	L.F.	.14	.72		.86	
Minimum Charge	Job		163		163	
2″ x 8″						
Demolish	L.F.		.39		.39	Cost includes material and labor to
Install	L.F.	.91	2.90		3.81	install 2″ x 8″ hem-fir exterior trim
Demolish and Install	L.F.	.91	3.29		4.20	board.
Reinstall	L.F.		2.32		2.32	
Clean	L.F.	.02	.20		.22	
Paint	L.F.	.14	.72		.86	
Minimum Charge	Job		163		163	
2″ x 10″						
Demolish	L.F.		.45		.45	Cost includes material and labor to
Install	L.F.	1.30	3.62		4.92	install 2″ x 10″ hem-fir exterior trim
Demolish and Install	L.F.	1.30	4.07		5.37	board.
Reinstall	L.F.		2.90		2.90	
Clean	L.F.	.03	.21		.24	
Paint	L.F.	.14	.72		.86	
Minimum Charge	Job		163		163	

Barge Rafter

	Unit	Material	Labor	Equip.	Total	Specification
2″ x 6″						
Demolish	L.F.		.70		.70	Cost includes material and labor to
Install	L.F.	.59	.36		.95	install 2″ x 6″ barge rafters.
Demolish and Install	L.F.	.59	1.06		1.65	
Reinstall	L.F.		.29		.29	
Clean	L.F.	.01	.20		.21	
Paint	L.F.	.14	.72		.86	
Minimum Charge	Job		163		163	
2″ x 8″						
Demolish	L.F.		.71		.71	Cost includes material and labor to
Install	L.F.	.91	.38		1.29	install 2″ x 8″ barge rafters.
Demolish and Install	L.F.	.91	1.09		2	
Reinstall	L.F.		.31		.31	
Clean	L.F.	.02	.20		.22	
Paint	L.F.	.14	.72		.86	
Minimum Charge	Job		163		163	

Exterior Trim

Fascia

	Unit	Material	Labor	Equip.	Total	Specification
2″ x 10″						
Demolish	L.F.		.72		.72	Cost includes material and labor to
Install	L.F.	1.30	.40		1.70	install 2″ x 10″ barge rafters.
Demolish and Install	L.F.	1.30	1.12		2.42	
Reinstall	L.F.		.32		.32	
Clean	L.F.	.03	.21		.24	
Paint	L.F.	.14	.72		.86	
Minimum Charge	Job		163		163	

Soffit

	Unit	Material	Labor	Equip.	Total	Specification
Aluminum						
12″ w / Fascia						
Demolish	L.F.		.46		.46	Cost includes material and labor to
Install	L.F.	2.70	1.86		4.56	install 12″ aluminum soffit with 6″
Demolish and Install	L.F.	2.70	2.32		5.02	fascia and J-channel.
Reinstall	L.F.		1.49		1.49	
Clean	L.F.		.29		.29	
Paint	L.F.	.18	.72		.90	
Minimum Charge	Job		163		163	
18″ w / Fascia						
Demolish	L.F.		.46		.46	Cost includes material and labor to
Install	L.F.	3.21	2.17		5.38	install 18″ aluminum soffit with 6″
Demolish and Install	L.F.	3.21	2.63		5.84	fascia and J-channel.
Reinstall	L.F.		1.74		1.74	
Clean	L.F.		.29		.29	
Paint	L.F.	.18	.72		.90	
Minimum Charge	Job		163		163	
24″ w / Fascia						
Demolish	L.F.		.46		.46	Cost includes material and labor to
Install	L.F.	3.76	2.41		6.17	install 24″ aluminum soffit with 6″
Demolish and Install	L.F.	3.76	2.87		6.63	fascia and J-channel.
Reinstall	L.F.		1.93		1.93	
Clean	L.F.		.43		.43	
Paint	L.F.	.18	.72		.90	
Minimum Charge	Job		163		163	
Vinyl						
Demolish	L.F.		.46		.46	Cost includes material and labor to
Install	L.F.	2.34	1.86		4.20	install 6″ vinyl fascia with 12″ vinyl
Demolish and Install	L.F.	2.34	2.32		4.66	soffit and J-channel.
Reinstall	L.F.		1.49		1.49	
Clean	L.F.		.29		.29	
Paint	L.F.	.18	.72		.90	
Minimum Charge	Job		163		163	
Plywood						
Demolish	L.F.		.50		.50	Cost includes material and labor to
Install	L.F.	1.13	.93		2.06	install 12″ wide plywood soffit.
Demolish and Install	L.F.	1.13	1.43		2.56	
Reinstall	L.F.		.75		.75	
Clean	L.F.	.03	.20		.23	
Paint	L.F.	.14	.72		.86	
Minimum Charge	Job		163		163	

Exterior Trim

Soffit		Unit	Material	Labor	Equip.	Total	Specification
Redwood							
	Demolish	L.F.		.50		.50	Cost includes material and labor to
	Install	L.F.	4.82	.93		5.75	install 12″ wide redwood soffit.
	Demolish and Install	L.F.	4.82	1.43		6.25	
	Reinstall	L.F.		.75		.75	
	Clean	L.F.	.03	.20		.23	
	Paint	L.F.	.14	.72		.86	
	Minimum Charge	Job		163		163	
Cedar							
	Demolish	L.F.		.50		.50	Cost includes material and labor to
	Install	L.F.	1.77	.93		2.70	install 12″ wide cedar soffit.
	Demolish and Install	L.F.	1.77	1.43		3.20	
	Reinstall	L.F.		.75		.75	
	Clean	L.F.	.03	.20		.23	
	Paint	L.F.	.14	.72		.86	
	Minimum Charge	Job		163		163	
Douglas Fir							
	Demolish	L.F.		.50		.50	Cost includes material and labor to
	Install	L.F.	1.95	.93		2.88	install 12″ wide Douglas fir soffit.
	Demolish and Install	L.F.	1.95	1.43		3.38	
	Reinstall	L.F.		.75		.75	
	Clean	L.F.	.03	.20		.23	
	Paint	L.F.	.14	.72		.86	
	Minimum Charge	Job		163		163	

Roofing and Flashing

Tile Roofing

Metal Flashing	Unit	Material	Labor	Equip.	Total	Specification
Galvanized						
6″						
Demolish	L.F.		.17		.17	Includes material and labor to install
Install	L.F.	.41	.75		1.16	flat galvanized steel flashing, 6″ wide.
Demolish and Install	L.F.	.41	.92		1.33	
Paint	L.F.	.10	.36		.46	
Minimum Charge	Job		150		150	
14″						
Demolish	L.F.		.17		.17	Includes material and labor to install
Install	L.F.	.95	.75		1.70	flat galvanized steel flashing, 14″
Demolish and Install	L.F.	.95	.92		1.87	wide.
Paint	L.F.	.10	.36		.46	
Minimum Charge	Job		150		150	
20″						
Demolish	L.F.		.17		.17	Includes material and labor to install
Install	L.F.	1.34	.75		2.09	flat galvanized steel flashing, 20″
Demolish and Install	L.F.	1.34	.92		2.26	wide.
Paint	L.F.	.10	.36		.46	
Minimum Charge	Job		150		150	
Aluminum						
6″						
Demolish	L.F.		.17		.17	Includes material and labor to install
Install	L.F.	.25	.75		1	flat aluminum flashing, 6″ wide.
Demolish and Install	L.F.	.25	.92		1.17	
Paint	L.F.	.10	.36		.46	
Minimum Charge	Job		150		150	
14″						
Demolish	L.F.		.17		.17	Includes material and labor to install
Install	L.F.	.50	.75		1.25	flat aluminum flashing, 14″ wide.
Demolish and Install	L.F.	.50	.92		1.42	
Paint	L.F.	.10	.36		.46	
Minimum Charge	Job		150		150	
20″						
Demolish	L.F.		.17		.17	Includes material and labor to install
Install	L.F.	.65	.75		1.40	flat aluminum flashing, 20″ wide.
Demolish and Install	L.F.	.65	.92		1.57	
Paint	L.F.	.10	.36		.46	
Minimum Charge	Job		150		150	

Roofing

Metal Flashing	Unit	Material	Labor	Equip.	Total	Specification
Copper						
6"						
Demolish	L.F.		.17		.17	Includes material and labor to install
Install	L.F.	1.72	1		2.72	flat copper flashing, 6" wide.
Demolish and Install	L.F.	1.72	1.17		2.89	
Minimum Charge	Job		178		178	
14"						
Demolish	L.F.		.17		.17	Includes material and labor to install
Install	L.F.	3.98	1		4.98	flat copper flashing, 14" wide.
Demolish and Install	L.F.	3.98	1.17		5.15	
Minimum Charge	Job		178		178	
18"						
Demolish	L.F.		.17		.17	Includes material and labor to install
Install	L.F.	5.70	1		6.70	flat copper flashing, 18" wide.
Demolish and Install	L.F.	5.70	1.17		6.87	
Minimum Charge	Job		178		178	
Aluminum Valley						
Demolish	L.F.		.17		.17	Includes material and labor to install
Install	L.F.	2.53	1.92		4.45	aluminum valley flashing, .024" thick.
Demolish and Install	L.F.	2.53	2.09		4.62	
Paint	L.F.	.10	.36		.46	
Minimum Charge	Job		163		163	
Gravel Stop						
Demolish	L.F.		.16		.16	Includes material and labor to install
Install	L.F.	4.20	2.46		6.66	aluminum gravel stop, .050" thick, 4"
Demolish and Install	L.F.	4.20	2.62		6.82	high, mill finish.
Reinstall	L.F.		1.97		1.97	
Paint	L.F.	.15	.36		.51	
Minimum Charge	Job		163		163	
Drip Edge						
Demolish	L.F.		.16		.16	Includes material and labor to install
Install	L.F.	.28	.82		1.10	aluminum drip edge, .016" thick, 5"
Demolish and Install	L.F.	.28	.98		1.26	wide, mill finish.
Reinstall	L.F.		.65		.65	
Paint	L.F.	.15	.36		.51	
Minimum Charge	Job		163		163	
End Wall						
Demolish	L.F.		.81		.81	Includes material and labor to install
Install	L.F.	1.52	1.92		3.44	aluminum end wall flashing, .024"
Demolish and Install	L.F.	1.52	2.73		4.25	thick.
Minimum Charge	Job		163		163	
Side Wall						
Demolish	L.F.		.16		.16	Includes material and labor to install
Install	L.F.	1.51	1.92		3.43	aluminum side wall flashing, .024"
Demolish and Install	L.F.	1.51	2.08		3.59	thick.
Minimum Charge	Job		163		163	
Coping and Wall Cap						
Demolish	L.F.		1.09		1.09	Includes material and labor to install
Install	L.F.	1.60	2.07		3.67	aluminum flashing, mill finish, 0.40
Demolish and Install	L.F.	1.60	3.16		4.76	thick, 12" wide.
Clean	L.F.	.08	.74		.82	
Minimum Charge	Job		150		150	

Roofing

Roof Shingles

	Unit	Material	Labor	Equip.	Total	Specification
Underlayment						
15#						
Demolish	Sq.		8.75		8.75	Includes material and labor to install
Install	Sq.	3.75	4.68		8.43	#15 felt underlayment.
Demolish and Install	Sq.	3.75	13.43		17.18	
Minimum Charge	Job		150		150	
30#						
Demolish	Sq.		8.75		8.75	Includes material and labor to install
Install	Sq.	7.65	5.15		12.80	#30 felt underlayment.
Demolish and Install	Sq.	7.65	13.90		21.55	
Minimum Charge	Job		150		150	
Self Adhering						
Demolish	Sq.		22		22	Includes material and labor to install
Install	Sq.	46	13.60		59.60	self adhering ice barrier roofing
Demolish and Install	Sq.	46	35.60		81.60	underlayment.
Minimum Charge	Job		150		150	

Rolled Roofing

	Unit	Material	Labor	Equip.	Total	Specification
90 lb						
Demolish	Sq.		22		22	Cost includes material and labor to
Install	Sq.	16.90	19.95		36.85	install 90 lb mineral surface rolled
Demolish and Install	Sq.	16.90	41.95		58.85	roofing.
Minimum Charge	Job		150		150	
110 lb						
Demolish	Sq.		22		22	Cost includes material and labor to
Install	Sq.	44.50	19.95		64.45	install 110 lb mineral surface rolled
Demolish and Install	Sq.	44.50	41.95		86.45	roofing.
Minimum Charge	Job		150		150	
140 lb						
Demolish	Sq.		22		22	Cost includes material and labor to
Install	Sq.	44.50	30		74.50	install 140 lb mineral surface rolled
Demolish and Install	Sq.	44.50	52		96.50	roofing.
Minimum Charge	Job		150		150	

Composition Shingles

	Unit	Material	Labor	Equip.	Total	Specification
Fiberglass (3-tab)						
225 lb, 20 Year						
Demolish	Sq.		23.50		23.50	Cost includes material and labor to
Install	Sq.	28	58.50		86.50	install 225 lb, 20 year fiberglass
Demolish and Install	Sq.	28	82		110	shingles.
Minimum Charge	Job		150		150	
300 lb, 25 Year						
Demolish	Sq.		23.50		23.50	Cost includes material and labor to
Install	Sq.	28.50	58.50		87	install 300 lb, 25 year fiberglass
Demolish and Install	Sq.	28.50	82		110.50	shingles.
Minimum Charge	Job		150		150	
Emergency Repairs						
Install	Ea.	72	150		222	Includes labor and material for
Minimum Charge	Job		150		150	emergency repairs.

Roofing

Composition Shingles	Unit	Material	Labor	Equip.	Total	Specification
Architectural						
Laminated, 25 Year						
Demolish	Sq.		23.50		23.50	Cost includes material and labor to
Install	Sq.	51	58.50		109.50	install 25 year fiberglass roof shingles.
Demolish and Install	Sq.	51	82		133	
Minimum Charge	Job		150		150	
Laminated, 30 Year						
Demolish	Sq.		23.50		23.50	Cost includes material and labor to
Install	Sq.	56.50	58.50		115	install 30 year fiberglass roof shingles.
Demolish and Install	Sq.	56.50	82		138.50	
Minimum Charge	Job		150		150	
Laminated, 40 Year						
Demolish	Sq.		23.50		23.50	Cost includes material and labor to
Install	Sq.	73.50	58.50		132	install 40 year fiberglass roof shingles.
Demolish and Install	Sq.	73.50	82		155.50	
Minimum Charge	Job		150		150	
Laminated Shake, 40 Year						
Demolish	Sq.		23.50		23.50	Cost includes material and labor to
Install	Sq.	83	78		161	install 40 year fiberglass roof shakes.
Demolish and Install	Sq.	83	101.50		184.50	
Minimum Charge	Job		150		150	
Emergency Repairs						
Install	Ea.	72	150		222	Includes labor and material for
Minimum Charge	Job		150		150	emergency repairs.
Roof Jack						
Demolish	Ea.		4.76		4.76	Includes material and labor to install
Install	Ea.	13.55	32.50		46.05	residential roof jack, w/bird screen,
Demolish and Install	Ea.	13.55	37.26		50.81	backdraft damper, 3" & 4" dia. round
Paint	Ea.	.06	3.58		3.64	duct.
Minimum Charge	Job		178		178	
Ridge Vent						
Demolish	L.F.		.84		.84	Includes material and labor to install a
Install	L.F.	2.78	2.30		5.08	mill finish aluminum ridge vent strip.
Demolish and Install	L.F.	2.78	3.14		5.92	
Minimum Charge	Job		163		163	
Shingle Molding						
1" x 2"						
Demolish	L.F.		.16		.16	Cost includes material and labor to
Install	L.F.	.36	.99		1.35	install 1" x 2" pine trim.
Demolish and Install	L.F.	.36	1.15		1.51	
Paint	L.F.	.02	.45		.47	
Minimum Charge	Job		163		163	
1" x 3"						
Demolish	L.F.		.16		.16	Cost includes material and labor to
Install	L.F.	.44	1.12		1.56	install 1" x 3" pine trim.
Demolish and Install	L.F.	.44	1.28		1.72	
Paint	L.F.	.02	.45		.47	
Minimum Charge	Job		163		163	
1" x 4"						
Demolish	L.F.		.16		.16	Cost includes material and labor to
Install	L.F.	.75	1.30		2.05	install 1" x 4" pine trim.
Demolish and Install	L.F.	.75	1.46		2.21	
Paint	L.F.	.02	.45		.47	
Minimum Charge	Job		163		163	

Roofing

Composition Shingles

Composition Shingles		Unit	Material	Labor	Equip.	Total	Specification
Add for Steep Pitch	Install	Sq.		18.75		18.75	Includes labor and material added for 2nd story or steep roofs.
Add for Additional Story	Install	Sq.		18.75		18.75	Includes labor and material added for 2nd story or steep roofs.

Wood Shingles

Wood Shingles		Unit	Material	Labor	Equip.	Total	Specification
Cedar Wood Shingles							
16" Red Label (#2)							
	Demolish	Sq.		58		58	Includes material and labor to install
	Install	Sq.	133	130		263	wood shingles/shakes.
	Demolish and Install	Sq.	133	188		321	
	Minimum Charge	Job		163		163	
16" Blue Label (#1)							
	Demolish	Sq.		58		58	Includes material and labor to install
	Install	Sq.	179	100		279	wood shingles/shakes.
	Demolish and Install	Sq.	179	158		337	
	Minimum Charge	Job		163		163	
18" Red Label (#2)							
	Demolish	Sq.		58		58	Includes material and labor to install
	Install	Sq.	219	83.50		302.50	wood shingles/shakes.
	Demolish and Install	Sq.	219	141.50		360.50	
	Minimum Charge	Job		163		163	
18" Blue Label (#1)							
	Demolish	Sq.		58		58	Includes material and labor to install
	Install	Sq.	176	91.50		267.50	wood shingles/shakes.
	Demolish and Install	Sq.	176	149.50		325.50	
	Minimum Charge	Job		163		163	
Replace Shingle							
	Install	Ea.	13.20	8.15		21.35	Includes labor and material to replace
	Minimum Charge	Job		163		163	individual wood shingle/shake. Cost is per shingle.
Cedar Shake							
18" Medium Handsplit							
	Demolish	Sq.		58		58	Includes material and labor to install
	Install	Sq.	107	125		232	wood shingles/shakes.
	Demolish and Install	Sq.	107	183		290	
	Minimum Charge	Job		163		163	
18" Heavy Handsplit							
	Demolish	Sq.		58		58	Includes material and labor to install
	Install	Sq.	107	139		246	wood shingles/shakes.
	Demolish and Install	Sq.	107	197		304	
	Minimum Charge	Job		163		163	
24" Heavy Handsplit							
	Demolish	Sq.		58		58	Includes material and labor to install
	Install	Sq.	152	112		264	wood shingles/shakes.
	Demolish and Install	Sq.	152	170		322	
	Minimum Charge	Job		163		163	
24" Medium Handsplit							
	Demolish	Sq.		58		58	Includes material and labor to install
	Install	Sq.	152	100		252	wood shingles/shakes.
	Demolish and Install	Sq.	152	158		310	
	Minimum Charge	Job		163		163	

Roofing

Wood Shingles

Wood Shingles		Unit	Material	Labor	Equip.	Total	Specification
24″ Straight Split							
	Demolish	Sq.		58		58	Includes material and labor to install
	Install	Sq.	163	148		311	wood shingles/shakes.
	Demolish and Install	Sq.	163	206		369	
	Minimum Charge	Job		163		163	
24″ Tapersplit							
	Demolish	Sq.		58		58	Includes material and labor to install
	Install	Sq.	163	148		311	wood shingles/shakes.
	Demolish and Install	Sq.	163	206		369	
	Minimum Charge	Job		163		163	
18″ Straight Split							
	Demolish	Sq.		58		58	Includes material and labor to install
	Install	Sq.	121	148		269	wood shingles/shakes.
	Demolish and Install	Sq.	121	206		327	
	Minimum Charge	Job		163		163	
Replace Shingle							
	Install	Ea.	13.20	8.15		21.35	Includes labor and material to replace
	Minimum Charge	Job		163		163	individual wood shingle/shake. Cost is per shingle.

Slate Tile Roofing

Slate Tile Roofing		Unit	Material	Labor	Equip.	Total	Specification
Unfading Green							
	Demolish	Sq.		52.50		52.50	Includes material and labor to install
	Install	Sq.	380	173		553	clear Vermont slate tile.
	Demolish and Install	Sq.	380	225.50		605.50	
	Reinstall	Sq.		138.24		138.24	
	Minimum Charge	Job		151		151	
Unfading Purple							
	Demolish	Sq.		52.50		52.50	Includes material and labor to install
	Install	Sq.	405	173		578	clear Vermont slate tile.
	Demolish and Install	Sq.	405	225.50		630.50	
	Reinstall	Sq.		138.24		138.24	
	Minimum Charge	Job		151		151	
Replace Slate Tiles							
	Install	Ea.	5.95	15.90		21.85	Includes labor and material to replace
	Minimum Charge	Job		151		151	individual slate tiles. Cost is per tile.
Variegated Purple							
	Demolish	Sq.		52.50		52.50	Includes material and labor to install
	Install	Sq.	425	173		598	clear Vermont slate tile.
	Demolish and Install	Sq.	425	225.50		650.50	
	Reinstall	Sq.		138.24		138.24	
	Minimum Charge	Job		151		151	
Unfading Grey / Black							
	Demolish	Sq.		52.50		52.50	Includes material and labor to install
	Install	Sq.	375	173		548	clear Vermont slate tile.
	Demolish and Install	Sq.	375	225.50		600.50	
	Reinstall	Sq.		138.24		138.24	
	Minimum Charge	Job		151		151	
Unfading Red							
	Demolish	Sq.		52.50		52.50	Includes material and labor to install
	Install	Sq.	1200	173		1373	clear Vermont slate tile.
	Demolish and Install	Sq.	1200	225.50		1425.50	
	Reinstall	Sq.		138.24		138.24	
	Minimum Charge	Job		151		151	

Roofing

Slate Tile Roofing

Slate Tile Roofing		Unit	Material	Labor	Equip.	Total	Specification
Weathering Black							
	Demolish	Sq.		52.50		52.50	Includes material and labor to install
	Install	Sq.	495	173		668	clear Pennsylvania slate tile.
	Demolish and Install	Sq.	495	225.50		720.50	
	Reinstall	Sq.		138.24		138.24	
	Minimum Charge	Job		151		151	
Weathering Green							
	Demolish	Sq.		52.50		52.50	Includes material and labor to install
	Install	Sq.	365	173		538	clear Vermont slate tile.
	Demolish and Install	Sq.	365	225.50		590.50	
	Reinstall	Sq.		138.24		138.24	
	Minimum Charge	Job		151		151	

Flat Clay Tile Roofing

Flat Clay Tile Roofing		Unit	Material	Labor	Equip.	Total	Specification
Glazed							
	Demolish	Sq.		58		58	Includes material and labor to install
	Install	Sq.	495	121		616	tile roofing. Cost based on terra cotta
	Demolish and Install	Sq.	495	179		674	red flat clay tile and includes ridge
	Reinstall	Sq.		96.77		96.77	and rake tiles.
	Minimum Charge	Job		151		151	
Terra Cotta Red							
	Demolish	Sq.		58		58	Includes material and labor to install
	Install	Sq.	495	121		616	tile roofing. Cost based on terra cotta
	Demolish and Install	Sq.	495	179		674	red flat clay tile and includes ridge
	Reinstall	Sq.		96.77		96.77	and rake tiles.
	Minimum Charge	Job		151		151	

Mission Tile Roofing

Mission Tile Roofing		Unit	Material	Labor	Equip.	Total	Specification
Glazed Red							
	Demolish	Sq.		58		58	Includes material and labor to install
	Install	Sq.	715	202		917	tile roofing. Cost based on glazed red
	Demolish and Install	Sq.	715	260		975	tile and includes birdstop, ridge and
	Reinstall	Sq.		161.28		161.28	rake tiles.
	Minimum Charge	Job		151		151	
Unglazed Red							
	Demolish	Sq.		58		58	Includes material and labor to install
	Install	Sq.	715	202		917	tile roofing. Cost based on unglazed
	Demolish and Install	Sq.	715	260		975	red tile and includes birdstop, ridge
	Reinstall	Sq.		161.28		161.28	and rake tiles.
	Minimum Charge	Job		151		151	
Glazed Blue							
	Demolish	Sq.		58		58	Includes material and labor to install
	Install	Sq.	715	202		917	tile roofing. Cost based on glazed
	Demolish and Install	Sq.	715	260		975	blue tile and includes birdstop, ridge
	Reinstall	Sq.		161.28		161.28	and rake tiles.
	Minimum Charge	Job		151		151	
Glazed White							
	Demolish	Sq.		58		58	Includes material and labor to install
	Install	Sq.	715	202		917	tile roofing. Cost based on glazed
	Demolish and Install	Sq.	715	260		975	white tile and includes birdstop, ridge
	Reinstall	Sq.		161.28		161.28	and rake tiles.
	Minimum Charge	Job		151		151	

Roofing

Mission Tile Roofing		Unit	Material	Labor	Equip.	Total	Specification
Unglazed White							
	Demolish	Sq.		58		58	Includes material and labor to install
	Install	Sq.	715	202		917	tile roofing. Cost based on unglazed
	Demolish and Install	Sq.	715	260		975	white tile and includes birdstop, ridge
	Reinstall	Sq.		161.28		161.28	and rake tiles.
	Minimum Charge	Job		151		151	
Color Blend							
	Demolish	Sq.		58		58	Includes material and labor to install
	Install	Sq.	715	202		917	tile roofing. Cost based on color blend
	Demolish and Install	Sq.	715	260		975	tile and includes birdstop, ridge and
	Reinstall	Sq.		161.28		161.28	rake tiles.
	Minimum Charge	Job		151		151	
Accessories / Extras							
Add for Cap Furring Strips							
	Install	Sq.	78	189		267	Includes labor and material to install vertically placed wood furring strips under cap tiles.
Add for Tile Adhesive							
	Install	Sq.	13.35	11.05		24.40	Includes material and labor for adhesive material placed between tiles.
Add for Wire Attachment							
	Install	Sq.	13.20	131		144.20	Includes labor and material to install single wire anchors for attaching tile.
Add for Braided Runners							
	Install	Sq.	23.50	126		149.50	Includes labor and material to install braided wire runners for secure attachment in high wind areas.
Add for Stainless Steel Nails							
	Install	Sq.	18.20			18.20	Includes material only for stainless steel nails.
Add for Brass Nails							
	Install	Sq.	24			24	Includes material only for brass nails.
Add for Hurricane Clips							
	Install	Sq.	15.95	178		193.95	Includes material and labor for Hurricane or wind clips for high wind areas.
Add for Copper Nails							
	Install	Sq.	24			24	Includes material only for copper nails.
Add for Vertical Furring Strips							
	Install	Sq.	15.40	35		50.40	Includes labor and material to install vertically placed wood furring strips under tiles.
Add for Horizontal Furring Strips							
	Install	Sq.	9.90	15.75		25.65	Includes labor and material to install horizontally placed wood furring strips under tiles.

Roofing

Spanish Tile Roofing		Unit	Material	Labor	Equip.	Total	Specification
Glazed Red							
	Demolish	Sq.		58		58	Includes material and labor to install
	Install	Sq.	495	126		621	tile roofing. Cost based on glazed red
	Demolish and Install	Sq.	495	184		679	tile and includes birdstop, ridge and
	Reinstall	Sq.		100.80		100.80	rake tiles.
	Minimum Charge	Job		151		151	
Glazed White							
	Demolish	Sq.		58		58	Includes material and labor to install
	Install	Sq.	495	126		621	tile roofing. Cost based on glazed
	Demolish and Install	Sq.	495	184		679	white tile and includes birdstop, ridge
	Reinstall	Sq.		100.80		100.80	and rake tiles.
	Minimum Charge	Job		151		151	
Glazed Blue							
	Demolish	Sq.		58		58	Includes material and labor to install
	Install	Sq.	495	126		621	tile roofing. Cost based on glazed
	Demolish and Install	Sq.	495	184		679	blue tile and includes birdstop, ridge
	Reinstall	Sq.		100.80		100.80	and rake tiles.
	Minimum Charge	Job		151		151	
Unglazed Red							
	Demolish	Sq.		58		58	Includes material and labor to install
	Install	Sq.	495	126		621	tile roofing. Cost based on unglazed
	Demolish and Install	Sq.	495	184		679	red tile and includes birdstop, ridge
	Reinstall	Sq.		100.80		100.80	and rake tiles.
	Minimum Charge	Job		151		151	
Unglazed White							
	Demolish	Sq.		58		58	Includes material and labor to install
	Install	Sq.	495	126		621	tile roofing. Cost based on unglazed
	Demolish and Install	Sq.	495	184		679	white tile and includes birdstop, ridge
	Reinstall	Sq.		100.80		100.80	and rake tiles.
	Minimum Charge	Job		151		151	
Color Blend							
	Demolish	Sq.		58		58	Includes material and labor to install
	Install	Sq.	495	126		621	tile roofing. Cost based on color blend
	Demolish and Install	Sq.	495	184		679	tile and includes birdstop, ridge and
	Reinstall	Sq.		100.80		100.80	rake tiles.
	Minimum Charge	Job		151		151	
Accessories / Extras							
Add for Vertical Furring Strips							
	Install	Sq.	15.40	35		50.40	Includes labor and material to install vertically placed wood furring strips under tiles.
Add for Wire Attachment							
	Install	Sq.	11	101		112	Includes labor and material to install single wire ties.
Add for Braided Runners							
	Install	Sq.	19.45	86.50		105.95	Includes labor and material to install braided wire runners for secure attachment in high wind areas.
Add for Stainless Steel Nails							
	Install	Sq.	15.40			15.40	Includes material only for stainless steel nails.
Add for Copper Nails							
	Install	Sq.	20.50			20.50	Includes material only for copper nails.

Roofing

Spanish Tile Roofing

Spanish Tile Roofing		Unit	Material	Labor	Equip.	Total	Specification
Add for Brass Nails							
	Install	Sq.	20.50			20.50	Includes material only for brass nails.
Add for Hurricane Clips							
	Install	Sq.	12.20	104		116.20	Includes material and labor to install hurricane or wind clips for high wind areas.
Add for Mortar Set Tiles							
	Install	Sq.	38	151		189	Includes material and labor to install setting tiles by using mortar material placed between tiles.
Add for Tile Adhesive							
	Install	Sq.	12.40	9.80		22.20	Includes material and labor for adhesive placed between tiles.
Add for Cap Furring Strips							
	Install	Sq.	78	189		267	Includes labor and material to install vertically placed wood furring strips under cap tiles.
Add for Horizontal Furring Strips							
	Install	Sq.	9.90	15.75		25.65	Includes labor and material to install horizontally placed wood furring strips under tiles.

Concrete Tile Roofing

Concrete Tile Roofing		Unit	Material	Labor	Equip.	Total	Specification
Corrugated Red / Brown							
	Demolish	Sq.		58		58	Includes material and labor to install tile roofing. Cost includes birdstop, booster, ridge and rake tiles.
	Install	Sq.	121	224		345	
	Demolish and Install	Sq.	121	282		403	
	Reinstall	Sq.		179.20		179.20	
	Minimum Charge	Job		151		151	
Corrugated Gray							
	Demolish	Sq.		58		58	Includes material and labor to install tile roofing. Cost includes birdstop, booster, ridge and rake tiles.
	Install	Sq.	121	224		345	
	Demolish and Install	Sq.	121	282		403	
	Reinstall	Sq.		179.20		179.20	
	Minimum Charge	Job		151		151	
Corrugated Bright Red							
	Demolish	Sq.		58		58	Includes material and labor to install tile roofing. Cost includes birdstop, booster, ridge and rake tiles.
	Install	Sq.	121	224		345	
	Demolish and Install	Sq.	121	282		403	
	Reinstall	Sq.		179.20		179.20	
	Minimum Charge	Job		151		151	
Corrugated Black							
	Demolish	Sq.		58		58	Includes material and labor to install tile roofing. Cost includes birdstop, booster, ridge and rake tiles.
	Install	Sq.	121	224		345	
	Demolish and Install	Sq.	121	282		403	
	Reinstall	Sq.		179.20		179.20	
	Minimum Charge	Job		151		151	
Corrugated Green							
	Demolish	Sq.		58		58	Includes material and labor to install tile roofing. Cost includes birdstop, booster, ridge and rake tiles.
	Install	Sq.	121	224		345	
	Demolish and Install	Sq.	121	282		403	
	Reinstall	Sq.		179.20		179.20	
	Minimum Charge	Job		151		151	

Roofing

Concrete Tile Roofing

	Unit	Material	Labor	Equip.	Total	Specification
Corrugated Blue						
Demolish	Sq.		58		58	Includes material and labor to install
Install	Sq.	121	224		345	tile roofing. Cost includes birdstop,
Demolish and Install	Sq.	121	282		403	booster, ridge and rake tiles.
Reinstall	Sq.		179.20		179.20	
Minimum Charge	Job		151		151	
Flat Natural Gray						
Demolish	Sq.		58		58	Includes material and labor to install
Install	Sq.	105	224		329	tile roofing. Cost includes birdstop,
Demolish and Install	Sq.	105	282		387	booster, ridge and rake tiles.
Reinstall	Sq.		179.20		179.20	
Minimum Charge	Job		151		151	
Flat Red / Brown						
Demolish	Sq.		58		58	Includes material and labor to install
Install	Sq.	105	224		329	tile roofing. Cost includes birdstop,
Demolish and Install	Sq.	105	282		387	booster, ridge and rake tiles.
Reinstall	Sq.		179.20		179.20	
Minimum Charge	Job		151		151	
Flat Bright Red						
Demolish	Sq.		58		58	Includes material and labor to install
Install	Sq.	105	224		329	tile roofing. Cost includes birdstop,
Demolish and Install	Sq.	105	282		387	booster, ridge and rake tiles.
Reinstall	Sq.		179.20		179.20	
Minimum Charge	Job		151		151	
Flat Green						
Demolish	Sq.		58		58	Includes material and labor to install
Install	Sq.	105	224		329	tile roofing. Cost includes birdstop,
Demolish and Install	Sq.	105	282		387	booster, ridge and rake tiles.
Minimum Charge	Job		151		151	
Flat Blue						
Demolish	Sq.		58		58	Includes material and labor to install
Install	Sq.	105	224		329	tile roofing. Cost includes birdstop,
Demolish and Install	Sq.	105	282		387	booster, ridge and rake tiles.
Reinstall	Sq.		179.20		179.20	
Minimum Charge	Job		151		151	
Flat Black						
Demolish	Sq.		58		58	Includes material and labor to install
Install	Sq.	105	224		329	tile roofing. Cost includes birdstop,
Demolish and Install	Sq.	105	282		387	booster, ridge and rake tiles.
Reinstall	Sq.		179.20		179.20	
Minimum Charge	Job		151		151	

Aluminum Sheet

	Unit	Material	Labor	Equip.	Total	Specification
Corrugated						
.019" Thick Natural						
Demolish	S.F.		.62		.62	Includes material and labor to install
Install	S.F.	.88	1.48		2.36	aluminum sheet metal roofing.
Demolish and Install	S.F.	.88	2.10		2.98	
Reinstall	S.F.		1.19		1.19	
Minimum Charge	Job		355		355	
.016" Thick Natural						
Demolish	S.F.		.62		.62	Includes material and labor to install
Install	S.F.	.68	1.48		2.16	aluminum sheet metal roofing.
Demolish and Install	S.F.	.68	2.10		2.78	
Reinstall	S.F.		1.19		1.19	
Minimum Charge	Job		355		355	

Roofing

Aluminum Sheet	Unit	Material	Labor	Equip.	Total	Specification
.016" Colored Finish						
Demolish	S.F.		.62		.62	Includes material and labor to install
Install	S.F.	.98	1.48		2.46	aluminum sheet metal roofing.
Demolish and Install	S.F.	.98	2.10		3.08	
Reinstall	S.F.		1.19		1.19	
Minimum Charge	Job		355		355	
.019" Colored Finish						
Demolish	S.F.		.62		.62	Includes material and labor to install
Install	S.F.	1.09	1.48		2.57	aluminum sheet metal roofing.
Demolish and Install	S.F.	1.09	2.10		3.19	
Reinstall	S.F.		1.19		1.19	
Minimum Charge	Job		355		355	
Ribbed						
.019" Thick Natural						
Demolish	S.F.		.62		.62	Includes material and labor to install
Install	S.F.	.88	1.48		2.36	aluminum sheet metal roofing.
Demolish and Install	S.F.	.88	2.10		2.98	
Reinstall	S.F.		1.19		1.19	
Minimum Charge	Job		355		355	
.016" Thick Natural						
Demolish	S.F.		.62		.62	Includes material and labor to install
Install	S.F.	.68	1.48		2.16	aluminum sheet metal roofing.
Demolish and Install	S.F.	.68	2.10		2.78	
Reinstall	S.F.		1.19		1.19	
Minimum Charge	Job		355		355	
.050" Colored Finish						
Demolish	S.F.		.62		.62	Includes material and labor to install
Install	S.F.	3.05	1.48		4.53	aluminum sheet metal roofing.
Demolish and Install	S.F.	3.05	2.10		5.15	
Reinstall	S.F.		1.19		1.19	
Minimum Charge	Job		355		355	
.032" Thick Natural						
Demolish	S.F.		.62		.62	Includes material and labor to install
Install	S.F.	1.71	1.48		3.19	aluminum sheet metal roofing.
Demolish and Install	S.F.	1.71	2.10		3.81	
Reinstall	S.F.		1.19		1.19	
Minimum Charge	Job		355		355	
.040" Thick Natural						
Demolish	S.F.		.62		.62	Includes material and labor to install
Install	S.F.	2.08	1.48		3.56	aluminum sheet metal roofing.
Demolish and Install	S.F.	2.08	2.10		4.18	
Reinstall	S.F.		1.19		1.19	
Minimum Charge	Job		355		355	
.050" Thick Natural						
Demolish	S.F.		.62		.62	Includes material and labor to install
Install	S.F.	2.52	1.48		4	aluminum sheet metal roofing.
Demolish and Install	S.F.	2.52	2.10		4.62	
Reinstall	S.F.		1.19		1.19	
Minimum Charge	Job		355		355	
.016" Colored Finish						
Demolish	S.F.		.62		.62	Includes material and labor to install
Install	S.F.	.98	1.48		2.46	aluminum sheet metal roofing.
Demolish and Install	S.F.	.98	2.10		3.08	
Reinstall	S.F.		1.19		1.19	
Minimum Charge	Job		355		355	

Roofing

Aluminum Sheet

Aluminum Sheet	Unit	Material	Labor	Equip.	Total	Specification
.019″ Colored Finish						
Demolish	S.F.		.62		.62	Includes material and labor to install
Install	S.F.	1.09	1.48		2.57	aluminum sheet metal roofing.
Demolish and Install	S.F.	1.09	2.10		3.19	
Reinstall	S.F.		1.19		1.19	
Minimum Charge	Job		355		355	
.032″ Colored Finish						
Demolish	S.F.		.62		.62	Includes material and labor to install
Install	S.F.	2.12	1.48		3.60	aluminum sheet metal roofing.
Demolish and Install	S.F.	2.12	2.10		4.22	
Reinstall	S.F.		1.19		1.19	
Minimum Charge	Job		355		355	
.040″ Colored Finish						
Demolish	S.F.		.62		.62	Includes material and labor to install
Install	S.F.	2.53	1.48		4.01	aluminum sheet metal roofing.
Demolish and Install	S.F.	2.53	2.10		4.63	
Reinstall	S.F.		1.19		1.19	
Minimum Charge	Job		355		355	

Fiberglass Sheet

Fiberglass Sheet	Unit	Material	Labor	Equip.	Total	Specification
12 Ounce Corrugated						
Demolish	S.F.		1.12		1.12	Cost includes material and labor to
Install	S.F.	3.55	1.18		4.73	install 12 ounce corrugated fiberglass
Demolish and Install	S.F.	3.55	2.30		5.85	sheet roofing.
Reinstall	S.F.		.95		.95	
Minimum Charge	Job		150		150	
8 Ounce Corrugated						
Demolish	S.F.		1.12		1.12	Cost includes material and labor to
Install	S.F.	2.46	1.18		3.64	install 8 ounce corrugated fiberglass
Demolish and Install	S.F.	2.46	2.30		4.76	roofing.
Reinstall	S.F.		.95		.95	
Minimum Charge	Job		150		150	

Galvanized Steel

Galvanized Steel	Unit	Material	Labor	Equip.	Total	Specification
Corrugated						
24 Gauge						
Demolish	S.F.		.62		.62	Includes material and labor to install
Install	S.F.	1.16	1.25		2.41	galvanized steel sheet metal roofing.
Demolish and Install	S.F.	1.16	1.87		3.03	Cost based on 24 gauge corrugated
Reinstall	S.F.		1		1	or ribbed steel roofing.
Minimum Charge	Job		150		150	
26 Gauge						
Demolish	S.F.		.62		.62	Includes material and labor to install
Install	S.F.	1.07	1.18		2.25	galvanized steel sheet metal roofing.
Demolish and Install	S.F.	1.07	1.80		2.87	Cost based on 26 gauge corrugated
Reinstall	S.F.		.95		.95	or ribbed steel roofing.
Minimum Charge	Job		150		150	
28 Gauge						
Demolish	S.F.		.62		.62	Includes material and labor to install
Install	S.F.	.89	1.13		2.02	galvanized steel sheet metal roofing.
Demolish and Install	S.F.	.89	1.75		2.64	Cost based on 28 gauge corrugated
Reinstall	S.F.		.90		.90	or ribbed steel roofing.
Minimum Charge	Job		150		150	

Roofing

Galvanized Steel		Unit	Material	Labor	Equip.	Total	Specification
30 Gauge							
	Demolish	S.F.		.62		.62	Includes material and labor to install
	Install	S.F.	.85	1.08		1.93	galvanized steel sheet metal roofing.
	Demolish and Install	S.F.	.85	1.70		2.55	Cost based on 30 gauge corrugated
	Reinstall	S.F.		.86		.86	or ribbed steel roofing.
	Minimum Charge	Job		150		150	
Ribbed							
24 Gauge							
	Demolish	S.F.		.62		.62	Includes material and labor to install
	Install	S.F.	1.16	1.25		2.41	galvanized steel sheet metal roofing.
	Demolish and Install	S.F.	1.16	1.87		3.03	Cost based on 24 gauge corrugated
	Reinstall	S.F.		1		1	or ribbed steel roofing.
	Minimum Charge	Job		150		150	
26 Gauge							
	Demolish	S.F.		.62		.62	Includes material and labor to install
	Install	S.F.	1.07	1.18		2.25	galvanized steel sheet metal roofing.
	Demolish and Install	S.F.	1.07	1.80		2.87	Cost based on 26 gauge corrugated
	Reinstall	S.F.		.95		.95	or ribbed steel roofing.
	Minimum Charge	Job		150		150	
28 Gauge							
	Demolish	S.F.		.62		.62	Includes material and labor to install
	Install	S.F.	.89	1.13		2.02	galvanized steel sheet metal roofing.
	Demolish and Install	S.F.	.89	1.75		2.64	Cost based on 28 gauge corrugated
	Reinstall	S.F.		.90		.90	or ribbed steel roofing.
	Minimum Charge	Job		150		150	
30 Gauge							
	Demolish	S.F.		.62		.62	Includes material and labor to install
	Install	S.F.	.85	1.08		1.93	galvanized steel sheet metal roofing.
	Demolish and Install	S.F.	.85	1.70		2.55	Cost based on 30 gauge corrugated
	Reinstall	S.F.		.86		.86	or ribbed steel roofing.
	Minimum Charge	Job		150		150	

Standing Seam		Unit	Material	Labor	Equip.	Total	Specification
Copper							
16 Ounce							
	Demolish	Sq.		62.50		62.50	Includes material and labor to install
	Install	Sq.	430	274		704	copper standing seam metal roofing.
	Demolish and Install	Sq.	430	336.50		766.50	Cost based on 16 ounce copper
	Reinstall	Sq.		219.32		219.32	roofing.
	Minimum Charge	Job		178		178	
18 Ounce							
	Demolish	Sq.		62.50		62.50	Includes material and labor to install
	Install	Sq.	480	297		777	copper standing seam metal roofing.
	Demolish and Install	Sq.	480	359.50		839.50	Cost based on 18 ounce copper
	Reinstall	Sq.		237.60		237.60	roofing.
	Minimum Charge	Job		178		178	
20 Ounce							
	Demolish	Sq.		62.50		62.50	Includes material and labor to install
	Install	Sq.	510	325		835	copper standing seam metal roofing.
	Demolish and Install	Sq.	510	387.50		897.50	Cost based on 20 ounce copper
	Reinstall	Sq.		259.20		259.20	roofing.

Roofing

Standing Seam

Standing Seam	Unit	Material	Labor	Equip.	Total	Specification
Stainless Steel						
26 Gauge						
Demolish	Sq.		62.50		62.50	Cost includes material and labor to
Install	Sq.	420	310		730	install 26 gauge stainless steel
Demolish and Install	Sq.	420	372.50		792.50	standing seam metal roofing.
Reinstall	Sq.		247.93		247.93	
Minimum Charge	Job		178		178	
28 Gauge						
Demolish	Sq.		62.50		62.50	Includes material and labor to install
Install	Sq.	335	297		632	standing seam metal roofing. Cost
Demolish and Install	Sq.	335	359.50		694.50	based on 28 gauge stainless steel
Reinstall	Sq.		237.60		237.60	roofing.
Minimum Charge	Job		178		178	

Elastomeric Roofing

Elastomeric Roofing	Unit	Material	Labor	Equip.	Total	Specification
Neoprene						
1/16"						
Demolish	Sq.		21.50		21.50	Includes material and labor to install
Install	Sq.	163	143	24	330	neoprene membrane, fully adhered,
Demolish and Install	Sq.	163	164.50	24	351.50	1/16" thick.
Minimum Charge	Job		685	80.50	765.50	
GRM Membrane						
Demolish	Sq.		21.50		21.50	Includes material and labor to install
Install	Sq.	239	228	27	494	coal tar based elastomeric roofing
Demolish and Install	Sq.	239	249.50	27	515.50	membrane, polyester fiber reinforced.
Minimum Charge	Job		685	80.50	765.50	
Cant Strips						
3"						
Demolish	L.F.		.67		.67	Includes material and labor to install
Install	L.F.	.44	.92		1.36	cants 3" x 3", treated timber, cut
Demolish and Install	L.F.	.44	1.59		2.03	diagonally.
Minimum Charge	Job		150		150	
4"						
Demolish	L.F.		.67		.67	Includes material and labor to install
Install	L.F.	1.49	.92		2.41	cants 4" x 4", treated timber, cut
Demolish and Install	L.F.	1.49	1.59		3.08	diagonally.
Minimum Charge	Job		150		150	
Polyvinyl Chloride (PVC)						
45 mil (loose laid)						
Demolish	Sq.		21.50		21.50	Cost includes material and labor to
Install	Sq.	89.50	27	3.16	119.66	install 45 mil PVC elastomeric roofing,
Demolish and Install	Sq.	89.50	48.50	3.16	141.16	loose-laid with stone.
Minimum Charge	Job		685	80.50	765.50	
48 mil (loose laid)						
Demolish	Sq.		21.50		21.50	Cost includes material and labor to
Install	Sq.	104	27	3.16	134.16	install 48 mil PVC elastomeric roofing,
Demolish and Install	Sq.	104	48.50	3.16	155.66	loose-laid with stone.
Minimum Charge	Job		685	80.50	765.50	
60 mil (loose laid)						
Demolish	Sq.		21.50		21.50	Cost includes material and labor to
Install	Sq.	96.50	27	3.16	126.66	install 60 mil PVC elastomeric roofing,
Demolish and Install	Sq.	96.50	48.50	3.16	148.16	loose-laid with stone.
Minimum Charge	Job		685	80.50	765.50	

Elastomeric Roofing	Unit	Material	Labor	Equip.	Total	Specification
45 mil (attached)						
Demolish	Sq.		32		32	Cost includes material and labor to
Install	Sq.	78.50	52.50	6.20	137.20	install 45 mil PVC elastomeric roofing
Demolish and Install	Sq.	78.50	84.50	6.20	169.20	fully attached to roof deck.
Minimum Charge	Job		685	80.50	765.50	
48 mil (attached)						
Demolish	Sq.		32		32	Cost includes material and labor to
Install	Sq.	127	52.50	6.20	185.70	install 48 mil PVC elastomeric roofing
Demolish and Install	Sq.	127	84.50	6.20	217.70	fully attached to roof deck.
Minimum Charge	Job		685	80.50	765.50	
60 mil (attached)						
Demolish	Sq.		32		32	Cost includes material and labor to
Install	Sq.	119	52.50	6.20	177.70	install 60 mil PVC elastomeric roofing
Demolish and Install	Sq.	119	84.50	6.20	209.70	fully attached to roof deck.
Minimum Charge	Job		685	80.50	765.50	
CSPE Type						
45 mil (loose laid w / stone)						
Demolish	Sq.		21.50		21.50	Cost includes material and labor to
Install	Sq.	143	27	3.16	173.16	install 45 mil CSPE (chlorosulfonated
Demolish and Install	Sq.	143	48.50	3.16	194.66	polyethylene-hypalon) roofing material
Minimum Charge	Job		685	80.50	765.50	loose laid on deck and ballasted with
						stone.
45 mil (attached at seams)						
Demolish	Sq.		51.50		51.50	Cost includes material and labor to
Install	Sq.	138	39	4.61	181.61	install 45 mil CSPE (chlorosulfonated
Demolish and Install	Sq.	138	90.50	4.61	233.11	polyethylene-hypalon) roofing material
Minimum Charge	Job		685	80.50	765.50	attached to deck at seams with batten
						strips.
45 mil (fully attached)						
Demolish	Sq.		32		32	Cost includes material and labor to
Install	Sq.	135	52.50	6.20	193.70	install 45 mil CSPE (chlorosulfonated
Demolish and Install	Sq.	135	84.50	6.20	225.70	polyethylene-hypalon) roofing material
Minimum Charge	Job		685	80.50	765.50	fully attached to deck.
EPDM Type						
45 mil (attached at seams)						
Demolish	Sq.		51.50		51.50	Cost includes material and labor to
Install	Sq.	52.50	39	4.61	96.11	install 45 mil EPDM (ethylene
Demolish and Install	Sq.	52.50	90.50	4.61	147.61	propylene diene monomer) roofing
Minimum Charge	Job		685	80.50	765.50	material attached to deck at seams.
45 mil (fully attached)						
Demolish	Sq.		32		32	Cost includes material and labor to
Install	Sq.	74.50	52.50	6.20	133.20	install 45 mil EPDM (ethylene
Demolish and Install	Sq.	74.50	84.50	6.20	165.20	propylene diene monomer) roofing
Minimum Charge	Job		685	80.50	765.50	material fully attached to deck.
45 mil (loose laid w / stone)						
Demolish	Sq.		21.50		21.50	Cost includes material and labor to
Install	Sq.	61	27	3.16	91.16	install 45 mil EPDM (ethylene
Demolish and Install	Sq.	61	48.50	3.16	112.66	propylene diene monomer) roofing
Minimum Charge	Job		685	80.50	765.50	material loose laid on deck and
						ballasted with stones (approx. 1/2 ton
						of stone per square).
60 mil (attached at seams)						
Demolish	Sq.		51.50		51.50	Cost includes material and labor to
Install	Sq.	64.50	39	4.61	108.11	install 60 mil EPDM (ethylene
Demolish and Install	Sq.	64.50	90.50	4.61	159.61	propylene diene monomer) roofing
Minimum Charge	Job		685	80.50	765.50	material attached to deck at seams.

Roofing

Elastomeric Roofing

Elastomeric Roofing	Unit	Material	Labor	Equip.	Total	Specification
60 mil (fully attached)						Cost includes material and labor to
Demolish	Sq.		32		32	install 60 mil EPDM (ethylene
Install	Sq.	87	52.50	6.20	145.70	propylene diene monomer) roofing
Demolish and Install	Sq.	87	84.50	6.20	177.70	material fully attached to deck.
Minimum Charge	Job		685	80.50	765.50	
60 mil (loose laid w / stone)						Cost includes material and labor to
Demolish	Sq.		21.50		21.50	install 60 mil EPDM (ethylene
Install	Sq.	74.50	27	3.16	104.66	propylene diene monomer) roofing
Demolish and Install	Sq.	74.50	48.50	3.16	126.16	material loose laid on deck and
Minimum Charge	Job		685	80.50	765.50	ballasted with stones (approx. 1/2 ton of stone per square).
Silicone-urethane Foam						
Prep Roof for Foam System						
Install	Sq.	5.25	120		125.25	Includes labor and material to removal
Minimum Charge	Job		685	80.50	765.50	of any gravel and prepare existing built-up roof for application of silicone-urethane foam system.

Modified Bitumen Roofing

Modified Bitumen Roofing	Unit	Material	Labor	Equip.	Total	Specification
Fully Attached						
120 mil						
Demolish	Sq.		51.50		51.50	Cost includes material and labor to
Install	Sq.	78	6.65	1.11	85.76	install 120 mil modified bitumen
Demolish and Install	Sq.	78	58.15	1.11	137.26	roofing, fully attached to deck by
Minimum Charge	Job		150		150	torch.
150 mil						
Demolish	Sq.		51.50		51.50	Cost includes material and labor to
Install	Sq.	112	16.40	2.73	131.13	install 150 mil modified bitumen
Demolish and Install	Sq.	112	67.90	2.73	182.63	roofing, fully attached to deck by
Minimum Charge	Job		985	164	1149	torch.
160 mil						
Demolish	Sq.		51.50		51.50	Cost includes material and labor to
Install	Sq.	89.50	17.10	2.85	109.45	install 160 mil modified bitumen
Demolish and Install	Sq.	89.50	68.60	2.85	160.95	roofing, fully attached to deck by
Minimum Charge	Job		985	164	1149	torch.

Built-up Roofing

Built-up Roofing	Unit	Material	Labor	Equip.	Total	Specification
Emergency Repairs						
Install	Ea.	142	1400	234	1776	Minimum charge for work requiring
Minimum Charge	Job		985	164	1149	hot asphalt.
Asphalt						
3-ply Asphalt						
Demolish	Sq.		53.50		53.50	Cost includes material and labor to
Install	Sq.	39.50	82	13.65	135.15	install 3-ply asphalt built-up roof
Demolish and Install	Sq.	39.50	135.50	13.65	188.65	including one 30 lb and two 15 lb
Minimum Charge	Job		985	164	1149	layers of felt.

Roofing

Built-up Roofing		Unit	Material	Labor	Equip.	Total	Specification
4-ply Asphalt							Cost includes material and labor to
	Demolish	Sq.		64		64	install 4-ply asphalt built-up roof
	Install	Sq.	53.50	89.50	14.90	157.90	including one 30 lb and three 15 lb
	Demolish and Install	Sq.	53.50	153.50	14.90	221.90	layers of felt.
	Minimum Charge	Job		985	164	1149	
5-ply Asphalt							Cost includes material and labor to
	Demolish	Sq.		83.50		83.50	install 5-ply asphalt built-up roof
	Install	Sq.	76	98.50	16.40	190.90	including one 30 lb and four 15 lb
	Demolish and Install	Sq.	76	182	16.40	274.40	layers of felt.
	Minimum Charge	Job		985	164	1149	
Components / Accessories							
Flood Coat							Includes labor and material to install
	Install	Sq.	13.20	46		59.20	flood coat.
R / R Gravel and Flood Coat							Includes labor and material to install
	Install	Sq.	17.75	150		167.75	flood coat and gravel.
Aluminum UV Coat							Includes labor and material to install
	Install	Sq.	14.15	171	20	205.15	aluminum-based UV coating.
Gravel Coat							Includes labor and material to install
	Demolish	Sq.		27		27	gravel coat.
	Install	Sq.	4.59	25		29.59	
	Demolish and Install	Sq.	4.59	52		56.59	
Mineral Surface Cap Sheet							Cost includes labor and material to
	Install	Sq.	18.15	5.15		23.30	install 50 # mineral surfaced cap sheet.
Fibrated Aluminum UV Coating							Includes labor and material to install
	Install	Sq.	16.10	46		62.10	fibered aluminum-based UV coating.
Impregnated Walk							
1"							Cost includes labor and material to
	Install	S.F.	2.37	1.60		3.97	install 1" layer of asphalt coating.
1/2"							Cost includes labor and material to
	Install	S.F.	1.60	1.26		2.86	install 1/2" layer of asphalt coating.
3/4"							Cost includes labor and material to
	Install	S.F.	1.98	1.33		3.31	install 3/4" layer of asphalt coating.

Exterior Doors and Windows

Doors	Roof Window	Double Hung Window

Exterior Type Door		Unit	Material	Labor	Equip.	Total	Specification
Entry							
Flush (1-3/4")							
	Demolish	Ea.		30		30	Cost includes material and labor to
	Install	Ea.	320	17.15		337.15	install 3' x 6' 8" pre-hung 1-3/4"
	Demolish and Install	Ea.	320	47.15		367.15	door, including casing and stop,
	Reinstall	Ea.		17.16		17.16	hinges, jamb, aluminum sill,
	Clean	Ea.	2.92	9.80		12.72	weatherstripped.
	Paint	Ea.	7.75	41		48.75	
	Minimum Charge	Job		163		163	
Raised Panel							
	Demolish	Ea.		30		30	Cost includes material and labor to
	Install	Ea.	375	18.10		393.10	install 3' x 6' 8" pre-hung 1-3/4"
	Demolish and Install	Ea.	375	48.10		423.10	door, jamb, including casing and
	Reinstall	Ea.		18.11		18.11	stop, hinges, aluminum sill,
	Clean	Ea.	2.92	9.80		12.72	weatherstripping.
	Paint	Ea.	7.75	41		48.75	
	Minimum Charge	Job		163		163	
Entry w / Lights							
Good Quality							
	Demolish	Ea.		30		30	Cost includes material and labor to
	Install	Ea.	345	19.20		364.20	install 3' x 6' 8" pre-hung 1-3/4"
	Demolish and Install	Ea.	345	49.20		394.20	door, solid wood lower panel, with up
	Reinstall	Ea.		19.18		19.18	to 9-light glass panels, casing, sill,
	Clean	Ea.	2.92	9.80		12.72	weatherstripped.
	Paint	Ea.	5.70	19.05		24.75	
	Minimum Charge	Job		163		163	
Better Quality							
	Demolish	Ea.		30		30	Cost includes material and labor to
	Install	Ea.	375	19.20		394.20	install 3' x 6' 8" pre-hung 1-3/4" Ash
	Demolish and Install	Ea.	375	49.20		424.20	door, solid wood, raised panel door
	Reinstall	Ea.		19.18		19.18	without glass panels.
	Clean	Ea.	2.92	9.80		12.72	
	Paint	Ea.	7.75	41		48.75	
	Minimum Charge	Job		163		163	
Premium Quality							
	Demolish	Ea.		30		30	Cost includes material and labor to
	Install	Ea.	610	19.20		629.20	install 3' x 6' 8" pre-hung 1-3/4" Oak
	Demolish and Install	Ea.	610	49.20		659.20	door, solid wood, raised panel door
	Reinstall	Ea.		19.18		19.18	without glass panels.
	Clean	Ea.	2.92	9.80		12.72	
	Paint	Ea.	7.75	41		48.75	
	Minimum Charge	Job		163		163	

Exterior Doors and Windows

Exterior Type Door		Unit	Material	Labor	Equip.	Total	Specification
Custom Quality							
	Demolish	Ea.		30		30	Cost includes material and labor to
	Install	Ea.	1200	19.20		1219.20	install 3' x 6' 8" pre-hung 1-3/4"
	Demolish and Install	Ea.	1200	49.20		1249.20	hand-made, solid wood, raised panel
	Reinstall	Ea.		19.18		19.18	door with custom stained glass.
	Clean	Ea.	.20	9.80		10	
	Paint	Ea.	2.15	95.50		97.65	
	Minimum Charge	Job		163		163	
Miami Style							
	Demolish	Ea.		30		30	Cost includes material and labor to
	Install	Ea.	204	16.30		220.30	install 3' x 6' 8" pre-hung 1-3/4"
	Demolish and Install	Ea.	204	46.30		250.30	lauan door of solid wood.
	Reinstall	Ea.		16.30		16.30	
	Clean	Ea.	.20	9.80		10	
	Paint	Ea.	7.75	41		48.75	
	Minimum Charge	Job		163		163	
French Style							
1-Light							
	Demolish	Ea.		30		30	Cost includes material and labor to
	Install	Ea.	340	23.50		363.50	install 3' x 6' 8" pre-hung 1-3/4"
	Demolish and Install	Ea.	340	53.50		393.50	french style door with 1 light.
	Reinstall	Ea.		23.29		23.29	
	Clean	Ea.	1.83	14.75		16.58	
	Paint	Ea.	9.15	71.50		80.65	
	Minimum Charge	Job		163		163	
5-Light							
	Demolish	Ea.		30		30	Cost includes material and labor to
	Install	Ea.	340	23.50		363.50	install 3' x 6' 8" pre-hung 1-3/4"
	Demolish and Install	Ea.	340	53.50		393.50	french style door with 5 lights.
	Reinstall	Ea.		23.29		23.29	
	Clean	Ea.	.40	19.65		20.05	
	Paint	Ea.	9.15	71.50		80.65	
	Minimum Charge	Job		163		163	
9-Light							
	Demolish	Ea.		30		30	Cost includes material and labor to
	Install	Ea.	325	23.50		348.50	install 3' x 6' 8" pre-hung 1-3/4"
	Demolish and Install	Ea.	325	53.50		378.50	french style door with 9 lights.
	Reinstall	Ea.		23.29		23.29	
	Clean	Ea.	.40	19.65		20.05	
	Paint	Ea.	9.15	71.50		80.65	
	Minimum Charge	Job		163		163	
10-Light							
	Demolish	Ea.		32.50		32.50	Cost includes material and labor to
	Install	Ea.	550	41		591	install 3' x 6' 8" pre-hung 1-3/4"
	Demolish and Install	Ea.	550	73.50		623.50	french style door with 10 lights.
	Reinstall	Ea.		40.75		40.75	
	Clean	Ea.	.40	19.65		20.05	
	Paint	Ea.	9.15	71.50		80.65	
	Minimum Charge	Job		163		163	
15-Light							
	Demolish	Ea.		32.50		32.50	Cost includes material and labor to
	Install	Ea.	615	46.50		661.50	install 3' x 6' 8" pre-hung 1-3/4"
	Demolish and Install	Ea.	615	79		694	french style door with 15 lights.
	Reinstall	Ea.		46.57		46.57	
	Clean	Ea.	.40	19.65		20.05	
	Paint	Ea.	9.15	71.50		80.65	
	Minimum Charge	Job		163		163	

Exterior Doors and Windows

Exterior Type Door		Unit	Material	Labor	Equip.	Total	Specification
Jamb							
Wood							
	Demolish	Ea.		18.10		18.10	Includes material and labor to install
	Install	Ea.	80	8.80		88.80	flat pine jamb with square cut heads
	Demolish and Install	Ea.	80	26.90		106.90	and rabbeted sides for 6' 8" high and
	Clean	Ea.	.09	5.90		5.99	3-9/16" door including trim sets for
	Paint	Ea.	5.70	19.05		24.75	both sides.
	Minimum Charge	Job		163		163	
Metal 6' 8" High							
	Demolish	Ea.		18.10		18.10	Cost includes material and labor to
	Install	Ea.	85.50	20.50		106	install 18 gauge hollow metal door
	Demolish and Install	Ea.	85.50	38.60		124.10	frame to fit a 4-1/2" jamb 6' 8" high
	Reinstall	Ea.		20.38		20.38	and up to 3' 6" opening.
	Clean	Ea.	.61	4.91		5.52	
	Paint	Ea.	1.07	5.70		6.77	
	Minimum Charge	Job		163		163	
Metal 7' High							
	Demolish	Ea.		18.10		18.10	Cost includes material and labor to
	Install	Ea.	87.50	21.50		109	install 18 gauge hollow metal door
	Demolish and Install	Ea.	87.50	39.60		127.10	frame to fit a 4-1/2" jamb 7' high and
	Reinstall	Ea.		21.73		21.73	up to 3' 6" opening.
	Clean	Ea.	.61	4.91		5.52	
	Paint	Ea.	1.07	5.70		6.77	
	Minimum Charge	Job		163		163	
Metal 8' High							
	Demolish	Ea.		18.10		18.10	Cost includes material and labor to
	Install	Ea.	106	27		133	install 18 gauge hollow metal door
	Demolish and Install	Ea.	106	45.10		151.10	frame to fit a 4-1/2" jamb 8' high and
	Reinstall	Ea.		27.17		27.17	up to 3' 6" opening.
	Clean	Ea.	.86	5.90		6.76	
	Paint	Ea.	1.20	6.35		7.55	
	Minimum Charge	Job		163		163	
Metal 9' High							
	Demolish	Ea.		18.10		18.10	Cost includes material and labor to
	Install	Ea.	121	32.50		153.50	install 18 gauge hollow metal door
	Demolish and Install	Ea.	121	50.60		171.60	frame to fit a 4-1/2" jamb 9' high and
	Reinstall	Ea.		32.60		32.60	up to 3' 6" opening.
	Clean	Ea.	.86	5.90		6.76	
	Paint	Ea.	1.20	6.35		7.55	
	Minimum Charge	Job		163		163	
Casing Trim							
Single Width							
	Demolish	Ea.		5.75		5.75	Cost includes material and labor to
	Install	Opng.	15.90	55.50		71.40	install 11/16" x 2-1/2" pine ranch
	Demolish and Install	Opng.	15.90	61.25		77.15	style casing for one side of a standard
	Reinstall	Opng.		44.20		44.20	door opening.
	Clean	Opng.	.42	2.36		2.78	
	Paint	Ea.	3.15	5.70		8.85	
	Minimum Charge	Job		163		163	
Double Width							
	Demolish	Ea.		6.85		6.85	Cost includes material and labor to
	Install	Opng.	18.70	65		83.70	install 11/16" x 2-1/2" pine ranch
	Demolish and Install	Opng.	18.70	71.85		90.55	style door casing for one side of a
	Reinstall	Opng.		52.16		52.16	double door opening.
	Clean	Opng.	.48	4.71		5.19	
	Paint	Ea.	3.15	6.50		9.65	
	Minimum Charge	Job		163		163	

Exterior Doors and Windows

Residential Metal Door		Unit	Material	Labor	Equip.	Total	Specification
Pre-hung, Steel-clad							
2' 8" x 6' 8"							
	Demolish	Ea.		30		30	Cost includes material and labor to
	Install	Ea.	211	41		252	install 2' 6" x 6' 8" pre-hung 1-3/4"
	Demolish and Install	Ea.	211	71		282	24 gauge steel clad door with 4-1/2"
	Reinstall	Ea.		40.75		40.75	wood frame, trims, hinges, aluminum
	Clean	Ea.	2.92	9.80		12.72	sill, weatherstripped.
	Paint	Ea.	9.15	41		50.15	
	Minimum Charge	Job		163		163	
3' x 6' 8"							
	Demolish	Ea.		30		30	Cost includes material and labor to
	Install	Ea.	215	43.50		258.50	install 3' x 6' 8" pre-hung 1-3/4" 24
	Demolish and Install	Ea.	215	73.50		288.50	gauge steel clad door with 4-1/2"
	Reinstall	Ea.		43.47		43.47	wood frame, trims, hinges, aluminum
	Clean	Ea.	2.92	9.80		12.72	sill, weatherstripped.
	Paint	Ea.	9.15	41		50.15	
	Minimum Charge	Job		163		163	

Metal Door		Unit	Material	Labor	Equip.	Total	Specification
2' 8" x 6' 8", 18 Gauge							
	Demolish	Ea.		6.45		6.45	Cost includes material and labor to
	Install	Ea.	217	20.50		237.50	install 18 gauge 2' 8" x 6' 8" steel
	Demolish and Install	Ea.	217	26.95		243.95	door only, 1-3/4" thick, hinges.
	Reinstall	Ea.		20.38		20.38	
	Clean	Ea.	1.21	7.35		8.56	
	Paint	Ea.	3.87	24		27.87	
	Minimum Charge	Job		163		163	
3' x 6' 8", 18 Gauge							
	Demolish	Ea.		6.45		6.45	Cost includes material and labor to
	Install	Ea.	211	21.50		232.50	install 18 gauge 3' x 6' 8" steel door
	Demolish and Install	Ea.	211	27.95		238.95	only, 1-3/4" thick, hinges.
	Reinstall	Ea.		21.73		21.73	
	Clean	Ea.	1.21	7.35		8.56	
	Paint	Ea.	3.87	24		27.87	
	Minimum Charge	Job		163		163	
2' 8" x 7', 18 Gauge							
	Demolish	Ea.		6.45		6.45	Cost includes material and labor to
	Install	Ea.	254	21.50		275.50	install 18 gauge 2' 8" x 7' steel door
	Demolish and Install	Ea.	254	27.95		281.95	only, 1-3/4" thick, hinges.
	Reinstall	Ea.		21.73		21.73	
	Clean	Ea.	1.21	7.35		8.56	
	Paint	Ea.	3.87	24		27.87	
	Minimum Charge	Job		163		163	
3' x 7', 18 Gauge							
	Demolish	Ea.		6.45		6.45	Cost includes material and labor to
	Install	Ea.	248	23.50		271.50	install 18 gauge 3' x 7' steel door
	Demolish and Install	Ea.	248	29.95		277.95	only, 1-3/4" thick, hinges.
	Reinstall	Ea.		23.29		23.29	
	Clean	Ea.	1.21	7.35		8.56	
	Paint	Ea.	3.87	24		27.87	
	Minimum Charge	Job		163		163	
3' 6" x 7', 18 Gauge							
	Demolish	Ea.		6.45		6.45	Cost includes material and labor to
	Install	Ea.	276	25		301	install 18 gauge 3' 6" x 7' steel door
	Demolish and Install	Ea.	276	31.45		307.45	only, 1-3/4" thick, hinges.
	Reinstall	Ea.		25.08		25.08	
	Clean	Ea.	1.21	7.35		8.56	
	Paint	Ea.	3.87	24		27.87	
	Minimum Charge	Job		163		163	

Exterior Doors and Windows

Metal Door

Metal Door	Unit	Material	Labor	Equip.	Total	Specification
Casing Trim						
Single Width						Cost includes material and labor to
Demolish	Ea.		5.75		5.75	install 11/16" x 2-1/2" pine ranch
Install	Opng.	15.90	55.50		71.40	style casing for one side of a standard
Demolish and Install	Opng.	15.90	61.25		77.15	door opening.
Reinstall	Opng.		44.20		44.20	
Clean	Opng.	.42	2.36		2.78	
Paint	Ea.	3.15	5.70		8.85	
Minimum Charge	Job		163		163	
Double Width						
Demolish	Ea.		6.85		6.85	Cost includes material and labor to
Install	Opng.	18.70	65		83.70	install 11/16" x 2-1/2" pine ranch
Demolish and Install	Opng.	18.70	71.85		90.55	style door casing for one side of a
Reinstall	Opng.		52.16		52.16	double door opening.
Clean	Opng.	.48	4.71		5.19	
Paint	Ea.	3.15	6.50		9.65	
Minimum Charge	Job		163		163	

Solid Core Door

Solid Core Door	Unit	Material	Labor	Equip.	Total	Specification
Exterior Type						
Flush (1-3/4")						
Demolish	Ea.		30		30	Cost includes material and labor to
Install	Ea.	305	41		346	install 3' x 6' 8" pre-hung 1-3/4"
Demolish and Install	Ea.	305	71		376	door, including casing and stop,
Reinstall	Ea.		40.75		40.75	hinges, jamb, aluminum sill,
Clean	Ea.	2.92	9.80		12.72	weatherstripped.
Paint	Ea.	9.15	41		50.15	
Minimum Charge	Job		163		163	
Raised Panel						
Demolish	Ea.		30		30	Cost includes material and labor to
Install	Ea.	375	18.10		393.10	install 3' x 6' 8" pre-hung 1-3/4"
Demolish and Install	Ea.	375	48.10		423.10	door, jamb, including casing and
Reinstall	Ea.		18.11		18.11	stop, hinges, aluminum sill,
Clean	Ea.	2.92	9.80		12.72	weatherstripping.
Paint	Ea.	9.15	95.50		104.65	
Minimum Charge	Job		163		163	
Entry w / Lights						
Demolish	Ea.		30		30	Cost includes material and labor to
Install	Ea.	345	19.20		364.20	install 3' x 6' 8" pre-hung 1-3/4"
Demolish and Install	Ea.	345	49.20		394.20	door, solid wood lower panel, with up
Reinstall	Ea.		19.18		19.18	to 9-light glass panels, casing, sill,
Clean	Ea.	2.92	9.80		12.72	weatherstripped.
Paint	Ea.	2.15	95.50		97.65	
Minimum Charge	Job		163		163	
Custom Entry						
Demolish	Ea.		30		30	Includes material and labor to install
Install	Ea.	1200	19.20		1219.20	prehung, fir, single glazed with lead
Demolish and Install	Ea.	1200	49.20		1249.20	caming, 1-3/4" x 6' 8" x 3'.
Reinstall	Ea.		19.18		19.18	
Clean	Ea.	2.92	9.80		12.72	
Paint	Ea.	3.87	24		27.87	
Minimum Charge	Job		163		163	

Solid Core Door		Unit	Material	Labor	Equip.	Total	Specification
Dutch Style							
	Demolish	Ea.		30		30	Includes material and labor to install
	Install	Ea.	655	54.50		709.50	dutch door, pine, 1-3/4" x 6' 8" x 2'
	Demolish and Install	Ea.	655	84.50		739.50	8" wide.
	Reinstall	Ea.		54.33		54.33	
	Clean	Ea.	2.92	9.80		12.72	
	Paint	Ea.	2.15	95.50		97.65	
	Minimum Charge	Job		163		163	
Miami Style							
	Demolish	Ea.		30		30	Cost includes material and labor to
	Install	Ea.	204	16.30		220.30	install 3' x 6' 8" pre-hung 1-3/4"
	Demolish and Install	Ea.	204	46.30		250.30	lauan door of solid wood.
	Reinstall	Ea.		16.30		16.30	
	Clean	Ea.	2.92	9.80		12.72	
	Paint	Ea.	7.75	41		48.75	
	Minimum Charge	Job		163		163	
French Style							
	Demolish	Ea.		32.50		32.50	Cost includes material and labor to
	Install	Ea.	615	46.50		661.50	install 3' x 6' 8" pre-hung 1-3/4"
	Demolish and Install	Ea.	615	79		694	french style door with 15 lights.
	Reinstall	Ea.		46.57		46.57	
	Clean	Ea.	.20	9.80		10	
	Paint	Ea.	5.80	143		148.80	
	Minimum Charge	Job		163		163	
Double							
	Demolish	Ea.		22		22	Includes material and labor to install a
	Install	Ea.	1050	81.50		1131.50	pre-hung double entry door including
	Demolish and Install	Ea.	1050	103.50		1153.50	frame and exterior trim.
	Clean	Ea.	.11	34		34.11	
	Paint	Ea.	18.25	143		161.25	
	Minimum Charge	Job		81.50		81.50	
Commercial Grade							
3' x 6' 8"							
	Demolish	Ea.		30		30	Includes material and labor to install
	Install	Ea.	273	19.20		292.20	flush solid core 1-3/4" hardwood
	Demolish and Install	Ea.	273	49.20		322.20	veneer 3' x 6' 8' door only.
	Reinstall	Ea.		19.18		19.18	
	Clean	Ea.	2.92	9.80		12.72	
	Paint	Ea.	6.45	24		30.45	
	Minimum Charge	Job		163		163	
3' 6" x 6' 8"							
	Demolish	Ea.		30		30	Includes material and labor to install
	Install	Ea.	370	19.20		389.20	flush solid core 1-3/4" hardwood
	Demolish and Install	Ea.	370	49.20		419.20	veneer 3' 6" x 6' 8' door only.
	Reinstall	Ea.		19.18		19.18	
	Clean	Ea.	2.92	9.80		12.72	
	Paint	Ea.	6.45	24		30.45	
	Minimum Charge	Job		163		163	
3' x 7'							
	Demolish	Ea.		30		30	Includes material and labor to install
	Install	Ea.	294	46.50		340.50	flush solid core 1-3/4" hardwood
	Demolish and Install	Ea.	294	76.50		370.50	veneer 3' x 7' door only.
	Reinstall	Ea.		46.57		46.57	
	Clean	Ea.	2.92	9.80		12.72	
	Paint	Ea.	6.45	24		30.45	
	Minimum Charge	Job		163		163	

Exterior Doors and Windows

Solid Core Door		Unit	Material	Labor	Equip.	Total	Specification
3' 6" x 7'							
	Demolish	Ea.		30		30	Includes material and labor to install
	Install	Ea.	395	23.50		418.50	flush solid core 1-3/4" hardwood
	Demolish and Install	Ea.	395	53.50		448.50	veneer 3' 6" x 7' door only.
	Reinstall	Ea.		23.29		23.29	
	Clean	Ea.	2.92	9.80		12.72	
	Paint	Ea.	6.45	24		30.45	
	Minimum Charge	Job		163		163	
3' x 8'							
	Demolish	Ea.		30		30	Includes material and labor to install
	Install	Ea.	340	41		381	flush solid core 1-3/4" hardwood
	Demolish and Install	Ea.	340	71		411	veneer 3' x 8' door only.
	Reinstall	Ea.		40.75		40.75	
	Clean	Ea.	3.50	14.75		18.25	
	Paint	Ea.	6.45	24		30.45	
	Minimum Charge	Job		163		163	
3' 6" x 8'							
	Demolish	Ea.		30		30	Includes material and labor to install
	Install	Ea.	420	41		461	flush solid core 1-3/4" hardwood
	Demolish and Install	Ea.	420	71		491	veneer 3' 6" x 8' door only.
	Reinstall	Ea.		40.75		40.75	
	Clean	Ea.	3.50	14.75		18.25	
	Paint	Ea.	6.45	24		30.45	
	Minimum Charge	Job		163		163	
3' x 9'							
	Demolish	Ea.		30		30	Includes material and labor to install
	Install	Ea.	440	46.50		486.50	flush solid core 1-3/4" hardwood
	Demolish and Install	Ea.	440	76.50		516.50	veneer 3' x 9' door only.
	Reinstall	Ea.		46.57		46.57	
	Clean	Ea.	3.50	14.75		18.25	
	Paint	Ea.	6.45	24		30.45	
	Minimum Charge	Job		163		163	
Stain							
	Install	Ea.	5.90	32		37.90	Includes labor and material to stain
	Minimum Charge	Job		143		143	exterior type door and trim on both sides.
Door Only (SC)							
2' 4" x 6' 8"							
	Demolish	Ea.		6.45		6.45	Cost includes material and labor to
	Install	Ea.	184	18.10		202.10	install 2' 4" x 6' 8" lauan solid core
	Demolish and Install	Ea.	184	24.55		208.55	door slab.
	Reinstall	Ea.		18.11		18.11	
	Clean	Ea.	1.21	7.35		8.56	
	Paint	Ea.	2.41	36		38.41	
	Minimum Charge	Job		163		163	
2' 8" x 6' 8"							
	Demolish	Ea.		6.45		6.45	Cost includes material and labor to
	Install	Ea.	188	18.10		206.10	install 2' 8" x 6' 8" lauan solid core
	Demolish and Install	Ea.	188	24.55		212.55	door slab.
	Reinstall	Ea.		18.11		18.11	
	Clean	Ea.	1.21	7.35		8.56	
	Paint	Ea.	2.41	36		38.41	
	Minimum Charge	Job		163		163	

Exterior Doors and Windows

Solid Core Door		Unit	Material	Labor	Equip.	Total	Specification
3' x 6' 8"							
	Demolish	Ea.		6.45		6.45	Cost includes material and labor to
	Install	Ea.	196	19.20		215.20	install 3' x 6' 8" lauan solid core door
	Demolish and Install	Ea.	196	25.65		221.65	slab.
	Reinstall	Ea.		19.18		19.18	
	Clean	Ea.	1.21	7.35		8.56	
	Paint	Ea.	2.41	36		38.41	
	Minimum Charge	Job		163		163	
3' 6" x 6' 8"							
	Demolish	Ea.		6.45		6.45	Cost includes material and labor to
	Install	Ea.	222	20.50		242.50	install 3' 6" x 6' 8" lauan solid core
	Demolish and Install	Ea.	222	26.95		248.95	door slab.
	Reinstall	Ea.		20.38		20.38	
	Clean	Ea.	1.21	7.35		8.56	
	Paint	Ea.	2.41	36		38.41	
	Minimum Charge	Job		163		163	
Panel							
	Demolish	Ea.		6.45		6.45	Cost includes material and labor to
	Install	Ea.	163	38.50		201.50	install 3' x 6' 8" six (6) panel interior
	Demolish and Install	Ea.	163	44.95		207.95	door.
	Reinstall	Ea.		38.35		38.35	
	Clean	Ea.	1.46	7.35		8.81	
	Paint	Ea.	2.41	36		38.41	
	Minimum Charge	Job		163		163	
Casing Trim							
Single Width							
	Demolish	Ea.		5.75		5.75	Cost includes material and labor to
	Install	Opng.	15.90	55.50		71.40	install 11/16" x 2-1/2" pine ranch
	Demolish and Install	Opng.	15.90	61.25		77.15	style casing for one side of a standard
	Reinstall	Opng.		44.20		44.20	door opening.
	Clean	Opng.	.42	2.36		2.78	
	Paint	Ea.	3.15	5.70		8.85	
	Minimum Charge	Job		163		163	
Double Width							
	Demolish	Ea.		6.85		6.85	Cost includes material and labor to
	Install	Opng.	18.70	65		83.70	install 11/16" x 2-1/2" pine ranch
	Demolish and Install	Opng.	18.70	71.85		90.55	style door casing for one side of a
	Reinstall	Opng.		52.16		52.16	double door opening.
	Clean	Opng.	.48	4.71		5.19	
	Paint	Ea.	3.15	6.50		9.65	
	Minimum Charge	Job		163		163	

Components		Unit	Material	Labor	Equip.	Total	Specification
Hardware							
Doorknob							
	Demolish	Ea.		12.05		12.05	Includes material and labor to install
	Install	Ea.	18.20	20.50		38.70	residential passage lockset, keyless
	Demolish and Install	Ea.	18.20	32.55		50.75	bored type.
	Reinstall	Ea.		16.30		16.30	
	Clean	Ea.	.30	7.35		7.65	
	Minimum Charge	Job		163		163	

Exterior Doors and Windows

Components		Unit	Material	Labor	Equip.	Total	Specification
Doorknob w / Lock							
	Demolish	Ea.		12.05		12.05	Includes material and labor to install
	Install	Ea.	21	20.50		41.50	privacy lockset.
	Demolish and Install	Ea.	21	32.55		53.55	
	Reinstall	Ea.		16.30		16.30	
	Clean	Ea.	.30	7.35		7.65	
	Minimum Charge	Job		163		163	
Lever Handle							
	Demolish	Ea.		12.05		12.05	Includes material and labor to install
	Install	Ea.	27	32.50		59.50	residential passage lockset, keyless
	Demolish and Install	Ea.	27	44.55		71.55	bored type with lever handle.
	Reinstall	Ea.		26.08		26.08	
	Clean	Ea.	.30	7.35		7.65	
	Minimum Charge	Job		163		163	
Deadbolt							
	Demolish	Ea.		11.30		11.30	Includes material and labor to install a
	Install	Ea.	38.50	23.50		62	deadbolt.
	Demolish and Install	Ea.	38.50	34.80		73.30	
	Reinstall	Ea.		18.63		18.63	
	Clean	Ea.	.30	7.35		7.65	
	Minimum Charge	Job		163		163	
Grip Handle Entry							
	Demolish	Ea.		10.35		10.35	Includes material and labor to install
	Install	Ea.	123	36		159	residential keyed entry lock set with
	Demolish and Install	Ea.	123	46.35		169.35	grip type handle.
	Reinstall	Ea.		28.98		28.98	
	Clean	Ea.	.30	7.35		7.65	
	Minimum Charge	Job		163		163	
Panic Device							
	Demolish	Ea.		26		26	Includes material and labor to install
	Install	Ea.	340	54.50		394.50	touch bar, low profile, exit only.
	Demolish and Install	Ea.	340	80.50		420.50	
	Reinstall	Ea.		43.47		43.47	
	Clean	Ea.	.14	13.10		13.24	
	Minimum Charge	Job		163		163	
Closer							
	Demolish	Ea.		6.80		6.80	Includes material and labor to install
	Install	Ea.	162	54.50		216.50	pneumatic heavy duty closer for
	Demolish and Install	Ea.	162	61.30		223.30	exterior type doors.
	Reinstall	Ea.		43.47		43.47	
	Clean	Ea.	.14	9.80		9.94	
	Minimum Charge	Job		163		163	
Peep-hole							
	Demolish	Ea.		5.80		5.80	Includes material and labor to install
	Install	Ea.	14.65	10.20		24.85	peep-hole for door.
	Demolish and Install	Ea.	14.65	16		30.65	
	Reinstall	Ea.		8.15		8.15	
	Clean	Ea.	.02	2.61		2.63	
	Minimum Charge	Job		163		163	
Exit Sign							
	Demolish	Ea.		17.95		17.95	Includes material and labor to install a
	Install	Ea.	48.50	44.50		93	wall mounted interior electric exit sign.
	Demolish and Install	Ea.	48.50	62.45		110.95	
	Reinstall	Ea.		35.56		35.56	
	Clean	Ea.	.02	5.30		5.32	
	Minimum Charge	Job		163		163	

Exterior Doors and Windows

Components		Unit	Material	Labor	Equip.	Total	Specification
Kickplate							
	Demolish	Ea.		10.05		10.05	Includes material and labor to install
	Install	Ea.	20.50	21.50		42	aluminum kickplate, 10" x 28".
	Demolish and Install	Ea.	20.50	31.55		52.05	
	Reinstall	Ea.		17.39		17.39	
	Clean	Ea.	.30	7.35		7.65	
	Minimum Charge	Job		163		163	
Threshold							
	Install	L.F.	2.56	3.26		5.82	Includes labor and material to install
	Minimum Charge	Job		81.50		81.50	oak threshold.
Weatherstripping							
	Install	L.F.	.29	.82		1.11	Includes labor and material to install
	Minimum Charge	Job		81.50		81.50	rubber weatherstripping for door.
Metal Jamb							
6' 8" High							
	Demolish	Ea.		18.10		18.10	Cost includes material and labor to
	Install	Ea.	85.50	20.50		106	install 18 gauge hollow metal door
	Demolish and Install	Ea.	85.50	38.60		124.10	frame to fit a 4-1/2" jamb 6' 8" high
	Reinstall	Ea.		20.38		20.38	and up to 3' 6" opening.
	Clean	Ea.	.61	4.91		5.52	
	Paint	Ea.	1.07	5.70		6.77	
	Minimum Charge	Job		163		163	
7' High							
	Demolish	Ea.		18.10		18.10	Cost includes material and labor to
	Install	Ea.	87.50	21.50		109	install 18 gauge hollow metal door
	Demolish and Install	Ea.	87.50	39.60		127.10	frame to fit a 4-1/2" jamb 7' high and
	Reinstall	Ea.		21.73		21.73	up to 3' 6" opening.
	Clean	Ea.	.61	4.91		5.52	
	Paint	Ea.	1.07	5.70		6.77	
	Minimum Charge	Job		163		163	
8' High							
	Demolish	Ea.		18.10		18.10	Cost includes material and labor to
	Install	Ea.	106	27		133	install 18 gauge hollow metal door
	Demolish and Install	Ea.	106	45.10		151.10	frame to fit a 4-1/2" jamb 8' high and
	Reinstall	Ea.		27.17		27.17	up to 3' 6" opening.
	Clean	Ea.	.86	5.90		6.76	
	Paint	Ea.	1.20	6.35		7.55	
	Minimum Charge	Job		163		163	
9' High							
	Demolish	Ea.		18.10		18.10	Cost includes material and labor to
	Install	Ea.	121	32.50		153.50	install 18 gauge hollow metal door
	Demolish and Install	Ea.	121	50.60		171.60	frame to fit a 4-1/2" jamb 9' high and
	Reinstall	Ea.		32.60		32.60	up to 3' 6" opening.
	Clean	Ea.	.86	5.90		6.76	
	Paint	Ea.	1.20	6.35		7.55	
	Minimum Charge	Job		163		163	
Wood Jamb							
	Demolish	Ea.		22.50		22.50	Includes material and labor to install
	Install	Ea.	80	8.80		88.80	flat pine jamb with square cut heads
	Demolish and Install	Ea.	80	31.30		111.30	and rabbeted sides for 6' 8" high and
	Clean	Ea.	.09	5.90		5.99	3-9/16" door including trim sets for
	Paint	Ea.	5.70	19.05		24.75	both sides.
	Minimum Charge	Job		81.50		81.50	
Stain							
	Install	Ea.	4.05	24		28.05	Includes labor and material to stain
	Minimum Charge	Job		143		143	single door and trim on both sides.

Exterior Doors and Windows

Components	Unit	Material	Labor	Equip.	Total	Specification
Shave & Refit						
Install	Ea.		27		27	Includes labor to shave and rework
Minimum Charge	Job		81.50		81.50	door to fit opening at the job site.

Sliding Patio Door	Unit	Material	Labor	Equip.	Total	Specification
Aluminum						
6' Single Pane						
Demolish	Ea.		45.50		45.50	Cost includes material and labor to
Install	Ea.	400	163		563	install 6' x 6' 8" sliding anodized
Demolish and Install	Ea.	400	208.50		608.50	aluminum patio door with single pane
Reinstall	Ea.		130.40		130.40	tempered safety glass, anodized
Clean	Ea.	1.83	14.75		16.58	frame, screen, weatherstripped.
Minimum Charge	Job		163		163	
8' Single Pane						
Demolish	Ea.		45.50		45.50	Cost includes material and labor to
Install	Ea.	455	217		672	install 8' x 6' 8" sliding aluminum
Demolish and Install	Ea.	455	262.50		717.50	patio door with single pane tempered
Reinstall	Ea.		173.87		173.87	safety glass, anodized frame, screen,
Clean	Ea.	1.83	14.75		16.58	weatherstripped.
Minimum Charge	Job		163		163	
10' Single Pane						
Demolish	Ea.		45.50		45.50	Cost includes material and labor to
Install	Ea.	595	325		920	install 10' x 6' 8" sliding aluminum
Demolish and Install	Ea.	595	370.50		965.50	patio door with single pane tempered
Reinstall	Ea.		260.80		260.80	safety glass, anodized frame, screen,
Clean	Ea.	1.83	14.75		16.58	weatherstripped.
Minimum Charge	Job		163		163	
6' Insulated						
Demolish	Ea.		45.50		45.50	Cost includes material and labor to
Install	Ea.	535	163		698	install 6' x 6' 8" sliding aluminum
Demolish and Install	Ea.	535	208.50		743.50	patio door with double insulated
Reinstall	Ea.		130.40		130.40	tempered glass, baked-on enamel
Clean	Ea.	1.83	14.75		16.58	finish, screen, weatherstripped.
Minimum Charge	Job		163		163	
8' Insulated						
Demolish	Ea.		45.50		45.50	Cost includes material and labor to
Install	Ea.	625	217		842	install 8' x 6' 8" sliding aluminum
Demolish and Install	Ea.	625	262.50		887.50	patio door with insulated tempered
Reinstall	Ea.		173.87		173.87	glass, baked-on enamel finish, screen,
Clean	Ea.	1.83	14.75		16.58	weatherstripped.
Minimum Charge	Job		163		163	
10' Insulated						
Demolish	Ea.		45.50		45.50	Cost includes material and labor to
Install	Ea.	705	325		1030	install 10' x 6' 8" sliding aluminum
Demolish and Install	Ea.	705	370.50		1075.50	patio door with double insulated
Reinstall	Ea.		260.80		260.80	tempered glass, baked-on enamel
Clean	Ea.	1.83	14.75		16.58	finish, screen, weatherstripped.
Minimum Charge	Job		163		163	

Sliding Patio Door	Unit	Material	Labor	Equip.	Total	Specification
Vinyl						
6' Single Pane						
Demolish	Ea.		45.50		45.50	Cost includes material and labor to
Install	Ea.	380	163		543	install 6' x 6' 8" sliding vinyl patio
Demolish and Install	Ea.	380	208.50		588.50	door with single pane tempered safety
Reinstall	Ea.		130.40		130.40	glass, anodized frame, screen,
Clean	Ea.	1.83	14.75		16.58	weatherstripped.
Minimum Charge	Job		163		163	
8' Single Pane						
Demolish	Ea.		45.50		45.50	Cost includes material and labor to
Install	Ea.	435	217		652	install 8' x 6' 8" sliding vinyl patio
Demolish and Install	Ea.	435	262.50		697.50	door with single pane tempered safety
Reinstall	Ea.		173.87		173.87	glass, anodized frame, screen,
Clean	Ea.	1.83	14.75		16.58	weatherstripped.
Minimum Charge	Job		163		163	
10' Single Pane						
Demolish	Ea.		45.50		45.50	Cost includes material and labor to
Install	Ea.	565	325		890	install 10' x 6' 8" sliding vinyl patio
Demolish and Install	Ea.	565	370.50		935.50	door with single pane tempered safety
Reinstall	Ea.		260.80		260.80	glass, anodized frame, screen,
Clean	Ea.	1.83	14.75		16.58	weatherstripped.
Minimum Charge	Job		163		163	
6' Insulated						
Demolish	Ea.		45.50		45.50	Cost includes material and labor to
Install	Ea.	605	163		768	install 6' x 6' 8" sliding vinyl patio
Demolish and Install	Ea.	605	208.50		813.50	door with double insulated tempered
Reinstall	Ea.		130.40		130.40	glass, anodized frame, screen,
Clean	Ea.	1.83	14.75		16.58	weatherstripped.
Minimum Charge	Job		163		163	
8' Insulated						
Demolish	Ea.		45.50		45.50	Cost includes material and labor to
Install	Ea.	775	217		992	install 8' x 6' 8" sliding vinyl patio
Demolish and Install	Ea.	775	262.50		1037.50	door with double insulated tempered
Reinstall	Ea.		173.87		173.87	glass, anodized frame, screen,
Clean	Ea.	1.83	14.75		16.58	weatherstripped.
Minimum Charge	Job		163		163	
10' Insulated						
Demolish	Ea.		45.50		45.50	Cost includes material and labor to
Install	Ea.	865	325		1190	install 10' x 6' 8" sliding vinyl patio
Demolish and Install	Ea.	865	370.50		1235.50	door with double insulated tempered
Reinstall	Ea.		260.80		260.80	glass, anodized frame, screen,
Clean	Ea.	1.83	14.75		16.58	weatherstripped.
Minimum Charge	Job		163		163	
Wood Framed						
6' Premium Insulated						
Demolish	Ea.		45.50		45.50	Cost includes material and labor to
Install	Ea.	1075	163		1238	install 6' x 6' 8" sliding wood patio
Demolish and Install	Ea.	1075	208.50		1283.50	door with pine frame, double insulated
Reinstall	Ea.		130.40		130.40	glass, screen, lock and dead bolt,
Clean	Ea.	1.83	14.75		16.58	weatherstripped. Add for casing and
Minimum Charge	Job		163		163	finishing.

Exterior Doors and Windows

Sliding Patio Door		Unit	Material	Labor	Equip.	Total	Specification
8' Premium Insulated							Cost includes material and labor to
	Demolish	Ea.		45.50		45.50	install 8' x 6' 8" sliding wood patio
	Install	Ea.	1275	217		1492	door with pine frame, double insulated
	Demolish and Install	Ea.	1275	262.50		1537.50	glass, screen, lock and dead bolt,
	Reinstall	Ea.		173.87		173.87	weatherstripped. Add for casing and
	Clean	Ea.	1.83	14.75		16.58	finishing.
	Minimum Charge	Job		163		163	
9' Premium Insulated							Cost includes material and labor to
	Demolish	Ea.		45.50		45.50	install 9' x 6' 8" sliding wood patio
	Install	Ea.	1275	217		1492	door with pine frame, double insulated
	Demolish and Install	Ea.	1275	262.50		1537.50	glass, screen, lock and dead bolt,
	Reinstall	Ea.		173.87		173.87	weatherstripped. Add for casing and
	Clean	Ea.	1.83	14.75		16.58	finishing.
	Minimum Charge	Job		163		163	
Stain							Includes labor and material to stain
	Install	Ea.	5.90	32		37.90	exterior type door and trim on both
	Minimum Charge	Job		143		143	sides.
Casing Trim							Cost includes material and labor to
	Demolish	Ea.		6.85		6.85	install 11/16" x 2-1/2" pine ranch
	Install	Opng.	18.70	65		83.70	style door casing for one side of a
	Demolish and Install	Opng.	18.70	71.85		90.55	double door opening.
	Reinstall	Opng.		52.16		52.16	
	Clean	Opng.	.48	4.71		5.19	
	Paint	Ea.	3.15	6.50		9.65	
	Minimum Charge	Job		163		163	
Sliding Screen							Includes material and labor to install
	Demolish	Ea.		6.45		6.45	sliding screen door, 6' 8" x 3'.
	Install	Ea.	100	16.30		116.30	
	Demolish and Install	Ea.	100	22.75		122.75	
	Reinstall	Ea.		13.04		13.04	
	Clean	Ea.	.61	4.91		5.52	
	Minimum Charge	Job		163		163	

Screen Door		Unit	Material	Labor	Equip.	Total	Specification
Aluminum							Includes material and labor to install
	Demolish	Ea.		30		30	aluminum frame screen door with
	Install	Ea.	103	46.50		149.50	fiberglass screen cloth, closer, hinges,
	Demolish and Install	Ea.	103	76.50		179.50	latch.
	Reinstall	Ea.		37.26		37.26	
	Clean	Ea.	2.92	9.80		12.72	
	Paint	Ea.	3.03	11.45		14.48	
	Minimum Charge	Job		163		163	
Wood							Includes material and labor to install
	Demolish	Ea.		30		30	wood screen door with aluminum cloth
	Install	Ea.	133	54.50		187.50	screen. Frame and hardware not
	Demolish and Install	Ea.	133	84.50		217.50	included.
	Reinstall	Ea.		43.47		43.47	
	Clean	Ea.	2.92	9.80		12.72	
	Paint	Ea.	3.03	11.45		14.48	
	Minimum Charge	Job		163		163	
Shave & Refit							Includes labor to shave and rework
	Install	Ea.		27		27	door to fit opening at the job site.
	Minimum Charge	Job		163		163	

Exterior Doors and Windows

Storm Door	Unit	Material	Labor	Equip.	Total	Specification
Good Grade						
Demolish	Ea.		30		30	Includes material and labor to install
Install	Ea.	160	46.50		206.50	aluminum frame storm door with
Demolish and Install	Ea.	160	76.50		236.50	fiberglass screen cloth, upper glass
Reinstall	Ea.		37.26		37.26	panel, closer, hinges, latch.
Clean	Ea.	2.92	9.80		12.72	
Paint	Ea.	3.03	11.45		14.48	
Minimum Charge	Job		163		163	
Better Grade						
Demolish	Ea.		30		30	Includes material and labor to install
Install	Ea.	210	46.50		256.50	aluminum frame storm door with
Demolish and Install	Ea.	210	76.50		286.50	fiberglass screen cloth, upper and
Reinstall	Ea.		37.26		37.26	lower glass panel, closer, latch.
Clean	Ea.	2.92	9.80		12.72	
Paint	Ea.	3.03	11.45		14.48	
Minimum Charge	Job		163		163	
Premium Grade						
Demolish	Ea.		30		30	Includes material and labor to install
Install	Ea.	256	46.50		302.50	aluminum frame storm door with
Demolish and Install	Ea.	256	76.50		332.50	fiberglass screen cloth, upper and
Reinstall	Ea.		37.26		37.26	lower glass panel, closer, latch.
Clean	Ea.	2.92	9.80		12.72	
Paint	Ea.	3.03	11.45		14.48	
Minimum Charge	Job		163		163	
Insulated						
Demolish	Ea.		30		30	Includes material and labor to install
Install	Ea.	237	23.50		260.50	aluminum frame storm door with
Demolish and Install	Ea.	237	53.50		290.50	fiberglass screen cloth, upper and
Reinstall	Ea.		18.63		18.63	lower glass panel, closer, latch.
Clean	Ea.	2.92	9.80		12.72	
Paint	Ea.	3.03	11.45		14.48	
Minimum Charge	Job		163		163	

Garage Door	Unit	Material	Labor	Equip.	Total	Specification
9' One Piece Metal						
Demolish	Ea.		76.50		76.50	Includes material and labor to install
Install	Ea.	350	81.50		431.50	one piece steel garage door 9' x 7'
Demolish and Install	Ea.	350	158		508	primed including hardware.
Reinstall	Ea.		65.20		65.20	
Clean	Ea.	3.64	29.50		33.14	
Paint	Ea.	6.20	71.50		77.70	
Minimum Charge	Job		163		163	
16' One Piece Metal						
Demolish	Ea.		90.50		90.50	Includes material and labor to install
Install	Ea.	555	109		664	one piece steel garage door 16' x 7'
Demolish and Install	Ea.	555	199.50		754.50	primed including hardware.
Reinstall	Ea.		86.93		86.93	
Clean	Ea.	7.30	39.50		46.80	
Paint	Ea.	12.40	81.50		93.90	
Minimum Charge	Job		163		163	

Exterior Doors and Windows

Garage Door		Unit	Material	Labor	Equip.	Total	Specification
9' Metal Sectional							
	Demolish	Ea.		76.50		76.50	Includes material and labor to install
	Install	Ea.	495	123		618	sectional metal garage door 9' x 7'
	Demolish and Install	Ea.	495	199.50		694.50	primed including hardware.
	Reinstall	Ea.		98.79		98.79	
	Clean	Ea.	3.73	33.50		37.23	
	Paint	Ea.	12.40	81.50		93.90	
	Minimum Charge	Job		163		163	
16' Metal Sectional							
	Demolish	Ea.		90.50		90.50	Includes material and labor to install
	Install	Ea.	625	217		842	sectional metal garage door 16' x 7'
	Demolish and Install	Ea.	625	307.50		932.50	primed including hardware.
	Reinstall	Ea.		173.87		173.87	
	Clean	Ea.	7.30	39.50		46.80	
	Paint	Ea.	12.40	81.50		93.90	
	Minimum Charge	Job		163		163	
Metal / Steel Roll Up							
8' x 8' Steel							
	Demolish	Ea.		103		103	Cost includes material and labor to
	Install	Ea.	845	495		1340	install 8' x 8' steel overhead roll-up,
	Demolish and Install	Ea.	845	598		1443	chain hoist operated service door.
	Reinstall	Ea.		394.40		394.40	
	Clean	Ea.	3.73	33.50		37.23	
	Paint	Ea.	13.20	95.50		108.70	
	Minimum Charge	Job		790		790	
10' x 10' Steel							
	Demolish	Ea.		121		121	Cost includes material and labor to
	Install	Ea.	1125	565		1690	install 10' x 10' steel overhead roll-up,
	Demolish and Install	Ea.	1125	686		1811	chain hoist operated service door.
	Reinstall	Ea.		450.74		450.74	
	Clean	Ea.	5.85	47		52.85	
	Paint	Ea.	20.50	95.50		116	
	Minimum Charge	Job		790		790	
12' x 12' Steel							
	Demolish	Ea.		145		145	Cost includes material and labor to
	Install	Ea.	1450	655		2105	install 12' x 12' steel overhead roll-up,
	Demolish and Install	Ea.	1450	800		2250	chain hoist operated service door.
	Reinstall	Ea.		525.87		525.87	
	Clean	Ea.	8.40	52.50		60.90	
	Paint	Ea.	29.50	114		143.50	
	Minimum Charge	Job		790		790	
14' x 14' Steel							
	Demolish	Ea.		181		181	Cost includes material and labor to
	Install	Ea.	1925	985		2910	install 14' x 14' steel overhead roll-up,
	Demolish and Install	Ea.	1925	1166		3091	chain hoist operated service door.
	Reinstall	Ea.		788.80		788.80	
	Clean	Ea.	11.45	67.50		78.95	
	Paint	Ea.	34.50	143		177.50	
	Minimum Charge	Job		790		790	
18' x 18' Steel							
	Demolish	Ea.		201		201	Cost includes material and labor to
	Install	Opng.	2525	1325		3850	install 18' x 18' steel overhead roll-up,
	Demolish and Install	Opng.	2525	1526		4051	chain hoist operated service door.
	Reinstall	Opng.		1051.73		1051.73	
	Clean	Ea.	11.45	39.50		50.95	
	Paint	Ea.	67	286		353	
	Minimum Charge	Job		790		790	
Automatic Motor							
	Install	Ea.	390	32.50		422.50	Includes labor and material to install
	Minimum Charge	Job		163		163	electric opener motor, 1/3 HP.

Exterior Doors and Windows

Garage Door		Unit	Material	Labor	Equip.	Total	Specification
9' One Piece Wood							
	Demolish	Ea.		76.50		76.50	Includes material and labor to install
	Install	Ea.	385	81.50		466.50	one piece wood garage door 9' x 7'
	Demolish and Install	Ea.	385	158		543	primed including hardware.
	Reinstall	Ea.		65.20		65.20	
	Clean	Ea.	3.64	29.50		33.14	
	Paint	Ea.	6.20	71.50		77.70	
	Minimum Charge	Job		163		163	
9' Wood Sectional							
	Demolish	Ea.		76.50		76.50	Includes material and labor to install
	Install	Ea.	485	81.50		566.50	sectional wood garage door 9' x 7'
	Demolish and Install	Ea.	485	158		643	unfinished including hardware.
	Reinstall	Ea.		65.20		65.20	
	Clean	Ea.	3.64	29.50		33.14	
	Paint	Ea.	6.20	71.50		77.70	
	Minimum Charge	Job		163		163	
16' Wood Sectional							
	Demolish	Ea.		90.50		90.50	Includes material and labor to install
	Install	Ea.	1050	109		1159	sectional fiberglass garage door 16' x
	Demolish and Install	Ea.	1050	199.50		1249.50	7' unfinished including hardware.
	Reinstall	Ea.		86.93		86.93	
	Clean	Ea.	7.30	39.50		46.80	
	Paint	Ea.	12.40	81.50		93.90	
	Minimum Charge	Job		163		163	
Electric Opener							
	Demolish	Ea.		52.50		52.50	Includes material and labor to install
	Install	Ea.	264	61		325	electric opener, 1/3 HP economy.
	Demolish and Install	Ea.	264	113.50		377.50	
	Reinstall	Ea.		48.93		48.93	
	Clean	Ea.	.14	13.10		13.24	
	Minimum Charge	Job		163		163	
Springs							
	Install	Pr.	39	81.50		120.50	Includes labor and material to replace
	Minimum Charge	Job		163		163	springs for garage door.

Commercial Metal Door		Unit	Material	Labor	Equip.	Total	Specification
Pre-hung Steel Door							
2' 8" x 6' 8"							
	Demolish	Ea.		45.50		45.50	Cost includes material and labor to
	Install	Ea.	375	20.50		395.50	install 2' 8" x 6' 8" pre-hung 1-3/4"
	Demolish and Install	Ea.	375	66		441	18 gauge steel door with steel frame,
	Reinstall	Ea.		20.38		20.38	hinges, aluminum sill,
	Clean	Ea.	2.92	9.80		12.72	weather-stripped.
	Paint	Ea.	3.87	24		27.87	
	Minimum Charge	Job		163		163	
3' x 6' 8"							
	Demolish	Ea.		45.50		45.50	Cost includes material and labor to
	Install	Ea.	365	21.50		386.50	install 3' x 6' 8" pre-hung 1-3/4" 18
	Demolish and Install	Ea.	365	67		432	gauge steel door with steel frame,
	Reinstall	Ea.		21.73		21.73	hinges, aluminum sill,
	Clean	Ea.	2.92	9.80		12.72	weather-stripped.
	Paint	Ea.	3.87	24		27.87	
	Minimum Charge	Job		163		163	

Exterior Doors and Windows

Commercial Metal Door		Unit	Material	Labor	Equip.	Total	Specification
3' 6" x 6' 8"							
	Demolish	Ea.		45.50		45.50	Cost includes material and labor to
	Install	Ea.	425	25		450	install 3' 6" x 6' 8" pre-hung 1-3/4"
	Demolish and Install	Ea.	425	70.50		495.50	18 gauge steel door with steel frame,
	Reinstall	Ea.		25.08		25.08	hinges, aluminum sill,
	Clean	Ea.	2.92	9.80		12.72	weather-stripped.
	Paint	Ea.	3.87	24		27.87	
	Minimum Charge	Job		163		163	
4' x 6' 8"							
	Demolish	Ea.		45.50		45.50	Cost includes material and labor to
	Install	Ea.	455	32.50		487.50	install 4' x 6' 8" pre-hung 1-3/4" 18
	Demolish and Install	Ea.	455	78		533	gauge steel door with steel frame,
	Reinstall	Ea.		32.60		32.60	hinges, aluminum sill,
	Clean	Ea.	2.92	9.80		12.72	weather-stripped.
	Paint	Ea.	3.87	24		27.87	
	Minimum Charge	Job		163		163	
Metal Jamb							
6' 8" High							
	Demolish	Ea.		18.10		18.10	Cost includes material and labor to
	Install	Ea.	85.50	20.50		106	install 18 gauge hollow metal door
	Demolish and Install	Ea.	85.50	38.60		124.10	frame to fit a 4-1/2" jamb 6' 8" high
	Reinstall	Ea.		20.38		20.38	and up to 3' 6" opening.
	Clean	Ea.	.61	4.91		5.52	
	Paint	Ea.	1.07	5.70		6.77	
	Minimum Charge	Job		163		163	
7' High							
	Demolish	Ea.		18.10		18.10	Cost includes material and labor to
	Install	Ea.	87.50	21.50		109	install 18 gauge hollow metal door
	Demolish and Install	Ea.	87.50	39.60		127.10	frame to fit a 4-1/2" jamb 7' high and
	Reinstall	Ea.		21.73		21.73	up to 3' 6" opening.
	Clean	Ea.	.61	4.91		5.52	
	Paint	Ea.	1.07	5.70		6.77	
	Minimum Charge	Job		163		163	
8' High							
	Demolish	Ea.		18.10		18.10	Cost includes material and labor to
	Install	Ea.	106	27		133	install 18 gauge hollow metal door
	Demolish and Install	Ea.	106	45.10		151.10	frame to fit a 4-1/2" jamb 8' high and
	Reinstall	Ea.		27.17		27.17	up to 3' 6" opening.
	Clean	Ea.	.86	5.90		6.76	
	Paint	Ea.	1.20	6.35		7.55	
	Minimum Charge	Job		163		163	
9' High							
	Demolish	Ea.		18.10		18.10	Cost includes material and labor to
	Install	Ea.	121	32.50		153.50	install 18 gauge hollow metal door
	Demolish and Install	Ea.	121	50.60		171.60	frame to fit a 4-1/2" jamb 9' high and
	Reinstall	Ea.		32.60		32.60	up to 3' 6" opening.
	Clean	Ea.	.86	5.90		6.76	
	Paint	Ea.	1.20	6.35		7.55	
	Minimum Charge	Job		163		163	
Hardware							
Doorknob w / Lock							
	Demolish	Ea.		12.05		12.05	Includes material and labor to install
	Install	Ea.	102	27		129	keyed entry lock set, bored type
	Demolish and Install	Ea.	102	39.05		141.05	including knob trim, strike and strike
	Reinstall	Ea.		21.73		21.73	box for standard commercial
	Clean	Ea.	.30	7.35		7.65	application.
	Minimum Charge	Job		163		163	

Exterior Doors and Windows

Commercial Metal Door	Unit	Material	Labor	Equip.	Total	Specification
Lever Handle						
Demolish	Ea.		12.05		12.05	Includes material and labor to install
Install	Ea.	55	32.50		87.50	passage lockset, keyless bored type
Demolish and Install	Ea.	55	44.55		99.55	with lever handle, non-locking
Reinstall	Ea.		26.08		26.08	standard commercial latchset.
Clean	Ea.	.30	7.35		7.65	
Minimum Charge	Job		163		163	
Deadbolt						
Demolish	Ea.		11.30		11.30	Includes material and labor to install a
Install	Ea.	38.50	23.50		62	deadbolt.
Demolish and Install	Ea.	38.50	34.80		73.30	
Reinstall	Ea.		18.63		18.63	
Clean	Ea.	.30	7.35		7.65	
Minimum Charge	Job		163		163	
Panic Device						
Demolish	Ea.		26		26	Includes material and labor to install
Install	Ea.	340	54.50		394.50	touch bar, low profile, exit only.
Demolish and Install	Ea.	340	80.50		420.50	
Reinstall	Ea.		43.47		43.47	
Clean	Ea.	.14	13.10		13.24	
Minimum Charge	Job		163		163	
Closer						
Demolish	Ea.		6.80		6.80	Includes material and labor to install
Install	Ea.	172	54.50		226.50	pneumatic heavy duty model for
Demolish and Install	Ea.	172	61.30		233.30	exterior type doors.
Reinstall	Ea.		43.47		43.47	
Clean	Ea.	.14	9.80		9.94	
Minimum Charge	Job		163		163	
Peep-Hole						
Demolish	Ea.		5.80		5.80	Includes material and labor to install
Install	Ea.	14.65	10.20		24.85	peep-hole for door.
Demolish and Install	Ea.	14.65	16		30.65	
Reinstall	Ea.		8.15		8.15	
Clean	Ea.	.02	2.61		2.63	
Minimum Charge	Job		163		163	
Exit Sign						
Demolish	Ea.		17.95		17.95	Includes material and labor to install a
Install	Ea.	48.50	44.50		93	wall mounted interior electric exit sign.
Demolish and Install	Ea.	48.50	62.45		110.95	
Reinstall	Ea.		35.56		35.56	
Clean	Ea.	.02	5.30		5.32	
Minimum Charge	Job		163		163	
Kickplate						
Demolish	Ea.		10.05		10.05	Includes material and labor to install
Install	Ea.	46	21.50		67.50	stainless steel kickplate, 10" x 36".
Demolish and Install	Ea.	46	31.55		77.55	
Reinstall	Ea.		17.39		17.39	
Clean	Ea.	.30	7.35		7.65	
Minimum Charge	Job		163		163	

Aluminum Window	Unit	Material	Labor	Equip.	Total	Specification
Single Hung Two Light						
2' x 2'						
Demolish	Ea.		16.35		16.35	Cost includes material and labor to
Install	Ea.	59.50	71.50		131	install 2' x 2' single hung, two light,
Demolish and Install	Ea.	59.50	87.85		147.35	double glazed aluminum window
Reinstall	Ea.		57.37		57.37	including screen.
Clean	Ea.	.03	6.05		6.08	
Minimum Charge	Job		197		197	

Exterior Doors and Windows

Aluminum Window	Unit	Material	Labor	Equip.	Total	Specification
2' x 2' 6"						
Demolish	Ea.		16.35		16.35	Cost includes material and labor to
Install	Ea.	62.50	71.50		134	install 2' x 2' 6" single hung, two
Demolish and Install	Ea.	62.50	87.85		150.35	light, double glazed aluminum
Reinstall	Ea.		57.37		57.37	window including screen.
Clean	Ea.	.03	6.05		6.08	
Minimum Charge	Job		197		197	
3' x 1' 6"						
Demolish	Ea.		16.35		16.35	Cost includes material and labor to
Install	Ea.	59	79		138	install 3' x 1' 6" single hung, two
Demolish and Install	Ea.	59	95.35		154.35	light, double glazed aluminum
Reinstall	Ea.		63.10		63.10	window including screen.
Clean	Ea.	.02	5.30		5.32	
Minimum Charge	Job		197		197	
3' x 2'						
Demolish	Ea.		16.35		16.35	Cost includes material and labor to
Install	Ea.	71	79		150	install 3' x 2' single hung, two light,
Demolish and Install	Ea.	71	95.35		166.35	double glazed aluminum window
Reinstall	Ea.		63.10		63.10	including screen.
Clean	Ea.	.02	5.30		5.32	
Minimum Charge	Job		197		197	
3' x 2' 6"						
Demolish	Ea.		16.35		16.35	Cost includes material and labor to
Install	Ea.	74.50	79		153.50	install 3' x 2' 6" single hung, two
Demolish and Install	Ea.	74.50	95.35		169.85	light, double glazed aluminum
Reinstall	Ea.		63.10		63.10	window including screen.
Clean	Ea.	.03	6.05		6.08	
Minimum Charge	Job		197		197	
3' x 3'						
Demolish	Ea.		16.35		16.35	Cost includes material and labor to
Install	Ea.	83	79		162	install 3' x 3' single hung, two light,
Demolish and Install	Ea.	83	95.35		178.35	double glazed aluminum window
Reinstall	Ea.		63.10		63.10	including screen.
Clean	Ea.	.03	6.05		6.08	
Minimum Charge	Job		197		197	
3' x 3' 6"						
Demolish	Ea.		16.35		16.35	Cost includes material and labor to
Install	Ea.	89	79		168	install 3' x 3' 6" single hung, two
Demolish and Install	Ea.	89	95.35		184.35	light, double glazed aluminum
Reinstall	Ea.		63.10		63.10	window including screen.
Clean	Ea.	.04	7.05		7.09	
Minimum Charge	Job		197		197	
3' x 4'						
Demolish	Ea.		16.35		16.35	Cost includes material and labor to
Install	Ea.	98	79		177	install 3' x 4' single hung, two light,
Demolish and Install	Ea.	98	95.35		193.35	double glazed aluminum window
Reinstall	Ea.		63.10		63.10	including screen.
Clean	Ea.	.04	7.05		7.09	
Minimum Charge	Job		197		197	
3' x 5'						
Demolish	Ea.		24		24	Cost includes material and labor to
Install	Ea.	106	79		185	install 3' x 5' single hung, two light,
Demolish and Install	Ea.	106	103		209	double glazed aluminum window
Reinstall	Ea.		63.10		63.10	including screen.
Clean	Ea.	.06	8.50		8.56	
Minimum Charge	Job		197		197	

Exterior Doors and Windows

Aluminum Window		Unit	Material	Labor	Equip.	Total	Specification
3' x 6'							Cost includes material and labor to
	Demolish	Ea.		24		24	install 3' x 6' single hung, two light,
	Install	Ea.	117	87.50		204.50	double glazed aluminum window
	Demolish and Install	Ea.	117	111.50		228.50	including screen.
	Reinstall	Ea.		70.12		70.12	
	Clean	Ea.	.07	10		10.07	
	Minimum Charge	Job		197		197	
Double Hung Two Light							
2' x 2'							Cost includes material and labor to
	Demolish	Ea.		16.35		16.35	install 2' x 2' double hung, two light,
	Install	Ea.	99.50	71.50		171	double glazed aluminum window
	Demolish and Install	Ea.	99.50	87.85		187.35	including screen.
	Reinstall	Ea.		57.37		57.37	
	Clean	Ea.	.02	5.30		5.32	
	Minimum Charge	Job		197		197	
2' x 2' 6"							Cost includes material and labor to
	Demolish	Ea.		16.35		16.35	install 2' x 2' 6" double hung, two
	Install	Ea.	108	71.50		179.50	light, double glazed aluminum
	Demolish and Install	Ea.	108	87.85		195.85	window including screen.
	Reinstall	Ea.		57.37		57.37	
	Clean	Ea.	.02	5.30		5.32	
	Minimum Charge	Job		197		197	
3' x 1' 6"							Cost includes material and labor to
	Demolish	Ea.		16.35		16.35	install 3' x 1' 6" double hung, two
	Install	Ea.	256	79		335	light, double glazed aluminum
	Demolish and Install	Ea.	256	95.35		351.35	window including screen.
	Reinstall	Ea.		63.10		63.10	
	Clean	Ea.	.02	5.30		5.32	
	Minimum Charge	Job		197		197	
3' x 2'							Cost includes material and labor to
	Demolish	Ea.		16.35		16.35	install 3' x 2' double hung, two light,
	Install	Ea.	263	79		342	double glazed aluminum window
	Demolish and Install	Ea.	263	95.35		358.35	including screen.
	Reinstall	Ea.		63.10		63.10	
	Clean	Ea.	.02	5.30		5.32	
	Minimum Charge	Job		197		197	
3' x 2' 6"							Cost includes material and labor to
	Demolish	Ea.		16.35		16.35	install 3' x 2' 6" double hung, two
	Install	Ea.	265	79		344	light, double glazed aluminum
	Demolish and Install	Ea.	265	95.35		360.35	window including screen.
	Reinstall	Ea.		63.10		63.10	
	Clean	Ea.	.03	6.05		6.08	
	Minimum Charge	Job		197		197	
3' x 3'							Cost includes material and labor to
	Demolish	Ea.		16.35		16.35	install 3' x 3' double hung, two light,
	Install	Ea.	286	79		365	double glazed aluminum window
	Demolish and Install	Ea.	286	95.35		381.35	including screen.
	Reinstall	Ea.		63.10		63.10	
	Clean	Ea.	.03	6.05		6.08	
	Minimum Charge	Job		197		197	
3' x 3' 6"							Cost includes material and labor to
	Demolish	Ea.		16.35		16.35	install 3' x 3' 6" double hung, two
	Install	Ea.	315	79		394	light, double glazed aluminum
	Demolish and Install	Ea.	315	95.35		410.35	window including screen.
	Reinstall	Ea.		63.10		63.10	
	Clean	Ea.	.04	7.05		7.09	
	Minimum Charge	Job		197		197	

Exterior Doors and Windows

Aluminum Window		Unit	Material	Labor	Equip.	Total	Specification
3' x 4'							
	Demolish	Ea.		16.35		16.35	Cost includes material and labor to
	Install	Ea.	325	79		404	install 3' x 4' double hung, two light,
	Demolish and Install	Ea.	325	95.35		420.35	double glazed aluminum window
	Reinstall	Ea.		63.10		63.10	including screen.
	Clean	Ea.	.04	7.05		7.09	
	Minimum Charge	Job		197		197	
3' x 5'							
	Demolish	Ea.		24		24	Cost includes material and labor to
	Install	Ea.	365	79		444	install 3' x 5' double hung, two light,
	Demolish and Install	Ea.	365	103		468	double glazed aluminum window
	Reinstall	Ea.		63.10		63.10	including screen.
	Clean	Ea.	.06	8.50		8.56	
	Minimum Charge	Job		197		197	
3' x 6'							
	Demolish	Ea.		24		24	Cost includes material and labor to
	Install	Ea.	470	87.50		557.50	install 3' x 6' double hung, two light,
	Demolish and Install	Ea.	470	111.50		581.50	double glazed aluminum window
	Reinstall	Ea.		70.12		70.12	including screen.
	Clean	Ea.	.07	10		10.07	
	Minimum Charge	Job		197		197	

Sliding Sash

		Unit	Material	Labor	Equip.	Total	Specification
1' 6" x 3'							
	Demolish	Ea.		16.35		16.35	Cost includes material and labor to
	Install	Ea.	56.50	71.50		128	install 1' 6" x 3' sliding sash, two
	Demolish and Install	Ea.	56.50	87.85		144.35	light, double glazed aluminum
	Reinstall	Ea.		57.37		57.37	window including screen.
	Clean	Ea.	.02	5.30		5.32	
	Minimum Charge	Job		197		197	
2' x 2'							
	Demolish	Ea.		16.35		16.35	Cost includes material and labor to
	Install	Ea.	50.50	79		129.50	install 2' x 2' sliding sash, two light,
	Demolish and Install	Ea.	50.50	95.35		145.85	double glazed aluminum window
	Reinstall	Ea.		63.10		63.10	including screen.
	Clean	Ea.	.02	5.30		5.32	
	Minimum Charge	Job		197		197	
3' x 2'							
	Demolish	Ea.		16.35		16.35	Cost includes material and labor to
	Install	Ea.	66	79		145	install 3' x 2' sliding sash, two light,
	Demolish and Install	Ea.	66	95.35		161.35	double glazed aluminum window
	Reinstall	Ea.		63.10		63.10	including screen.
	Clean	Ea.	.02	5.30		5.32	
	Minimum Charge	Job		197		197	
3' x 2' 6"							
	Demolish	Ea.		16.35		16.35	Cost includes material and labor to
	Install	Ea.	68	79		147	install 3' x 2' 6" sliding sash, two
	Demolish and Install	Ea.	68	95.35		163.35	light, double glazed aluminum
	Reinstall	Ea.		63.10		63.10	window including screen.
	Clean	Ea.	.03	6.05		6.08	
	Minimum Charge	Job		197		197	
3' x 3'							
	Demolish	Ea.		16.35		16.35	Cost includes material and labor to
	Install	Ea.	70	79		149	install 3' x 3' sliding sash, two light,
	Demolish and Install	Ea.	70	95.35		165.35	double glazed aluminum window
	Reinstall	Ea.		63.10		63.10	including screen.
	Clean	Ea.	.03	6.05		6.08	
	Minimum Charge	Job		197		197	

Aluminum Window

Aluminum Window	Unit	Material	Labor	Equip.	Total	Specification
3' x 3' 6"						
Demolish	Ea.		16.35		16.35	Cost includes material and labor to
Install	Ea.	81.50	79		160.50	install 3' x 3' 6" sliding sash, two
Demolish and Install	Ea.	81.50	95.35		176.85	light, double glazed aluminum
Reinstall	Ea.		63.10		63.10	window including screen.
Clean	Ea.	.04	7.05		7.09	
Minimum Charge	Job		197		197	
3' x 4'						
Demolish	Ea.		16.35		16.35	Cost includes material and labor to
Install	Ea.	89	87.50		176.50	install 3' x 4' sliding sash, two light,
Demolish and Install	Ea.	89	103.85		192.85	double glazed aluminum window
Reinstall	Ea.		70.12		70.12	including screen.
Clean	Ea.	.04	7.05		7.09	
Minimum Charge	Job		197		197	
3' x 5'						
Demolish	Ea.		24		24	Cost includes material and labor to
Install	Ea.	113	87.50		200.50	install 3' x 5' sliding sash, two light,
Demolish and Install	Ea.	113	111.50		224.50	double glazed aluminum window
Reinstall	Ea.		70.12		70.12	including screen.
Clean	Ea.	.06	8.50		8.56	
Minimum Charge	Job		197		197	
4' x 5'						
Demolish	Ea.		24		24	Cost includes material and labor to
Install	Ea.	114	113		227	install 4' x 5' sliding sash, two light,
Demolish and Install	Ea.	114	137		251	double glazed aluminum window
Reinstall	Ea.		90.15		90.15	including screen.
Clean	Ea.	.09	12.10		12.19	
Minimum Charge	Job		197		197	
5' x 5'						
Demolish	Ea.		24		24	Cost includes material and labor to
Install	Ea.	160	113		273	install 5' x 5' sliding sash, two light,
Demolish and Install	Ea.	160	137		297	double glazed aluminum window
Reinstall	Ea.		90.15		90.15	including screen.
Clean	Ea.	.09	12.10		12.19	
Minimum Charge	Job		197		197	

Awning Sash

Awning Sash	Unit	Material	Labor	Equip.	Total	Specification
26" x 26", 2 Light						
Demolish	Ea.		16.35		16.35	Cost includes material and labor to
Install	Ea.	248	87.50		335.50	install 26" x 26" two light awning,
Demolish and Install	Ea.	248	103.85		351.85	double glazed aluminum window
Reinstall	Ea.		70.12		70.12	including screen.
Clean	Ea.	.03	6.05		6.08	
Minimum Charge	Job		197		197	
26" x 37", 2 Light						
Demolish	Ea.		16.35		16.35	Cost includes material and labor to
Install	Ea.	282	87.50		369.50	install 26" x 37" two light awning,
Demolish and Install	Ea.	282	103.85		385.85	double glazed aluminum window
Reinstall	Ea.		70.12		70.12	including screen.
Clean	Ea.	.03	6.05		6.08	
Minimum Charge	Job		197		197	
38" x 26", 3 Light						
Demolish	Ea.		16.35		16.35	Cost includes material and labor to
Install	Ea.	315	87.50		402.50	install 38" x 26" three light awning,
Demolish and Install	Ea.	315	103.85		418.85	double glazed aluminum window
Reinstall	Ea.		70.12		70.12	including screen.
Clean	Ea.	.04	7.05		7.09	
Minimum Charge	Job		197		197	

Exterior Doors and Windows

Aluminum Window		Unit	Material	Labor	Equip.	Total	Specification
38" x 37", 3 Light							
	Demolish	Ea.		16.35		16.35	Cost includes material and labor to
	Install	Ea.	320	98.50		418.50	install 38" x 37" three light awning,
	Demolish and Install	Ea.	320	114.85		434.85	double glazed aluminum window
	Reinstall	Ea.		78.88		78.88	including screen.
	Clean	Ea.	.04	7.05		7.09	
	Minimum Charge	Job		197		197	
38" x 53", 3 Light							
	Demolish	Ea.		24		24	Cost includes material and labor to
	Install	Ea.	505	98.50		603.50	install 38" x 53" three light awning,
	Demolish and Install	Ea.	505	122.50		627.50	double glazed aluminum window
	Reinstall	Ea.		78.88		78.88	including screen.
	Clean	Ea.	.06	8.50		8.56	
	Minimum Charge	Job		197		197	
50" x 37", 4 Light							
	Demolish	Ea.		24		24	Cost includes material and labor to
	Install	Ea.	525	98.50		623.50	install 50" x 37" four light awning,
	Demolish and Install	Ea.	525	122.50		647.50	double glazed aluminum window
	Reinstall	Ea.		78.88		78.88	including screen.
	Clean	Ea.	.06	8.50		8.56	
	Minimum Charge	Job		197		197	
50" x 53", 4 Light							
	Demolish	Ea.		24		24	Cost includes material and labor to
	Install	Ea.	835	113		948	install 50" x 53" four light awning,
	Demolish and Install	Ea.	835	137		972	double glazed aluminum window
	Reinstall	Ea.		90.15		90.15	including screen.
	Clean	Ea.	.07	10		10.07	
	Minimum Charge	Job		197		197	
63" x 37", 5 Light							
	Demolish	Ea.		24		24	Cost includes material and labor to
	Install	Ea.	675	113		788	install 63" x 37" five light awning,
	Demolish and Install	Ea.	675	137		812	double glazed aluminum window
	Reinstall	Ea.		90.15		90.15	including screen.
	Clean	Ea.	.07	10		10.07	
	Minimum Charge	Job		197		197	
63" x 53", 5 Light							
	Demolish	Ea.		24		24	Cost includes material and labor to
	Install	Ea.	900	113		1013	install 63" x 53" five light awning,
	Demolish and Install	Ea.	900	137		1037	double glazed aluminum window
	Reinstall	Ea.		90.15		90.15	including screen.
	Clean	Ea.	.09	12.10		12.19	
	Minimum Charge	Job		197		197	
Mullion Bars							
	Install	Ea.	9.55	31.50		41.05	Includes labor and material to install
	Minimum Charge	Job		197		197	mullions for sliding aluminum window.
Glass Louvered							
2' x 2'							
	Demolish	Ea.		16.35		16.35	Cost includes material and labor to
	Install	Ea.	47	71.50		118.50	install 2' x 2' aluminum window
	Demolish and Install	Ea.	47	87.85		134.85	including 4" glass louvers, hardware,
	Reinstall	Ea.		57.37		57.37	screen.
	Clean	Ea.	.02	5.30		5.32	
	Minimum Charge	Job		197		197	

Exterior Doors and Windows

Aluminum Window		Unit	Material	Labor	Equip.	Total	Specification
2' x 2' 6"							
	Demolish	Ea.		16.35		16.35	Cost includes material and labor to
	Install	Ea.	57	71.50		128.50	install 2' x 2' 6" aluminum window
	Demolish and Install	Ea.	57	87.85		144.85	including 4" glass louvers, hardware,
	Reinstall	Ea.		57.37		57.37	screen.
	Clean	Ea.	.02	5.30		5.32	
	Minimum Charge	Job		197		197	
3' x 1' 6"							
	Demolish	Ea.		16.35		16.35	Cost includes material and labor to
	Install	Ea.	60	79		139	install 3' x 1' 6" aluminum window
	Demolish and Install	Ea.	60	95.35		155.35	including 4" glass louvers, hardware,
	Reinstall	Ea.		63.10		63.10	screen.
	Clean	Ea.	.02	5.30		5.32	
	Minimum Charge	Job		197		197	
3' x 2'							
	Demolish	Ea.		16.35		16.35	Cost includes material and labor to
	Install	Ea.	69	79		148	install 3' x 2' aluminum window
	Demolish and Install	Ea.	69	95.35		164.35	including 4" glass louvers, hardware,
	Reinstall	Ea.		63.10		63.10	screen.
	Clean	Ea.	.02	5.30		5.32	
	Minimum Charge	Job		197		197	
3' x 2' 6"							
	Demolish	Ea.		16.35		16.35	Cost includes material and labor to
	Install	Ea.	75	79		154	install 3' x 2' 6" aluminum window
	Demolish and Install	Ea.	75	95.35		170.35	including 4" glass louvers, hardware,
	Reinstall	Ea.		63.10		63.10	screen.
	Clean	Ea.	.03	6.05		6.08	
	Minimum Charge	Job		197		197	
3' x 3'							
	Demolish	Ea.		16.35		16.35	Cost includes material and labor to
	Install	Ea.	87.50	87.50		175	install 3' x 3' aluminum window
	Demolish and Install	Ea.	87.50	103.85		191.35	including 4" glass louvers, hardware,
	Reinstall	Ea.		70.12		70.12	screen.
	Clean	Ea.	.03	6.05		6.08	
	Minimum Charge	Job		197		197	
3' x 3' 6"							
	Demolish	Ea.		16.35		16.35	Cost includes material and labor to
	Install	Ea.	99.50	87.50		187	install 3' x 3' 6" aluminum window
	Demolish and Install	Ea.	99.50	103.85		203.35	including 4" glass louvers, hardware,
	Reinstall	Ea.		70.12		70.12	screen.
	Clean	Ea.	.04	7.05		7.09	
	Minimum Charge	Job		197		197	
3' x 4'							
	Demolish	Ea.		16.35		16.35	Cost includes material and labor to
	Install	Ea.	111	87.50		198.50	install 3' x 4' aluminum window
	Demolish and Install	Ea.	111	103.85		214.85	including 4" glass louvers, hardware,
	Reinstall	Ea.		70.12		70.12	screen.
	Clean	Ea.	.04	7.05		7.09	
	Minimum Charge	Job		197		197	
3' x 5'							
	Demolish	Ea.		24		24	Cost includes material and labor to
	Install	Ea.	142	98.50		240.50	install 3' x 5' aluminum window with
	Demolish and Install	Ea.	142	122.50		264.50	4" glass louvers, hardware, screen.
	Reinstall	Ea.		78.88		78.88	
	Clean	Ea.	.06	8.50		8.56	
	Minimum Charge	Job		197		197	

Exterior Doors and Windows

Aluminum Window		Unit	Material	Labor	Equip.	Total	Specification
3' x 6'							
	Demolish	Ea.		24		24	Cost includes material and labor to
	Install	Ea.	160	98.50		258.50	install 3' x 6' aluminum window with
	Demolish and Install	Ea.	160	122.50		282.50	4" glass louvers, hardware, screen.
	Reinstall	Ea.		78.88		78.88	
	Clean	Ea.	.07	10		10.07	
	Minimum Charge	Job		197		197	
Bay							
4' 8" x 6' 4"							
	Demolish	Ea.		17.45		17.45	Cost includes material and labor to
	Install	Ea.	1050	81.50		1131.50	install 4' 8" x 6' 4" angle bay double
	Demolish and Install	Ea.	1050	98.95		1148.95	hung unit, pine frame with aluminum
	Reinstall	Ea.		65.20		65.20	cladding, picture center sash,
	Clean	Ea.	.13	14.15		14.28	hardware, screens.
	Minimum Charge	Job		197		197	
4' 8" x 7' 2"							
	Demolish	Ea.		17.45		17.45	Cost includes material and labor to
	Install	Ea.	1075	81.50		1156.50	install 4' 8" x 7' 2" angle bay double
	Demolish and Install	Ea.	1075	98.95		1173.95	hung unit, pine frame with aluminum
	Reinstall	Ea.		65.20		65.20	cladding, picture center sash,
	Clean	Ea.	.13	14.15		14.28	hardware, screens.
	Minimum Charge	Job		197		197	
4' 8" x 7' 10"							
	Demolish	Ea.		17.45		17.45	Cost includes material and labor to
	Install	Ea.	1100	81.50		1181.50	install 4' 8" x 7' 10" angle bay
	Demolish and Install	Ea.	1100	98.95		1198.95	double hung unit, pine frame with
	Reinstall	Ea.		65.20		65.20	aluminum cladding, picture center
	Clean	Ea.	.13	14.15		14.28	sash, hardware, screens.
	Minimum Charge	Job		197		197	
4' 8" x 8' 6"							
	Demolish	Ea.		20		20	Cost includes material and labor to
	Install	Ea.	1250	93		1343	install 4' 8" x 8' 6" angle bay double
	Demolish and Install	Ea.	1250	113		1363	hung unit, pine frame with aluminum
	Reinstall	Ea.		74.51		74.51	cladding, picture center sash,
	Clean	Ea.	.02	18.85		18.87	hardware, screens.
	Minimum Charge	Job		197		197	
4' 8" x 8' 10"							
	Demolish	Ea.		20		20	Cost includes material and labor to
	Install	Ea.	1275	93		1368	install 4' 8" x 8' 10" angle bay
	Demolish and Install	Ea.	1275	113		1388	double hung unit, pine frame with
	Reinstall	Ea.		74.51		74.51	aluminum cladding, picture center
	Clean	Ea.	.02	18.85		18.87	sash, hardware, screens.
	Minimum Charge	Job		197		197	
4' 8" x 9' 2"							
	Demolish	Ea.		20		20	Cost includes material and labor to
	Install	Ea.	1325	163		1488	install 4' 8" x 9' 2" angle bay double
	Demolish and Install	Ea.	1325	183		1508	hung unit, pine frame with aluminum
	Reinstall	Ea.		130.40		130.40	cladding, picture center sash,
	Clean	Ea.	.02	18.85		18.87	hardware, screens.
	Minimum Charge	Job		197		197	
4' 8" x 9' 10"							
	Demolish	Ea.		20		20	Cost includes material and labor to
	Install	Ea.	1400	163		1563	install 4' 8" x 9' 10" angle bay
	Demolish and Install	Ea.	1400	183		1583	double hung unit, pine frame with
	Reinstall	Ea.		130.40		130.40	aluminum cladding, picture center
	Clean	Ea.	.02	18.85		18.87	sash, hardware, screens.
	Minimum Charge	Job		197		197	

Exterior Doors and Windows

Aluminum Window		Unit	Material	Labor	Equip.	Total	Specification
Casing							
	Demolish	Ea.		7.55		7.55	Cost includes material and labor to
	Install	Opng.	22	25		47	install 11/16" x 2-1/2" pine ranch
	Demolish and Install	Opng.	22	32.55		54.55	style window casing.
	Reinstall	Opng.		20.06		20.06	
	Clean	Opng.	.42	2.36		2.78	
	Paint	Ea.	3.15	5.70		8.85	
	Minimum Charge	Job		197		197	
Storm Window							
2' x 3' 6"							
	Demolish	Ea.		9.70		9.70	Includes material and labor to install
	Install	Ea.	72	21.50		93.50	residential aluminum storm window
	Demolish and Install	Ea.	72	31.20		103.20	with mill finish, 2' x 3' 5" high.
	Reinstall	Ea.		17.39		17.39	
	Clean	Ea.	.02	5.30		5.32	
	Minimum Charge	Job		197		197	
2' 6" x 5'							
	Demolish	Ea.		9.70		9.70	Includes material and labor to install
	Install	Ea.	81	23.50		104.50	residential aluminum storm window
	Demolish and Install	Ea.	81	33.20		114.20	with mill finish, 2' 6" x 5' high.
	Reinstall	Ea.		18.63		18.63	
	Clean	Ea.	.04	7.05		7.09	
	Minimum Charge	Job		197		197	
4' x 6'							
	Demolish	Ea.		18.70		18.70	Includes material and labor to install
	Install	Ea.	95.50	26		121.50	residential aluminum storm window, 4'
	Demolish and Install	Ea.	95.50	44.70		140.20	x 6' high.
	Reinstall	Ea.		20.86		20.86	
	Clean	Ea.	.09	10.60		10.69	
	Minimum Charge	Job		197		197	
Hardware							
Weatherstripping							
	Install	L.F.	.28	2.72		3	Includes labor and material to install
	Minimum Charge	Job		109		109	vinyl V strip weatherstripping for
							double-hung window.
Trim Set							
	Demolish	Ea.		7.55		7.55	Cost includes material and labor to
	Install	Opng.	22	25		47	install 11/16" x 2-1/2" pine ranch
	Demolish and Install	Opng.	22	32.55		54.55	style window casing.
	Reinstall	Opng.		20.06		20.06	
	Clean	Opng.	.42	2.36		2.78	
	Paint	Ea.	3.15	5.70		8.85	
	Minimum Charge	Job		163		163	

Vinyl Window		Unit	Material	Labor	Equip.	Total	Specification
Double Hung							
2' x 2' 6"							
	Demolish	Ea.		11.90		11.90	Cost includes material and labor to
	Install	Ea.	164	43.50		207.50	install 2' x 2' 6" double hung, two
	Demolish and Install	Ea.	164	55.40		219.40	light, vinyl window with screen.
	Reinstall	Ea.		34.77		34.77	
	Clean	Ea.	.02	5.30		5.32	
	Minimum Charge	Job		325		325	

Exterior Doors and Windows

Vinyl Window

Vinyl Window	Unit	Material	Labor	Equip.	Total	Specification
2′ x 3′ 6″						
Demolish	Ea.		11.90		11.90	Cost includes material and labor to
Install	Ea.	172	46.50		218.50	install 2′ x 3′ 6″ double hung, two
Demolish and Install	Ea.	172	58.40		230.40	light, vinyl window with screen.
Reinstall	Ea.		37.26		37.26	
Clean	Ea.	.03	6.05		6.08	
Minimum Charge	Job		325		325	
2′ 6″ x 4′ 6″						
Demolish	Ea.		11.90		11.90	Cost includes material and labor to
Install	Ea.	205	50		255	install 2′ 6″ x 4′ 6″ double hung, two
Demolish and Install	Ea.	205	61.90		266.90	light, vinyl window with screen.
Reinstall	Ea.		40.12		40.12	
Clean	Ea.	.04	7.05		7.09	
Minimum Charge	Job		325		325	
3′ x 4′						
Demolish	Ea.		11.90		11.90	Includes material and labor to install
Install	Ea.	208	65		273	medium size double hung, 3′ x 4′,
Demolish and Install	Ea.	208	76.90		284.90	two light, vinyl window with screen.
Reinstall	Ea.		52.16		52.16	
Clean	Ea.	.04	7.05		7.09	
Minimum Charge	Job		325		325	
3′ x 4′ 6″						
Demolish	Ea.		14.55		14.55	Cost includes material and labor to
Install	Ea.	230	72.50		302.50	install 3′ x 4′ 6″ double hung, two
Demolish and Install	Ea.	230	87.05		317.05	light, vinyl window with screen.
Reinstall	Ea.		57.96		57.96	
Clean	Ea.	.06	8.50		8.56	
Minimum Charge	Job		325		325	
4′ x 4′ 6″						
Demolish	Ea.		14.55		14.55	Cost includes material and labor to
Install	Ea.	253	81.50		334.50	install 4′ x 4′ 6″ double hung, two
Demolish and Install	Ea.	253	96.05		349.05	light, vinyl window with screen.
Reinstall	Ea.		65.20		65.20	
Clean	Ea.	.07	10		10.07	
Minimum Charge	Job		325		325	
4′ x 6′						
Demolish	Ea.		14.55		14.55	Includes material and labor to install
Install	Ea.	293	93		386	large double hung, 4′ x 6′, two light,
Demolish and Install	Ea.	293	107.55		400.55	vinyl window with screen.
Reinstall	Ea.		74.51		74.51	
Clean	Ea.	.09	12.10		12.19	
Minimum Charge	Job		325		325	
Trim Set						
Demolish	Ea.		7.55		7.55	Cost includes material and labor to
Install	Opng.	22	25		47	install 11/16″ x 2-1/2″ pine ranch
Demolish and Install	Opng.	22	32.55		54.55	style window casing.
Reinstall	Opng.		20.06		20.06	
Clean	Opng.	.42	2.36		2.78	
Paint	Ea.	3.15	5.70		8.85	
Minimum Charge	Job		197		197	

Wood Window

Wood Window	Unit	Material	Labor	Equip.	Total	Specification
2′ 2″ x 3′ 4″						
Demolish	Ea.		11.90		11.90	Cost includes material and labor to
Install	Ea.	178	43.50		221.50	install 2′ 2″ x 3′ 4″ double hung, two
Demolish and Install	Ea.	178	55.40		233.40	light, double glazed wood window
Reinstall	Ea.		34.77		34.77	with screen.
Clean	Ea.	.03	6.05		6.08	
Paint	Ea.	1.97	11.90		13.87	
Minimum Charge	Job		163		163	

Exterior Doors and Windows

Wood Window		Unit	Material	Labor	Equip.	Total	Specification
2' 2" x 4' 4"							
	Demolish	Ea.		11.90		11.90	Cost includes material and labor to
	Install	Ea.	200	46.50		246.50	install 2' 2" x 4' 4" double hung, two
	Demolish and Install	Ea.	200	58.40		258.40	light, double glazed wood window
	Reinstall	Ea.		37.26		37.26	with screen.
	Clean	Ea.	.03	6.05		6.08	
	Paint	Ea.	1.97	11.90		13.87	
	Minimum Charge	Job		163		163	
2' 6" x 3' 4"							
	Demolish	Ea.		11.90		11.90	Cost includes material and labor to
	Install	Ea.	188	50		238	install 2' 6" x 3' 4" double hung, two
	Demolish and Install	Ea.	188	61.90		249.90	light, double glazed wood window
	Reinstall	Ea.		40.12		40.12	with screen.
	Clean	Ea.	.04	7.05		7.09	
	Paint	Ea.	2.52	14.30		16.82	
	Minimum Charge	Job		163		163	
2' 6" x 4'							
	Demolish	Ea.		11.90		11.90	Cost includes material and labor to
	Install	Ea.	202	54.50		256.50	install 2' 6" x 4' double hung, two
	Demolish and Install	Ea.	202	66.40		268.40	light, double glazed wood window
	Reinstall	Ea.		43.47		43.47	with screen.
	Clean	Ea.	.04	7.05		7.09	
	Paint	Ea.	2.52	14.30		16.82	
	Minimum Charge	Job		163		163	
2' 6" x 4' 8"							
	Demolish	Ea.		11.90		11.90	Cost includes material and labor to
	Install	Ea.	223	54.50		277.50	install 2' 6" x 4' 8" double hung, two
	Demolish and Install	Ea.	223	66.40		289.40	light, double glazed wood window
	Reinstall	Ea.		43.47		43.47	with screen.
	Clean	Ea.	.04	7.05		7.09	
	Paint	Ea.	2.52	14.30		16.82	
	Minimum Charge	Job		163		163	
2' 10" x 3' 4"							
	Demolish	Ea.		11.90		11.90	Cost includes material and labor to
	Install	Ea.	198	65		263	install 2' 10" x 3' 4" double hung, two
	Demolish and Install	Ea.	198	76.90		274.90	light, double glazed wood window
	Reinstall	Ea.		52.16		52.16	with screen.
	Clean	Ea.	.04	7.05		7.09	
	Paint	Ea.	2.52	14.30		16.82	
	Minimum Charge	Job		163		163	
2' 10" x 4'							
	Demolish	Ea.		11.90		11.90	Cost includes material and labor to
	Install	Ea.	217	65		282	install 2' 10" x 4' double hung, two
	Demolish and Install	Ea.	217	76.90		293.90	light, double glazed wood window
	Reinstall	Ea.		52.16		52.16	with screen.
	Clean	Ea.	.04	7.05		7.09	
	Paint	Ea.	2.52	14.30		16.82	
	Minimum Charge	Job		163		163	
3' 7" x 3' 4"							
	Demolish	Ea.		11.90		11.90	Cost includes material and labor to
	Install	Ea.	223	72.50		295.50	install 3' 7" x 3' 4" double hung, two
	Demolish and Install	Ea.	223	84.40		307.40	light, double glazed wood window
	Reinstall	Ea.		57.96		57.96	with screen.
	Clean	Ea.	.04	7.05		7.09	
	Paint	Ea.	2.52	14.30		16.82	
	Minimum Charge	Job		163		163	

Exterior Doors and Windows

Wood Window		Unit	Material	Labor	Equip.	Total	Specification
3' 7" x 5' 4"							
	Demolish	Ea.		14.55		14.55	Cost includes material and labor to
	Install	Ea.	286	72.50		358.50	install 3' 7" x 5' 4" double hung, two
	Demolish and Install	Ea.	286	87.05		373.05	light, double glazed wood window
	Reinstall	Ea.		57.96		57.96	with screen.
	Clean	Ea.	.09	12.10		12.19	
	Paint	Ea.	4.20	24		28.20	
	Minimum Charge	Job		163		163	
4' 3" x 5' 4"							
	Demolish	Ea.		14.55		14.55	Cost includes material and labor to
	Install	Ea.	450	81.50		531.50	install 4' 3" x 5' 4" double hung, two
	Demolish and Install	Ea.	450	96.05		546.05	light, double glazed wood window
	Reinstall	Ea.		65.20		65.20	with screen.
	Clean	Ea.	.09	12.10		12.19	
	Paint	Ea.	2.70	28.50		31.20	
	Minimum Charge	Job		163		163	
Casement							
2' x 3' 4"							
	Demolish	Ea.		11.90		11.90	Cost includes material and labor to
	Install	Ea.	187	32.50		219.50	install 2' x 3' 4" casement window
	Demolish and Install	Ea.	187	44.40		231.40	with 3/4" insulated glass, screens,
	Reinstall	Ea.		26.08		26.08	weatherstripping, hardware.
	Clean	Ea.	.03	6.05		6.08	
	Paint	Ea.	1.97	11.90		13.87	
	Minimum Charge	Job		325		325	
2' x 4'							
	Demolish	Ea.		11.90		11.90	Cost includes material and labor to
	Install	Ea.	210	36		246	install 2' x 4' casement window with
	Demolish and Install	Ea.	210	47.90		257.90	3/4" insulated glass, screens,
	Reinstall	Ea.		28.98		28.98	weatherstripping, hardware.
	Clean	Ea.	.03	6.05		6.08	
	Paint	Ea.	1.97	11.90		13.87	
	Minimum Charge	Job		325		325	
2' x 5'							
	Demolish	Ea.		11.90		11.90	Cost includes material and labor to
	Install	Ea.	235	38.50		273.50	install 2' x 5' casement window with
	Demolish and Install	Ea.	235	50.40		285.40	3/4" insulated glass, screens,
	Reinstall	Ea.		30.68		30.68	weatherstripping, hardware.
	Clean	Ea.	.04	7.05		7.09	
	Paint	Ea.	2.52	14.30		16.82	
	Minimum Charge	Job		325		325	
2' x 6'							
	Demolish	Ea.		11.90		11.90	Cost includes material and labor to
	Install	Ea.	265	41		306	install 2' x 6' casement window with
	Demolish and Install	Ea.	265	52.90		317.90	3/4" insulated glass, screens,
	Reinstall	Ea.		32.60		32.60	weatherstripping, hardware.
	Clean	Ea.	.04	7.05		7.09	
	Paint	Ea.	2.52	14.30		16.82	
	Minimum Charge	Job		325		325	
4' x 3' 4"							
	Demolish	Ea.		14.55		14.55	Cost includes material and labor to
	Install	Ea.	540	43.50		583.50	install 4' x 3' 4" casement windows
	Demolish and Install	Ea.	540	58.05		598.05	with 3/4" insulated glass, screens,
	Reinstall	Ea.		34.77		34.77	weatherstripping, hardware.
	Clean	Ea.	.04	7.05		7.09	
	Paint	Ea.	1.34	14.30		15.64	
	Minimum Charge	Job		325		325	

Exterior Doors and Windows

Wood Window	Unit	Material	Labor	Equip.	Total	Specification
4' x 4'						
Demolish	Ea.		14.55		14.55	Cost includes material and labor to
Install	Ea.	595	43.50		638.50	install 4' x 4' casement windows with
Demolish and Install	Ea.	595	58.05		653.05	3/4" insulated glass, screens,
Reinstall	Ea.		34.77		34.77	weatherstripping, hardware.
Clean	Ea.	.07	10		10.07	
Paint	Ea.	1.60	17.90		19.50	
Minimum Charge	Job		325		325	
4' x 5'						
Demolish	Ea.		14.55		14.55	Cost includes material and labor to
Install	Ea.	670	46.50		716.50	install 4' x 5' casement windows with
Demolish and Install	Ea.	670	61.05		731.05	3/4" insulated glass, screens,
Reinstall	Ea.		37.26		37.26	weatherstripping, hardware.
Clean	Ea.	.09	12.10		12.19	
Paint	Ea.	2.08	24		26.08	
Minimum Charge	Job		325		325	
4' x 6'						
Demolish	Ea.		14.55		14.55	Cost includes material and labor to
Install	Ea.	760	54.50		814.50	install 4' x 6' casement windows with
Demolish and Install	Ea.	760	69.05		829.05	3/4" insulated glass, screens,
Reinstall	Ea.		43.47		43.47	weatherstripping, hardware.
Clean	Ea.	.09	12.10		12.19	
Paint	Ea.	2.45	28.50		30.95	
Minimum Charge	Job		325		325	
Casement Bow						
4' 8" x 8' 1"						
Demolish	Ea.		20		20	Includes material and labor to install
Install	Ea.	1425	122		1547	casement bow window unit with 4
Demolish and Install	Ea.	1425	142		1567	lights, pine frame, 1/2" insulated
Reinstall	Ea.		97.80		97.80	glass, weatherstripping, hardware,
Clean	Ea.	.13	14.15		14.28	screens.
Paint	Ea.	4.53	57		61.53	
Minimum Charge	Job		325		325	
4' 8" x 10'						
Demolish	Ea.		20		20	Includes material and labor to install
Install	Ea.	1525	163		1688	casement bow window unit with 5
Demolish and Install	Ea.	1525	183		1708	lights, pine frame, 1/2" insulated
Reinstall	Ea.		130.40		130.40	glass, weatherstripping, hardware,
Clean	Ea.	.02	18.85		18.87	screens.
Paint	Ea.	5.60	71.50		77.10	
Minimum Charge	Job		325		325	
5' 4" x 8' 1"						
Demolish	Ea.		20		20	Includes material and labor to install
Install	Ea.	1575	122		1697	casement bow window unit with 4
Demolish and Install	Ea.	1575	142		1717	lights, pine frame, 1/2" insulated
Reinstall	Ea.		97.80		97.80	glass, weatherstripping, hardware,
Clean	Ea.	.02	18.85		18.87	screens.
Paint	Ea.	5.15	57		62.15	
Minimum Charge	Job		325		325	
5' 4" x 10'						
Demolish	Ea.		20		20	Includes material and labor to install
Install	Ea.	1725	163		1888	casement bow window unit with 5
Demolish and Install	Ea.	1725	183		1908	lights, pine frame, 1/2" insulated
Reinstall	Ea.		130.40		130.40	glass, weatherstripping, hardware,
Clean	Ea.	.02	18.85		18.87	screens.
Paint	Ea.	6.50	71.50		78	
Minimum Charge	Job		325		325	

Exterior Doors and Windows

Wood Window		Unit	Material	Labor	Equip.	Total	Specification
6' x 8' 1"							
	Demolish	Ea.		20		20	Includes material and labor to install
	Install	Ea.	1525	163		1688	casement bow window unit with 5
	Demolish and Install	Ea.	1525	183		1708	lights, pine frame, 1/2" insulated
	Reinstall	Ea.		130.40		130.40	glass, weatherstripping, hardware,
	Clean	Ea.	.02	18.85		18.87	screens.
	Paint	Ea.	5.90	71.50		77.40	
	Minimum Charge	Job		325		325	
6' x 10'							
	Demolish	Ea.		20		20	Includes material and labor to install
	Install	Ea.	1825	163		1988	casement bow window unit with 5
	Demolish and Install	Ea.	1825	183		2008	lights, pine frame, 1/2" insulated
	Reinstall	Ea.		130.40		130.40	glass, weatherstripping, hardware,
	Clean	Ea.	.02	18.85		18.87	screens.
	Paint	Ea.	7.35	95.50		102.85	
	Minimum Charge	Job		325		325	
Casing							
	Demolish	Ea.		7.55		7.55	Cost includes material and labor to
	Install	Opng.	22	25		47	install 11/16" x 2-1/2" pine ranch
	Demolish and Install	Opng.	22	32.55		54.55	style window casing.
	Reinstall	Opng.		20.06		20.06	
	Clean	Opng.	.42	2.36		2.78	
	Paint	Ea.	3.15	5.70		8.85	
	Minimum Charge	Job		197		197	
Trim Set							
	Demolish	Ea.		5.75		5.75	Cost includes material and labor to
	Install	Opng.	22	25		47	install 11/16" x 2-1/2" pine ranch
	Demolish and Install	Opng.	22	30.75		52.75	style window casing.
	Reinstall	Opng.		20.06		20.06	
	Clean	Opng.	.42	2.36		2.78	
	Paint	Ea.	3.15	5.70		8.85	
	Minimum Charge	Job		163		163	
Wood Picture Window							
3' 6" x 4'							
	Demolish	Ea.		14.55		14.55	Includes material and labor to install a
	Install	Ea.	320	54.50		374.50	wood framed picture window
	Demolish and Install	Ea.	320	69.05		389.05	including exterior trim.
	Clean	Ea.	.09	12.10		12.19	
	Paint	Ea.	1.72	22		23.72	
	Minimum Charge	Job		325		325	
4' x 4' 6"							
	Demolish	Ea.		14.55		14.55	Includes material and labor to install a
	Install	Ea.	325	59.50		384.50	wood framed picture window
	Demolish and Install	Ea.	325	74.05		399.05	including exterior trim.
	Clean	Ea.	.09	12.10		12.19	
	Paint	Ea.	1.72	22		23.72	
	Minimum Charge	Job		325		325	
5' x 4'							
	Demolish	Ea.		14.55		14.55	Includes material and labor to install a
	Install	Ea.	405	59.50		464.50	wood framed picture window
	Demolish and Install	Ea.	405	74.05		479.05	including exterior trim.
	Clean	Ea.	.09	12.10		12.19	
	Paint	Ea.	3.67	36		39.67	
	Minimum Charge	Job		325		325	

Exterior Doors and Windows

Wood Window

	Unit	Material	Labor	Equip.	Total	Specification
6' x 4' 6"						
Demolish	Ea.		14.55		14.55	Includes material and labor to install a
Install	Ea.	525	65		590	wood framed picture window
Demolish and Install	Ea.	525	79.55		604.55	including exterior trim.
Clean	Ea.	.09	12.10		12.19	
Paint	Ea.	3.67	36		39.67	
Minimum Charge	Job		325		325	

Window Screens

	Unit	Material	Labor	Equip.	Total	Specification
Small						
Demolish	Ea.		2.73		2.73	Includes material and labor to install
Install	Ea.	15.95	9.90		25.85	small window screen, 4 S.F.
Demolish and Install	Ea.	15.95	12.63		28.58	
Reinstall	Ea.		7.90		7.90	
Clean	Ea.	.30	4.91		5.21	
Minimum Charge	Job		81.50		81.50	
Standard Size						
Demolish	Ea.		2.73		2.73	Includes material and labor to install
Install	Ea.	38.50	9.90		48.40	medium window screen, 12 S.F.
Demolish and Install	Ea.	38.50	12.63		51.13	
Reinstall	Ea.		7.90		7.90	
Clean	Ea.	.30	4.91		5.21	
Minimum Charge	Job		81.50		81.50	
Large						
Demolish	Ea.		2.73		2.73	Includes material and labor to install
Install	Ea.	51	9.90		60.90	large window screen, 16 S.F.
Demolish and Install	Ea.	51	12.63		63.63	
Reinstall	Ea.		7.90		7.90	
Clean	Ea.	.30	4.91		5.21	
Minimum Charge	Job		81.50		81.50	

Window Glass

	Unit	Material	Labor	Equip.	Total	Specification
Plate, Clear, 1/4"						
Demolish	S.F.		1.31		1.31	Includes material and labor to install
Install	S.F.	6.75	5.15		11.90	clear glass.
Demolish and Install	S.F.	6.75	6.46		13.21	
Reinstall	S.F.		4.13		4.13	
Clean	S.F.	.02	.14		.16	
Minimum Charge	Job		155		155	
Insulated						
Demolish	S.F.		1.31		1.31	Includes material and labor to install
Install	S.F.	11.20	8.85		20.05	insulated glass.
Demolish and Install	S.F.	11.20	10.16		21.36	
Reinstall	S.F.		7.09		7.09	
Clean	S.F.	.02	.14		.16	
Minimum Charge	Job		155		155	
Tinted						
Demolish	S.F.		1.31		1.31	Includes material and labor to install
Install	S.F.	6.90	4.77		11.67	tinted glass.
Demolish and Install	S.F.	6.90	6.08		12.98	
Reinstall	S.F.		3.82		3.82	
Clean	S.F.	.02	.14		.16	
Minimum Charge	Job		155		155	

Exterior Doors and Windows

Window Glass		Unit	Material	Labor	Equip.	Total	Specification
Obscure / Florentine							
	Demolish	S.F.		1.31		1.31	Includes material and labor to install
	Install	S.F.	7.70	4.43		12.13	obscure glass.
	Demolish and Install	S.F.	7.70	5.74		13.44	
	Reinstall	S.F.		3.54		3.54	
	Clean	S.F.	.02	.14		.16	
	Minimum Charge	Job		155		155	
Laminated Safety							
	Demolish	S.F.		1.31		1.31	Includes material and labor to install
	Install	S.F.	9.35	6.90		16.25	clear laminated glass.
	Demolish and Install	S.F.	9.35	8.21		17.56	
	Reinstall	S.F.		5.51		5.51	
	Clean	S.F.	.02	.14		.16	
	Minimum Charge	Job		155		155	
Wired							
	Demolish	S.F.		1.31		1.31	Includes material and labor to install
	Install	S.F.	12.25	4.59		16.84	rough obscure wire glass.
	Demolish and Install	S.F.	12.25	5.90		18.15	
	Reinstall	S.F.		3.67		3.67	
	Clean	S.F.	.02	.14		.16	
	Minimum Charge	Job		155		155	
Tempered							
	Demolish	S.F.		1.31		1.31	Includes material and labor to install
	Install	S.F.	6.75	5.15		11.90	clear glass.
	Demolish and Install	S.F.	6.75	6.46		13.21	
	Reinstall	S.F.		4.13		4.13	
	Clean	S.F.	.02	.14		.16	
	Minimum Charge	Job		155		155	
Stained							
	Demolish	S.F.		1.31		1.31	Includes material and labor to install
	Install	S.F.	48	5.15		53.15	stained glass.
	Demolish and Install	S.F.	48	6.46		54.46	
	Reinstall	S.F.		4.13		4.13	
	Clean	S.F.	.02	.14		.16	
	Minimum Charge	Job		155		155	
Etched							
	Demolish	S.F.		1.31		1.31	Includes material and labor to install
	Install	S.F.	39	5.15		44.15	etched glass.
	Demolish and Install	S.F.	39	6.46		45.46	
	Reinstall	S.F.		4.13		4.13	
	Clean	S.F.	.02	.14		.16	
	Minimum Charge	Job		155		155	
Wall Mirror							
	Demolish	S.F.		1.31		1.31	Includes material and labor to install
	Install	S.F.	7.35	4.96		12.31	distortion-free float glass mirror, 1/4"
	Demolish and Install	S.F.	7.35	6.27		13.62	thick, cut and polished edges.
	Reinstall	S.F.		3.97		3.97	
	Clean	S.F.	.02	.14		.16	
	Minimum Charge	Job		155		155	
Bullet Resistant							
	Demolish	S.F.		1.31		1.31	Includes material and labor to install
	Install	S.F.	89.50	62		151.50	bullet resistant clear laminated glass.
	Demolish and Install	S.F.	89.50	63.31		152.81	
	Reinstall	S.F.		49.60		49.60	
	Clean	S.F.	.02	.14		.16	
	Minimum Charge	Job		155		155	

Exterior Doors and Windows

Storefront System	Unit	Material	Labor	Equip.	Total	Specification
Clear Glass						
Stub Wall to 8'						
Demolish	S.F.		2.05		2.05	Includes material and labor to install
Install	S.F.	16.15	8.55		24.70	commercial grade 4-1/2" section,
Demolish and Install	S.F.	16.15	10.60		26.75	anodized aluminum with clear glass in
Reinstall	S.F.		6.84		6.84	upper section and safety glass in
Clean	S.F.		.28		.28	lower section.
Minimum Charge	Job		197		197	
Floor to 10'						
Demolish	S.F.		2.05		2.05	Includes material and labor to install
Install	S.F.	17.70	8.55		26.25	commercial grade 4-1/2" section,
Demolish and Install	S.F.	17.70	10.60		28.30	anodized aluminum with clear glass in
Reinstall	S.F.		6.84		6.84	upper section and safety glass in
Clean	S.F.		.28		.28	lower section.
Minimum Charge	Job		197		197	
Polished Plate Glass						
Stub Wall to 9'						
Demolish	S.F.		2.05		2.05	Cost includes material and labor to
Install	S.F.	12.25	8.55		20.80	install 4' x 9' polished plate glass
Demolish and Install	S.F.	12.25	10.60		22.85	window built on 8" stub wall.
Reinstall	S.F.		6.84		6.84	
Clean	S.F.		.28		.28	
Minimum Charge	Job		197		197	
Stub Wall to 13'						
Demolish	S.F.		2.05		2.05	Cost includes material and labor to
Install	S.F.	12.80	8.55		21.35	install 4' x 13' polished plate glass
Demolish and Install	S.F.	12.80	10.60		23.40	window built on 8" stub wall.
Reinstall	S.F.		6.84		6.84	
Clean	S.F.		.28		.28	
Minimum Charge	Job		197		197	
Floor to 9'						
Demolish	S.F.		2.05		2.05	Cost includes material and labor to
Install	S.F.	13.30	8.55		21.85	install 4' x 9' polished plate glass
Demolish and Install	S.F.	13.30	10.60		23.90	window built on floor.
Reinstall	S.F.		6.84		6.84	
Clean	S.F.		.28		.28	
Minimum Charge	Job		197		197	
Floor to 13'						
Demolish	S.F.		2.05		2.05	Cost includes material and labor to
Install	S.F.	14.30	8.55		22.85	install 4' x 13' polished plate glass
Demolish and Install	S.F.	14.30	10.60		24.90	window built on floor.
Reinstall	S.F.		6.84		6.84	
Clean	S.F.		.28		.28	
Minimum Charge	Job		197		197	
Clear Plate Glass						
Floor to 9'						
Demolish	S.F.		2.05		2.05	Cost includes material and labor to
Install	S.F.	19.20	8.55		27.75	install 4' x 9' clear plate glass window
Demolish and Install	S.F.	19.20	10.60		29.80	built on floor.
Reinstall	S.F.		6.84		6.84	
Clean	S.F.		.28		.28	
Minimum Charge	Job		197		197	

Exterior Doors and Windows

Storefront System		Unit	Material	Labor	Equip.	Total	Specification
Floor to 11'							
	Demolish	S.F.		2.05		2.05	Cost includes material and labor to
	Install	S.F.	18.65	8.55		27.20	install 4' x 11' clear plate glass
	Demolish and Install	S.F.	18.65	10.60		29.25	window built on floor.
	Reinstall	S.F.		6.84		6.84	
	Clean	S.F.		.28		.28	
	Minimum Charge	Job		197		197	
Floor to 13'							
	Demolish	S.F.		2.05		2.05	Cost includes material and labor to
	Install	S.F.	18.80	8.55		27.35	install 4' x 13' clear plate glass
	Demolish and Install	S.F.	18.80	10.60		29.40	window built on floor.
	Reinstall	S.F.		6.84		6.84	
	Clean	S.F.		.28		.28	
	Minimum Charge	Job		197		197	
Decorative Type							
	Demolish	S.F.		2.05		2.05	Includes material and labor to install
	Install	S.F.	14.65	6.20		20.85	decorative storefront system.
	Demolish and Install	S.F.	14.65	8.25		22.90	
	Reinstall	S.F.		4.96		4.96	
	Clean	S.F.		.28		.28	
	Minimum Charge	Job		155		155	
Decorative Curved							
	Demolish	S.F.		2.05		2.05	Includes material and labor to install
	Install	S.F.	27.50	12.40		39.90	decorative curved storefront system.
	Demolish and Install	S.F.	27.50	14.45		41.95	
	Reinstall	S.F.		9.92		9.92	
	Clean	S.F.		.28		.28	
	Minimum Charge	Job		155		155	
Entrance							
Narrow Stile 3' x 7"							
	Demolish	Ea.		18.10		18.10	Cost includes material and labor to
	Install	Ea.	730	217		947	install 3' x 7' narrow stile door, center
	Demolish and Install	Ea.	730	235.10		965.10	pivot concealed closer.
	Reinstall	Ea.		173.87		173.87	
	Clean	Ea.	3.64	29.50		33.14	
	Minimum Charge	Job		197		197	
Tempered Glass							
	Demolish	Ea.		18.10		18.10	Cost includes material and labor to
	Install	Ea.	2575	425		3000	install 3' x 7' tempered glass door,
	Demolish and Install	Ea.	2575	443.10		3018.10	center pivot concealed closer,.
	Reinstall	Ea.		341.10		341.10	
	Clean	Ea.	3.64	29.50		33.14	
	Minimum Charge	Job		197		197	
Heavy Section							
	Demolish	Ea.		18.10		18.10	Includes material and labor to install
	Install	Ea.	835	395		1230	wide stile door with 3' 6" x 7'
	Demolish and Install	Ea.	835	413.10		1248.10	opening.
	Reinstall	Ea.		315.52		315.52	
	Clean	Ea.	.73	9.80		10.53	
	Minimum Charge	Job		620		620	

Exterior Doors and Windows

Curtain Wall System		Unit	Material	Labor	Equip.	Total	Specification
Regular Weight							
Bronze Anodized							
	Demolish	S.F.		2.05		2.05	Includes material and labor to install
	Install	S.F.	30.50	8.55		39.05	structural curtain wall system with
	Demolish and Install	S.F.	30.50	10.60		41.10	bronze anodized frame of heavy
	Clean	S.F.		.28		.28	weight, float glass.
	Minimum Charge	Job		163		163	
Black Anodized							
	Demolish	S.F.		2.05		2.05	Includes material and labor to install
	Install	S.F.	32.50	8.55		41.05	structural curtain wall system with
	Demolish and Install	S.F.	32.50	10.60		43.10	black anodized frame of regular
	Clean	S.F.		.28		.28	weight, float glass.
	Minimum Charge	Job		163		163	
Heavy Weight							
Bronze Anodized							
	Demolish	S.F.		2.05		2.05	Includes material and labor to install
	Install	S.F.	30.50	8.55		39.05	structural curtain wall system with
	Demolish and Install	S.F.	30.50	10.60		41.10	bronze anodized frame of heavy
	Clean	S.F.		.28		.28	weight, float glass.
	Minimum Charge	Job		163		163	
Black Anodized							
	Demolish	S.F.		2.05		2.05	Includes material and labor to install
	Install	S.F.	31	8.55		39.55	structural curtain wall system with
	Demolish and Install	S.F.	31	10.60		41.60	black anodized frame of heavy
	Clean	S.F.		.28		.28	weight, float glass.
	Minimum Charge	Job		163		163	
Glass							
1″ Insulated							
	Install	S.F.	13.25	8.25		21.50	Cost includes labor and material to
	Minimum Charge	Job		155		155	install 1″ insulated glass.
Heat Absorbing							
	Install	S.F.	7.25	8.25		15.50	Includes labor an material to install
	Minimum Charge	Job		155		155	heat absorbing glass.
Low Transmission							
	Install	S.F.	4.77	8.25		13.02	Includes labor an material to install
	Minimum Charge	Job		155		155	low transmission glass.

Skylight		Unit	Material	Labor	Equip.	Total	Specification
Dome							
22″ x 22″							
	Demolish	Ea.		63		63	Cost includes material and labor to
	Install	Ea.	49	98.50		147.50	install 22″ x 22″ fixed dome skylight.
	Demolish and Install	Ea.	49	161.50		210.50	
	Reinstall	Ea.		78.93		78.93	
	Clean	Ea.	.33	8.40		8.73	
	Minimum Charge	Job		590		590	
22″ x 46″							
	Demolish	Ea.		63		63	Cost includes material and labor to
	Install	Ea.	110	118		228	install 22″ x 46″ fixed dome skylight.
	Demolish and Install	Ea.	110	181		291	
	Reinstall	Ea.		94.72		94.72	
	Clean	Ea.		13.10		13.10	
	Minimum Charge	Job		590		590	

Exterior Doors and Windows

Skylight

	Unit	Material	Labor	Equip.	Total	Specification
30″ x 30″						
Demolish	Ea.		63		63	Cost includes material and labor to
Install	Ea.	92	98.50		190.50	install 30″ x 30″ fixed dome skylight.
Demolish and Install	Ea.	92	161.50		253.50	
Reinstall	Ea.		78.93		78.93	
Clean	Ea.		13.10		13.10	
Minimum Charge	Job		590		590	
30″ x 46″						
Demolish	Ea.		63		63	Cost includes material and labor to
Install	Ea.	160	118		278	install 30″ x 46″ fixed dome skylight.
Demolish and Install	Ea.	160	181		341	
Reinstall	Ea.		94.72		94.72	
Clean	Ea.		13.10		13.10	
Minimum Charge	Job		590		590	

Fixed

	Unit	Material	Labor	Equip.	Total	Specification
22″ x 27″						
Demolish	Ea.		63		63	Includes material and labor to install
Install	Ea.	234	98.50		332.50	double glazed skylight, 22″ x 27″,
Demolish and Install	Ea.	234	161.50		395.50	fixed.
Reinstall	Ea.		78.93		78.93	
Clean	Ea.	.33	8.40		8.73	
Minimum Charge	Job		590		590	
22″ x 46″						
Demolish	Ea.		63		63	Includes material and labor to install
Install	Ea.	295	118		413	double glazed skylight, 22″ x 46″,
Demolish and Install	Ea.	295	181		476	fixed.
Reinstall	Ea.		94.72		94.72	
Clean	Ea.		13.10		13.10	
Minimum Charge	Job		590		590	
44″ x 46″						
Demolish	Ea.		63		63	Includes material and labor to install
Install	Ea.	420	118		538	double glazed skylight, 44″ x 46″,
Demolish and Install	Ea.	420	181		601	fixed.
Reinstall	Ea.		94.72		94.72	
Clean	Ea.		14.75		14.75	
Minimum Charge	Job		590		590	

Operable

	Unit	Material	Labor	Equip.	Total	Specification
22″ x 27″						
Demolish	Ea.		63		63	Includes material and labor to install
Install	Ea.	345	98.50		443.50	double glazed skylight, 22″ x 27″,
Demolish and Install	Ea.	345	161.50		506.50	operable.
Reinstall	Ea.		78.93		78.93	
Clean	Ea.	.33	8.40		8.73	
Minimum Charge	Job		590		590	
22″ x 46″						
Demolish	Ea.		63		63	Includes material and labor to install
Install	Ea.	405	118		523	double glazed skylight, 22″ x 46″,
Demolish and Install	Ea.	405	181		586	operable.
Reinstall	Ea.		94.72		94.72	
Clean	Ea.		13.10		13.10	
Minimum Charge	Job		590		590	
44″ x 46″						
Demolish	Ea.		63		63	Includes material and labor to install
Install	Ea.	565	118		683	double glazed skylight, 44″ x 46″,
Demolish and Install	Ea.	565	181		746	operable.
Reinstall	Ea.		94.72		94.72	
Clean	Ea.		14.75		14.75	
Minimum Charge	Job		590		590	

Exterior Doors and Windows

Industrial Steel Window	Unit	Material	Labor	Equip.	Total	Specification
100% Fixed						
Demolish	S.F.		1.68		1.68	Includes material and labor to install
Install	S.F.	27.50	3.94		31.44	commercial grade steel windows,
Demolish and Install	S.F.	27.50	5.62		33.12	including glazing.
Reinstall	S.F.		3.16		3.16	
Clean	Ea.	.13	.88		1.01	
Paint	S.F.	.11	.64		.75	
Minimum Charge	Job		197		197	
50% Vented						
Demolish	S.F.		1.68		1.68	Includes material and labor to install
Install	S.F.	39.50	3.94		43.44	commercial grade steel windows,
Demolish and Install	S.F.	39.50	5.62		45.12	including glazing.
Reinstall	S.F.		3.16		3.16	
Clean	Ea.	.13	.88		1.01	
Paint	S.F.	.11	.64		.75	
Minimum Charge	Job		197		197	

Brick Veneer

Concrete Masonry Units

Masonry Wall		Unit	Material	Labor	Equip.	Total	Specification
Unreinforced CMU							
4″ x 8″ x 16″							
	Demolish	S.F.		1.45		1.45	Includes material and labor to install
	Install	S.F.	1.08	3.23		4.31	normal weight 4″ x 8″ x 16″ block.
	Demolish and Install	S.F.	1.08	4.68		5.76	
	Clean	S.F.	.04	.36		.40	
	Paint	S.F.	.13	.65		.78	
	Minimum Charge	Job		164		164	
6″ x 8″ x 16″							
	Demolish	S.F.		1.54		1.54	Includes material and labor to install
	Install	S.F.	1.58	3.38		4.96	normal weight 6″ x 8″ x 16″ block.
	Demolish and Install	S.F.	1.58	4.92		6.50	
	Clean	S.F.	.04	.36		.40	
	Paint	S.F.	.13	.65		.78	
	Minimum Charge	Job		164		164	
8″ x 8″ x 16″							
	Demolish	S.F.		1.75		1.75	Includes material and labor to install
	Install	S.F.	1.72	3.72		5.44	normal weight 8″ x 8″ x 16″ block.
	Demolish and Install	S.F.	1.72	5.47		7.19	
	Clean	S.F.	.04	.36		.40	
	Paint	S.F.	.13	.65		.78	
	Minimum Charge	Job		164		164	
12″ x 8″ x 16″							
	Demolish	S.F.		1.75		1.75	Includes material and labor to install
	Install	S.F.	2.49	5.60		8.09	normal weight 12″ x 8″ x 16″ block.
	Demolish and Install	S.F.	2.49	7.35		9.84	
	Clean	S.F.	.04	.36		.40	
	Paint	S.F.	.13	.65		.78	
	Minimum Charge	Job		164		164	
Reinforced CMU							
4″ x 8″ x 16″							
	Demolish	S.F.		1.45	.78	2.23	Includes material and labor to install
	Install	S.F.	1.19	3.31		4.50	normal weight 4″ x 8″ x 16″ block,
	Demolish and Install	S.F.	1.19	4.76	.78	6.73	mortar and 1/2″ re-bar installed
	Clean	S.F.	.04	.36		.40	vertically every 24″ O.C.
	Paint	S.F.	.13	.65		.78	
	Minimum Charge	Job		290		290	
6″ x 8″ x 16″							
	Demolish	S.F.		1.53	.36	1.89	Includes material and labor to install
	Install	S.F.	1.71	3.46		5.17	normal weight 6″ x 8″ x 16″ block,
	Demolish and Install	S.F.	1.71	4.99	.36	7.06	mortar and 1/2″ re-bar installed
	Clean	S.F.	.04	.36		.40	vertically every 24″ O.C.
	Paint	S.F.	.13	.65		.78	
	Minimum Charge	Job		290		290	

Masonry

Masonry Wall		Unit	Material	Labor	Equip.	Total	Specification
8″ x 8″ x 16″							
	Demolish	S.F.		1.62	.38	2	Includes material and labor to install
	Install	S.F.	1.84	3.77		5.61	normal weight 8″ x 8″ x 16″ block,
	Demolish and Install	S.F.	1.84	5.39	.38	7.61	mortar and 1/2″ re-bar installed
	Clean	S.F.	.04	.36		.40	vertically every 24″ O.C.
	Paint	S.F.	.13	.65		.78	
	Minimum Charge	Job		290		290	
12″ x 8″ x 16″							
	Demolish	S.F.		1.67	.39	2.06	Includes material and labor to install
	Install	S.F.	2.62	5.80		8.42	normal weight 12″ x 8″ x 16″ block,
	Demolish and Install	S.F.	2.62	7.47	.39	10.48	mortar and 1/2″ re-bar installed
	Clean	S.F.	.04	.36		.40	vertically every 24″ O.C.
	Paint	S.F.	.13	.65		.78	
	Minimum Charge	Job		290		290	
Structural Brick							
4″ Single Row							
	Demolish	S.F.		2.09		2.09	Cost includes material and labor to
	Install	S.F.	3.11	6.90		10.01	install 4″ red brick wall, running bond
	Demolish and Install	S.F.	3.11	8.99		12.10	with 3/8″ concave joints, mortar and
	Clean	S.F.	.04	.31		.35	ties.
	Paint	S.F.	.13	.65		.78	
	Minimum Charge	Job		290		290	
8″ Double Row							
	Demolish	S.F.		2.09		2.09	Cost includes material and labor to
	Install	S.F.	5.20	11		16.20	install 8″ red brick wall, running bond,
	Demolish and Install	S.F.	5.20	13.09		18.29	with 3/8″ concave joints, mortar and
	Clean	S.F.	.04	.31		.35	wall ties.
	Paint	S.F.	.13	.65		.78	
	Minimum Charge	Job		290		290	
12″ Triple Row							
	Demolish	S.F.		2.91		2.91	Cost includes material and labor to
	Install	S.F.	7.80	15.65		23.45	install 12″ red brick wall, running
	Demolish and Install	S.F.	7.80	18.56		26.36	bond, with 3/8″ concave joints,
	Clean	S.F.	.04	.31		.35	mortar and wall ties.
	Paint	S.F.	.13	.65		.78	
	Minimum Charge	Job		290		290	
Brick Veneer							
4″ Red Common							
	Demolish	S.F.		2.09		2.09	Cost includes material and labor to
	Install	S.F.	2.48	6.45		8.93	install 4″ common red face brick,
	Demolish and Install	S.F.	2.48	8.54		11.02	running bond with 3/8″ concave
	Clean	S.F.	.04	.31		.35	joints, mortar and wall ties.
	Paint	S.F.	.13	.65		.78	
	Minimum Charge	Job		164		164	
4″ Used							
	Demolish	S.F.		2.09		2.09	Cost includes material and labor to
	Install	S.F.	3.11	6.90		10.01	install 4″ red brick wall, running bond
	Demolish and Install	S.F.	3.11	8.99		12.10	with 3/8″ concave joints, mortar and
	Clean	S.F.	.04	.31		.35	ties.
	Paint	S.F.	.13	.65		.78	
	Minimum Charge	Job		164		164	

Masonry

Masonry Wall		Unit	Material	Labor	Equip.	Total	Specification
6" Jumbo							
	Demolish	S.F.		2.09		2.09	Cost includes material and labor to
	Install	S.F.	3.98	3.42		7.40	install 6" red jumbo brick veneer,
	Demolish and Install	S.F.	3.98	5.51		9.49	running bond with 3/8" concave
	Clean	S.F.	.04	.31		.35	joints, mortar and ties.
	Paint	S.F.	.13	.65		.78	
	Minimum Charge	Job		164		164	
Norman (11-1/2")							
	Demolish	S.F.		2.09		2.09	Cost includes material and labor to
	Install	S.F.	4.32	4.65		8.97	install 4" norman red brick wall,
	Demolish and Install	S.F.	4.32	6.74		11.06	running bond with 3/8" concave
	Clean	S.F.	.04	.31		.35	joints, mortar and ties.
	Paint	S.F.	.13	.65		.78	
	Minimum Charge	Job		164		164	
Glazed							
	Demolish	S.F.		2.09		2.09	Includes material, labor and
	Install	S.F.	7.70	7.10		14.80	equipment to install glazed brick.
	Demolish and Install	S.F.	7.70	9.19		16.89	
	Clean	S.F.	.04	.31		.35	
	Paint	S.F.	.13	.65		.78	
	Minimum Charge	Job		164		164	
Slumpstone							
6" x 4"							
	Demolish	S.F.		1.45		1.45	Cost includes material and labor to
	Install	S.F.	4.60	3.62		8.22	install 6" x 4" x 16" concrete slump
	Demolish and Install	S.F.	4.60	5.07		9.67	block.
	Clean	S.F.	.04	.36		.40	
	Paint	S.F.	.13	.65		.78	
	Minimum Charge	Job		164		164	
6" x 6"							
	Demolish	S.F.		1.54		1.54	Cost includes material and labor to
	Install	S.F.	4.20	3.74		7.94	install 6" x 6" x 16" concrete slump
	Demolish and Install	S.F.	4.20	5.28		9.48	block.
	Clean	S.F.	.04	.36		.40	
	Paint	S.F.	.13	.65		.78	
	Minimum Charge	Job		164		164	
8" x 6"							
	Demolish	S.F.		1.54		1.54	Cost includes material and labor to
	Install	S.F.	5.50	3.86		9.36	install 8" x 6" x 16" concrete slump
	Demolish and Install	S.F.	5.50	5.40		10.90	block.
	Clean	S.F.	.04	.36		.40	
	Paint	S.F.	.13	.65		.78	
	Minimum Charge	Job		164		164	
12" x 4"							
	Demolish	S.F.		1.45		1.45	Cost includes material and labor to
	Install	S.F.	9.25	4.46		13.71	install 12" x 4" x 16" concrete slump
	Demolish and Install	S.F.	9.25	5.91		15.16	block.
	Clean	S.F.	.04	.36		.40	
	Paint	S.F.	.13	.65		.78	
	Minimum Charge	Job		164		164	
12" x 6"							
	Demolish	S.F.		1.54		1.54	Cost includes material and labor to
	Install	S.F.	8.75	4.83		13.58	install 12" x 6" x 16" concrete slump
	Demolish and Install	S.F.	8.75	6.37		15.12	block.
	Clean	S.F.	.04	.36		.40	
	Paint	S.F.	.13	.65		.78	
	Minimum Charge	Job		164		164	

Masonry

Masonry Wall		Unit	Material	Labor	Equip.	Total	Specification
Stucco							
Three Coat							
	Demolish	S.Y.		1.81		1.81	Includes material and labor to install
	Install	S.Y.	3.58	6.90	.57	11.05	stucco on exterior masonry walls.
	Demolish and Install	S.Y.	3.58	8.71	.57	12.86	
	Clean	S.Y.		3.27		3.27	
	Paint	S.F.	.12	.47		.59	
	Minimum Charge	Job		146		146	
Three Coat w / Wire Mesh							
	Demolish	S.Y.		3.27		3.27	Includes material and labor to install
	Install	S.Y.	5.90	26.50	1.80	34.20	stucco on wire mesh over wood
	Demolish and Install	S.Y.	5.90	29.77	1.80	37.47	framing.
	Clean	S.Y.		3.27		3.27	
	Paint	S.F.	.12	.47		.59	
	Minimum Charge	Job		146		146	
Fill Cracks							
	Install	S.F.	.40	2.34		2.74	Includes labor and material to fill-in
	Minimum Charge	Job		146		146	hairline cracks up to 1/8" with filler,
							including sand and fill per S.F.
Furring							
1" x 2"							
	Demolish	L.F.		.10		.10	Cost includes material and labor to
	Install	L.F.	.21	.66		.87	install 1" x 2" furring on masonry wall.
	Demolish and Install	L.F.	.21	.76		.97	
	Minimum Charge	Job		163		163	
2" x 2"							
	Demolish	L.F.		.11		.11	Cost includes material and labor to
	Install	L.F.	.22	.72		.94	install 2" x 2" furring on masonry wall.
	Demolish and Install	L.F.	.22	.83		1.05	
	Minimum Charge	Job		163		163	
2" x 4"							
	Demolish	L.F.		.12		.12	Cost includes material and labor to
	Install	L.F.	.37	.83		1.20	install 2" x 4" furring on masonry wall.
	Demolish and Install	L.F.	.37	.95		1.32	
	Reinstall	L.F.		.66		.66	
	Minimum Charge	Job		163		163	
Maintenance / Cleaning							
Repoint							
	Install	S.F.	.30	4.11		4.41	Includes labor and material to cut and
	Minimum Charge	Hr.		41		41	repoint brick, hard mortar, running
							bond.
Acid Etch							
	Install	S.F.	.02	.59		.61	Includes labor and material to acid
	Minimum Charge	Job		605	78	683	wash smooth brick.
Pressure Wash							
	Clean	S.F.		.80	.10	.90	Includes labor and material to high
	Minimum Charge	Job		605	78	683	pressure water wash masonry,
							average.
Waterproof							
	Install	S.F.	.18	.60		.78	Includes labor and material to install
	Minimum Charge	Job		150		150	masory waterproofing.

Masonry

Chimney

Chimney		Unit	Material	Labor	Equip.	Total	Specification
Demo & Haul Exterior							
	Demolish	C.F.		4.28		4.28	Includes minimum labor and
	Minimum Charge	Job		131		131	equipment to remove masonry and haul debris to truck or dumpster.
Demo & Haul Interior							
	Demolish	C.F.		5.90		5.90	Includes minimum labor and
	Minimum Charge	Job		131		131	equipment to remove masonry and haul debris to truck or dumpster.
Install New (w / flue)							
Brick							
	Install	V.L.F.	27.50	36		63.50	Includes material and labor to install
	Minimum Charge	Job		164		164	brick chimney with flue from ground level up. Foundation not included.
Natural Stone							
	Install	L.F.	69.50	58		127.50	Includes labor and material to install
	Minimum Charge	Job		164		164	natural stone chimney.
Install From Roof Up							
Brick							
	Install	V.L.F.	27.50	36		63.50	Includes material and labor to install
	Minimum Charge	Job		164		164	brick chimney with flue from ground level up. Foundation not included.
Natural Stone							
	Install	L.F.	69.50	48.50		118	Includes labor and material to install
	Minimum Charge	Job		164		164	natural stone chimney.
Install Base Pad							
	Install	Ea.	220	178	1.12	399.12	Includes labor and material to replace
	Minimum Charge	Job		750	26.50	776.50	spread footing, 5' square x 16" deep.
Metal Cap							
	Install	Ea.	39.50	20.50		60	Includes labor and materials to install
	Minimum Charge	Job		163		163	bird and squirrel screens in a chimney flue.

Fireplace

Fireplace		Unit	Material	Labor	Equip.	Total	Specification
Common Brick							
30"							
	Demolish	Ea.		935		935	Includes material and labor to install a
	Install	Ea.	1375	2900		4275	complete unit which includes common
	Demolish and Install	Ea.	1375	3835		5210	brick, foundation, damper, flue lining
	Minimum Charge	Job		164		164	stack to 15' high and a 30" fire box.
36"							
	Demolish	Ea.		935		935	Includes material and labor to install a
	Install	Ea.	1475	3225		4700	complete unit which includes common
	Demolish and Install	Ea.	1475	4160		5635	brick, foundation, damper, flue lining
	Minimum Charge	Job		164		164	stack to 15' high and a 36" fire box.
42"							
	Demolish	Ea.		935		935	Includes material and labor to install a
	Install	Ea.	1600	3625		5225	complete unit which includes common
	Demolish and Install	Ea.	1600	4560		6160	brick, foundation, damper, flue lining
	Minimum Charge	Job		164		164	stack to 15' high and a 42" fire box.

Masonry

Fireplace	Unit	Material	Labor	Equip.	Total	Specification
48"						
Demolish	Ea.		935		935	Includes material and labor to install a
Install	Ea.	1675	4125		5800	complete unit which includes common
Demolish and Install	Ea.	1675	5060		6735	brick, foundation, damper, flue lining
Minimum Charge	Job		164		164	stack to 15' high and a 48" fire box.
Raised Hearth						
Brick Hearth						
Demolish	S.F.		4.09		4.09	Includes material and labor to install a
Install	S.F.	12.80	24		36.80	brick fireplace hearth based on 30" x
Demolish and Install	S.F.	12.80	28.09		40.89	29" opening.
Clean	S.F.	.04	.31		.35	
Minimum Charge	Job		164		164	
Marble Hearth / Facing						
Demolish	Ea.		26		26	Includes material and labor to install
Install	Ea.	600	98		698	marble fireplace hearth.
Demolish and Install	Ea.	600	124		724	
Reinstall	Ea.		78.48		78.48	
Clean	Ea.	22	4.71		26.71	
Minimum Charge	Job		164		164	
Breast (face) Brick						
Demolish	S.F.		4.09		4.09	Includes material and labor to install
Install	S.F.	3.17	6.75		9.92	standard red brick, running bond.
Demolish and Install	S.F.	3.17	10.84		14.01	
Clean	S.F.		.80	.10	.90	
Minimum Charge	Hr.		41		41	
Fire Box						
Demolish	Ea.		209		209	Includes material and labor to install
Install	Ea.	138	290		428	fireplace box only (110 brick).
Demolish and Install	Ea.	138	499		637	
Clean	Ea.	.11	10.60		10.71	
Minimum Charge	Hr.		41		41	
Natural Stone						
Demolish	S.F.		4.09		4.09	Includes material and labor to install
Install	S.F.	14.15	12.40		26.55	high priced stone, split or rock face.
Demolish and Install	S.F.	14.15	16.49		30.64	
Clean	S.F.	.06	.72		.78	
Minimum Charge	Job		290		290	
Slumpstone Veneer						
Demolish	S.F.		1.54		1.54	Cost includes material and labor to
Install	S.F.	5.50	3.86		9.36	install 8" x 6" x 16" concrete slump
Demolish and Install	S.F.	5.50	5.40		10.90	block.
Clean	S.F.	.04	.36		.40	
Paint	S.F.	.13	.65		.78	
Minimum Charge	Job		164		164	
Prefabricated						
Firebox						
Demolish	Ea.		37.50		37.50	Includes material and labor to install
Install	Ea.	1550	296		1846	up to 43" radiant heat, zero
Demolish and Install	Ea.	1550	333.50		1883.50	clearance, prefabricated fireplace.
Reinstall	Ea.		237.09		237.09	Flues, doors or blowers not included.
Clean	Ea.	1.46	14.75		16.21	
Minimum Charge	Job		163		163	

Masonry

Fireplace		Unit	Material	Labor	Equip.	Total	Specification
Flue							
	Demolish	V.L.F.		4.36		4.36	Includes material and labor to install
	Install	V.L.F.	11.50	12.35		23.85	prefabricated chimney vent, 8"
	Demolish and Install	V.L.F.	11.50	16.71		28.21	exposed metal flue.
	Reinstall	V.L.F.		9.87		9.87	
	Clean	L.F.	.58	2.95		3.53	
	Minimum Charge	Job		320		320	
Blower							
	Demolish	Ea.		3.49		3.49	Includes material and labor to install
	Install	Ea.	108	22.50		130.50	blower for metal fireplace flue.
	Demolish and Install	Ea.	108	25.99		133.99	
	Reinstall	Ea.		17.82		17.82	
	Clean	Ea.	2.92	29.50		32.42	
	Minimum Charge	Job		320		320	
Accessories							
Wood Mantle							
	Demolish	L.F.		4.36		4.36	Includes material and labor to install
	Install	L.F.	6.80	9.30		16.10	wood fireplace mantel.
	Demolish and Install	L.F.	6.80	13.66		20.46	
	Reinstall	L.F.		9.31		9.31	
	Clean	Ea.	.02	.65		.67	
	Paint	L.F.	.07	7.75		7.82	
	Minimum Charge	Job		163		163	
Colonial Style							
	Demolish	Ea.		40.50		40.50	Includes material and labor to install
	Install	Opng.	2400	163		2563	wood fireplace mantel.
	Demolish and Install	Opng.	2400	203.50		2603.50	
	Reinstall	Opng.		163		163	
	Clean	Ea.		.47		.47	
	Paint	Ea.	11.60	107		118.60	
	Minimum Charge	Job		163		163	
Modern Design							
	Demolish	Ea.		40.50		40.50	Includes material and labor to install
	Install	Opng.	190	65		255	wood fireplace mantel.
	Demolish and Install	Opng.	190	105.50		295.50	
	Reinstall	Opng.		65.20		65.20	
	Clean	Ea.		.47		.47	
	Paint	Ea.	11.60	107		118.60	
	Minimum Charge	Job		163		163	
Glass Door Unit							
	Demolish	Ea.		5.45		5.45	Includes material and labor to install a
	Install	Ea.	128	6.85		134.85	fireplace glass door unit.
	Demolish and Install	Ea.	128	12.30		140.30	
	Reinstall	Ea.		6.85		6.85	
	Clean	Ea.	.35	9.05		9.40	
	Minimum Charge	Job		163		163	
Fire Screen							
	Demolish	Ea.		2.73		2.73	Includes material and labor to install a
	Install	Ea.	39	3.42		42.42	fireplace fire screen.
	Demolish and Install	Ea.	39	6.15		45.15	
	Reinstall	Ea.		3.42		3.42	
	Clean	Ea.	.18	7.85		8.03	
	Minimum Charge	Job		163		163	

Shingles	Board & Batten	Texture 1-11

Exterior Siding		Unit	Material	Labor	Equip.	Total	Specification
Trim							
1″ x 2″							
	Demolish	L.F.		.22		.22	Cost includes material and labor to
	Install	L.F.	.36	.99		1.35	install 1″ x 2″ pine trim.
	Demolish and Install	L.F.	.36	1.21		1.57	
	Clean	L.F.	.01	.16		.17	
	Paint	L.F.	.02	.45		.47	
	Minimum Charge	Job		163		163	
1″ x 3″							
	Demolish	L.F.		.22		.22	Cost includes material and labor to
	Install	L.F.	.31	1.09		1.40	install 1″ x 3″ pine trim.
	Demolish and Install	L.F.	.31	1.31		1.62	
	Clean	L.F.	.01	.16		.17	
	Paint	L.F.	.02	.45		.47	
	Minimum Charge	Job		163		163	
1″ x 4″							
	Demolish	L.F.		.22		.22	Cost includes material and labor to
	Install	L.F.	.75	1.30		2.05	install 1″ x 4″ pine trim.
	Demolish and Install	L.F.	.75	1.52		2.27	
	Clean	L.F.	.01	.16		.17	
	Paint	L.F.	.02	.45		.47	
	Minimum Charge	Job		163		163	
1″ x 6″							
	Demolish	L.F.		.22		.22	Cost includes material and labor to
	Install	L.F.	1.31	1.30		2.61	install 1″ x 6″ pine trim.
	Demolish and Install	L.F.	1.31	1.52		2.83	
	Clean	L.F.	.01	.16		.17	
	Paint	L.F.	.07	.45		.52	
	Minimum Charge	Job		163		163	
1″ x 8″							
	Demolish	L.F.		.24		.24	Cost includes material and labor to
	Install	L.F.	1.61	1.63		3.24	install 1″ x 8″ pine trim.
	Demolish and Install	L.F.	1.61	1.87		3.48	
	Clean	L.F.	.02	.17		.19	
	Paint	L.F.	.07	.45		.52	
	Minimum Charge	Job		163		163	
Aluminum							
8″ Non-Insulated							
	Demolish	S.F.		.59		.59	Includes material and labor to install
	Install	S.F.	1.35	1.29		2.64	horizontal, colored clapboard, 8″ to
	Demolish and Install	S.F.	1.35	1.88		3.23	10″ wide, plain.
	Reinstall	S.F.		1.29		1.29	
	Clean	S.F.		.24		.24	
	Minimum Charge	Job		163		163	

Siding

Exterior Siding	Unit	Material	Labor	Equip.	Total	Specification
8" Insulated						
Demolish	S.F.		.59		.59	Includes material and labor to install
Install	S.F.	1.53	1.29		2.82	horizontal, colored insulated
Demolish and Install	S.F.	1.53	1.88		3.41	clapboard, 8" wide, plain.
Reinstall	S.F.		1.29		1.29	
Clean	S.F.		.24		.24	
Minimum Charge	Job		163		163	
12" Non-Insulated						
Demolish	S.F.		.59		.59	Includes material and labor to install
Install	S.F.	1.42	1.11		2.53	horizontal, colored clapboard, plain.
Demolish and Install	S.F.	1.42	1.70		3.12	
Reinstall	S.F.		1.11		1.11	
Clean	S.F.		.24		.24	
Minimum Charge	Job		163		163	
12" Insulated						
Demolish	S.F.		.59		.59	Includes material and labor to install
Install	S.F.	1.67	1.11		2.78	horizontal, colored insulated
Demolish and Install	S.F.	1.67	1.70		3.37	clapboard, plain.
Reinstall	S.F.		1.11		1.11	
Clean	S.F.		.24		.24	
Minimum Charge	Job		163		163	
Clapboard 8" to 10"						
Demolish	S.F.		.59		.59	Includes material and labor to install
Install	S.F.	1.35	1.29		2.64	horizontal, colored clapboard, 8" to
Demolish and Install	S.F.	1.35	1.88		3.23	10" wide, plain.
Reinstall	S.F.		1.29		1.29	
Clean	S.F.		.24		.24	
Minimum Charge	Job		163		163	
Corner Strips						
Demolish	L.F.		.31		.31	Includes material and labor to install
Install	V.L.F.	1.87	1.29		3.16	corners for horizontal lapped
Demolish and Install	V.L.F.	1.87	1.60		3.47	aluminum siding.
Reinstall	V.L.F.		1.29		1.29	
Clean	L.F.		.16		.16	
Minimum Charge	Job		163		163	
Vinyl						
Non-Insulated						
Demolish	S.F.		.46		.46	Cost includes material and labor to
Install	S.F.	.76	1.28		2.04	install 10" non-insulated solid vinyl
Demolish and Install	S.F.	.76	1.74		2.50	siding.
Reinstall	S.F.		1.02		1.02	
Clean	S.F.		.24		.24	
Minimum Charge	Job		163		163	
Insulated						
Demolish	S.F.		.46		.46	Includes material and labor to install
Install	S.F.	.84	1.28		2.12	insulated solid vinyl siding panels.
Demolish and Install	S.F.	.84	1.74		2.58	
Reinstall	S.F.		1.02		1.02	
Clean	S.F.		.24		.24	
Minimum Charge	Job		163		163	
Corner Strips						
Demolish	L.F.		.29		.29	Includes material and labor to install
Install	L.F.	1.27	1.59		2.86	corner posts for vinyl siding panels.
Demolish and Install	L.F.	1.27	1.88		3.15	
Reinstall	L.F.		1.27		1.27	
Clean	L.F.		.16		.16	
Minimum Charge	Job		163		163	

132

Siding

Exterior Siding		Unit	Material	Labor	Equip.	Total	Specification
Hardboard							
4' x 8' Panel							
	Demolish	S.F.		.49		.49	Cost includes material and labor to
	Install	S.F.	1.02	.81		1.83	install 4' x 8' masonite siding 7/16"
	Demolish and Install	S.F.	1.02	1.30		2.32	thick.
	Clean	S.F.		.24		.24	
	Paint	S.F.	.24	.96		1.20	
	Minimum Charge	Job		163		163	
Lap							
	Demolish	S.F.		.48		.48	Cost includes material and labor to
	Install	S.F.	.93	.84		1.77	install 8" or 12" masonite lap siding
	Demolish and Install	S.F.	.93	1.32		2.25	7/16" thick with wood grain finish.
	Clean	S.F.		.24		.24	
	Paint	S.F.	.24	.96		1.20	
	Minimum Charge	Job		163		163	
Plywood Panel							
Douglas Fir (T-1-11)							
	Demolish	S.F.		.36		.36	Includes material and labor to install
	Install	S.F.	1.11	.98		2.09	Douglas fir 4' x 8' plywood exterior
	Demolish and Install	S.F.	1.11	1.34		2.45	siding plain or patterns.
	Reinstall	S.F.		.98		.98	
	Clean	S.F.		.24		.24	
	Paint	S.F.	.24	.96		1.20	
	Minimum Charge	Job		163		163	
Redwood							
	Demolish	S.F.		.36		.36	Includes material and labor to install
	Install	S.F.	1.98	.98		2.96	redwood 4' x 8' plywood exterior
	Demolish and Install	S.F.	1.98	1.34		3.32	siding plain or patterns.
	Reinstall	S.F.		.98		.98	
	Clean	S.F.		.24		.24	
	Paint	S.F.	.24	.96		1.20	
	Minimum Charge	Job		163		163	
Cedar							
	Demolish	S.F.		.36		.36	Cost includes material and labor to
	Install	S.F.	2.62	.98		3.60	install 4' x 8' cedar plywood exterior
	Demolish and Install	S.F.	2.62	1.34		3.96	siding plain or patterns.
	Reinstall	S.F.		.98		.98	
	Clean	S.F.		.24		.24	
	Paint	S.F.	.24	.96		1.20	
	Minimum Charge	Job		163		163	
Tongue and Groove							
Fir							
	Demolish	S.F.		.65		.65	Includes material and labor to install
	Install	S.F.	1.39	.87		2.26	fir tongue and groove exterior siding
	Demolish and Install	S.F.	1.39	1.52		2.91	1" x 8".
	Reinstall	S.F.		.70		.70	
	Clean	S.F.		.24		.24	
	Paint	S.F.	.24	.96		1.20	
	Minimum Charge	Job		163		163	
Pine							
	Demolish	S.F.		.65		.65	Includes material and labor to install
	Install	S.F.	.67	1.77		2.44	pine tongue and groove exterior
	Demolish and Install	S.F.	.67	2.42		3.09	siding 1" x 8".
	Reinstall	S.F.		1.77		1.77	
	Clean	S.F.		.24		.24	
	Paint	S.F.	.24	.96		1.20	
	Minimum Charge	Job		163		163	

Siding

Exterior Siding		Unit	Material	Labor	Equip.	Total	Specification
Cedar							
	Demolish	S.F.		.65		.65	Includes material and labor to install
	Install	S.F.	3.04	.87		3.91	cedar tongue and groove exterior
	Demolish and Install	S.F.	3.04	1.52		4.56	siding 1" x 8".
	Reinstall	S.F.		.70		.70	
	Clean	S.F.		.24		.24	
	Paint	S.F.	.24	.96		1.20	
	Minimum Charge	Job		163		163	
Redwood							
	Demolish	S.F.		.65		.65	Includes material and labor to install
	Install	S.F.	2.93	2.21		5.14	redwood tongue and groove exterior
	Demolish and Install	S.F.	2.93	2.86		5.79	siding.
	Reinstall	S.F.		2.21		2.21	
	Clean	S.F.		.24		.24	
	Paint	S.F.	.24	.96		1.20	
	Minimum Charge	Job		163		163	
Board and Batten							
Fir							
	Demolish	S.F.		.65		.65	Cost includes material and labor to
	Install	S.F.	1.49	.82		2.31	install 1" x 8" fir boards with 1" x 3"
	Demolish and Install	S.F.	1.49	1.47		2.96	battens.
	Reinstall	S.F.		.65		.65	
	Clean	S.F.		.24		.24	
	Paint	S.F.	.24	.96		1.20	
	Minimum Charge	Job		163		163	
Pine							
	Demolish	S.F.		.65		.65	Cost includes material and labor to
	Install	S.F.	.77	.82		1.59	install 1" x 10" pine boards with 1" x
	Demolish and Install	S.F.	.77	1.47		2.24	3" battens.
	Reinstall	S.F.		.65		.65	
	Clean	S.F.		.24		.24	
	Paint	S.F.	.24	.96		1.20	
	Minimum Charge	Job		163		163	
Cedar							
	Demolish	S.F.		.65		.65	Cost includes material and labor to
	Install	S.F.	2.18	.82		3	install 1" x 10" cedar boards with 1" x
	Demolish and Install	S.F.	2.18	1.47		3.65	3" battens.
	Reinstall	S.F.		.65		.65	
	Clean	S.F.		.24		.24	
	Paint	S.F.	.24	.96		1.20	
	Minimum Charge	Job		163		163	
Redwood							
	Demolish	S.F.		.65		.65	Cost includes material and labor to
	Install	S.F.	3.95	.82		4.77	install 1" x 10" redwood boards with
	Demolish and Install	S.F.	3.95	1.47		5.42	1" x 3" battens.
	Reinstall	S.F.		.65		.65	
	Clean	S.F.		.24		.24	
	Paint	S.F.	.24	.96		1.20	
	Minimum Charge	Job		163		163	
Wood Shingle							
	Demolish	Sq.		61		61	Includes material and labor to install
	Install	Sq.	130	145		275	wood shingles/shakes.
	Demolish and Install	Sq.	130	206		336	
	Clean	Sq.		25		25	
	Paint	Sq.	15.35	86		101.35	
	Minimum Charge	Job		163		163	

Siding

Exterior Siding	Unit	Material	Labor	Equip.	Total	Specification
Synthetic Stucco System						
Demolish	S.F.		2.18		2.18	Cost includes material and labor to
Install	S.F.	2.99	6.25	.51	9.75	install 2 coats of adhesive mixture,
Demolish and Install	S.F.	2.99	8.43	.51	11.93	glass fiber mesh, 1" insulation board,
Clean	S.F.		.21		.21	and textured finish coat.
Paint	S.F.	.12	.47		.59	
Minimum Charge	Job		690	56.50	746.50	

Site Work

Front-end Loader

Bulldozer

Grading		Unit	Material	Labor	Equip.	Total	Specification
Hand (fine)							
	Install	S.F.		.47		.47	Includes labor and material to install fine grade for slab on grade, hand grade.
	Minimum Charge	Job		118		118	
Machine (fine)							
	Install	S.Y.		.40	.37	.77	Includes labor and material to install fine grade for roadway, base or leveling course.
	Minimum Charge	Job		275	251	526	

Asphalt Paving		Unit	Material	Labor	Equip.	Total	Specification
Seal Coat							
	Install	S.F.	.06	.09		.15	Includes labor and material to install sealcoat for a small area.
	Minimum Charge	Job		118		118	
Wearing Course							
	Install	S.F.	.07	.25	.17	.49	Cost includes labor and material to install 1" thick wearing course.
	Minimum Charge	Job		118		118	

Masonry Pavers		Unit	Material	Labor	Equip.	Total	Specification
Natural Concrete							
Sand Base							
	Demolish	S.F.		2.62		2.62	Includes material and labor to install concrete pavers on sand base.
	Install	S.F.	2.13	5.50		7.63	
	Demolish and Install	S.F.	2.13	8.12		10.25	
	Reinstall	S.F.		4.38		4.38	
	Clean	S.F.		.94		.94	
	Minimum Charge	Job		164		164	
Mortar Base							
	Demolish	S.F.		4.29	1.32	5.61	Includes material and labor to install concrete pavers on mortar base.
	Install	S.F.	2.76	4.57		7.33	
	Demolish and Install	S.F.	2.76	8.86	1.32	12.94	
	Reinstall	S.F.		3.65		3.65	
	Clean	S.F.		.86		.86	
	Minimum Charge	Job		164		164	

Site Work

Masonry Pavers

Masonry Pavers	Unit	Material	Labor	Equip.	Total	Specification
Adobe						
Sand Base						
Demolish	S.F.		2.62		2.62	Includes material and labor to install
Install	S.F.	1.69	3.65		5.34	adobe pavers on sand base.
Demolish and Install	S.F.	1.69	6.27		7.96	
Reinstall	S.F.		2.92		2.92	
Clean	S.F.		.94		.94	
Minimum Charge	Job		164		164	
Mortar Base						
Demolish	S.F.		4.29	1.32	5.61	Includes material and labor to install
Install	S.F.	2.34	4.57		6.91	adobe pavers on mortar base.
Demolish and Install	S.F.	2.34	8.86	1.32	12.52	
Reinstall	S.F.		3.65		3.65	
Clean	S.F.		.86		.86	
Minimum Charge	Job		164		164	
Standard Brick						
Sand Base						
Demolish	S.F.		2.62		2.62	Includes material and labor to install
Install	S.F.	2.74	5.80		8.54	standard brick pavers on sand base.
Demolish and Install	S.F.	2.74	8.42		11.16	
Reinstall	S.F.		4.64		4.64	
Clean	S.F.		.94		.94	
Minimum Charge	Job		164		164	
Mortar Base						
Demolish	S.F.		4.29	1.32	5.61	Includes material and labor to install
Install	S.F.	5	8.20		13.20	standard brick pavers on mortar base.
Demolish and Install	S.F.	5	12.49	1.32	18.81	
Reinstall	S.F.		6.58		6.58	
Clean	S.F.		.86		.86	
Minimum Charge	Job		164		164	
Deluxe Brick						
Sand Base						
Demolish	S.F.		2.62		2.62	Includes material and labor to install
Install	S.F.	5	4.57		9.57	deluxe brick pavers on sand base.
Demolish and Install	S.F.	5	7.19		12.19	
Reinstall	S.F.		3.65		3.65	
Clean	S.F.		.94		.94	
Minimum Charge	Job		164		164	
Mortar Base						
Demolish	S.F.		4.29	1.32	5.61	Includes material and labor to install
Install	S.F.	5.40	8.20		13.60	deluxe brick pavers on mortar base.
Demolish and Install	S.F.	5.40	12.49	1.32	19.21	
Reinstall	S.F.		6.58		6.58	
Clean	S.F.		.86		.86	
Minimum Charge	Job		164		164	

Fencing	Unit	Material	Labor	Equip.	Total	Specification
Block w / Footing						
4″						
Demolish	S.F.		1.45		1.45	Cost includes material and labor to
Install	S.F.	1.03	3.46		4.49	install 4″ non-reinforced concrete
Demolish and Install	S.F.	1.03	4.91		5.94	block.
Clean	S.F.	.13	1	.13	1.26	
Paint	S.F.	.17	.38		.55	
Minimum Charge	Job		164		164	

Fencing		Unit	Material	Labor	Equip.	Total	Specification
4″ Reinforced							
	Demolish	S.F.		1.45	.78	2.23	Cost includes material and labor to
	Install	S.F.	1.14	3.50		4.64	install 4″ reinforced concrete block.
	Demolish and Install	S.F.	1.14	4.95	.78	6.87	
	Clean	S.F.	.13	.1	.13	1.26	
	Paint	S.F.	.17	.38		.55	
	Minimum Charge	Job		164		164	
6″							
	Demolish	S.F.		1.54		1.54	Cost includes material and labor to
	Install	S.F.	1.54	3.72		5.26	install 6″ non-reinforced concrete
	Demolish and Install	S.F.	1.54	5.26		6.80	block.
	Clean	S.F.	.13	1	.13	1.26	
	Paint	S.F.	.17	.38		.55	
	Minimum Charge	Job		164		164	
6″ Reinforced							
	Demolish	S.F.		1.53	.36	1.89	Cost includes material and labor to
	Install	S.F.	1.66	3.77		5.43	install 6″ reinforced concrete block.
	Demolish and Install	S.F.	1.66	5.30	.36	7.32	
	Clean	S.F.	.13	1	.13	1.26	
	Paint	S.F.	.17	.38		.55	
	Minimum Charge	Job		164		164	
Cap for Block Wall							
	Demolish	L.F.		1.64		1.64	Includes labor and equipment to install
	Install	L.F.	16.50	7.75		24.25	precast concrete coping.
	Demolish and Install	L.F.	16.50	9.39		25.89	
	Clean	L.F.	.01	2.45		2.46	
	Paint	S.F.	.17	.38		.55	
	Minimum Charge	Job		164		164	
Block Pilaster							
	Demolish	Ea.		112		112	Includes material and labor to install
	Install	Ea.	61	116		177	block pilaster for fence.
	Demolish and Install	Ea.	61	228		289	
	Reinstall	Ea.		115.94		115.94	
	Paint	Ea.	3.88	12.70		16.58	
	Minimum Charge	Job		164		164	
Block w / o Footing							
4″							
	Demolish	S.F.		1.45		1.45	Cost includes material and labor to
	Install	S.F.	1.23	2.64		3.87	install 4″ non-reinforced hollow
	Demolish and Install	S.F.	1.23	4.09		5.32	lightweight block.
	Clean	S.F.	.13	1	.13	1.26	
	Paint	S.F.	.17	.38		.55	
	Minimum Charge	Job		164		164	
4″ Reinforced							
	Demolish	S.F.		1.45	.78	2.23	Cost includes material and labor to
	Install	S.F.	1.42	3.42		4.84	install 4″ reinforced hollow lightweight
	Demolish and Install	S.F.	1.42	4.87	.78	7.07	block.
	Clean	S.F.	.13	1	.13	1.26	
	Paint	S.F.	.17	.38		.55	
	Minimum Charge	Job		164		164	
6″							
	Demolish	S.F.		1.54		1.54	Cost includes material and labor to
	Install	S.F.	1.69	2.90		4.59	install 6″ non-reinforced hollow
	Demolish and Install	S.F.	1.69	4.44		6.13	lightweight block.
	Clean	S.F.	.13	1	.13	1.26	
	Paint	S.F.	.17	.38		.55	
	Minimum Charge	Job		164		164	

Fencing	Unit	Material	Labor	Equip.	Total	Specification
6" Reinforced						
Demolish	S.F.		1.53	.36	1.89	Cost includes material and labor to
Install	S.F.	1.78	3.67		5.45	install 6" reinforced hollow lightweight
Demolish and Install	S.F.	1.78	5.20	.36	7.34	block.
Clean	S.F.	.13	1	.13	1.26	
Paint	S.F.	.17	.38		.55	
Minimum Charge	Job		164		164	
Cap for Block Wall						
Demolish	L.F.		1.64		1.64	Includes labor and equipment to install
Install	L.F.	16.50	7.75		24.25	precast concrete coping.
Demolish and Install	L.F.	16.50	9.39		25.89	
Clean	L.F.	.01	2.45		2.46	
Paint	S.F.	.17	.38		.55	
Minimum Charge	Job		164		164	
Block Pilaster						
Demolish	Ea.		112		112	Includes material and labor to install
Install	Ea.	61	116		177	block pilaster for fence.
Demolish and Install	Ea.	61	228		289	
Reinstall	Ea.		115.94		115.94	
Paint	Ea.	3.88	12.70		16.58	
Minimum Charge	Job		164		164	
Chain Link						
3' High w / o Top Rail						
Demolish	L.F.		2.02	.62	2.64	Includes material and labor to install
Install	L.F.	3.19	1.38	.28	4.85	residential galvanized fence 3' high
Demolish and Install	L.F.	3.19	3.40	.90	7.49	with 1-3/8" steel post 10' O.C. set
Reinstall	L.F.		1.10	.22	1.32	without concrete.
Clean	L.F.	.09	.86		.95	
Minimum Charge	Job		365		365	
4' High w / o Top Rail						
Demolish	L.F.		2.02	.62	2.64	Includes material and labor to install
Install	L.F.	3.39	1.71	.35	5.45	residential galvanized fence 4' high
Demolish and Install	L.F.	3.39	3.73	.97	8.09	with 1-3/8" steel post 10' O.C. set
Reinstall	L.F.		1.36	.28	1.64	without concrete.
Clean	L.F.	.12	.94		1.06	
Minimum Charge	Job		365		365	
5' High w / o Top Rail						
Demolish	L.F.		2.02	.62	2.64	Includes material and labor to install
Install	L.F.	3.97	2.23	.45	6.65	residential galvanized fence 5' high
Demolish and Install	L.F.	3.97	4.25	1.07	9.29	with 1-3/8" steel post 10' O.C. set
Reinstall	L.F.		1.78	.36	2.14	without concrete.
Clean	L.F.	.14	1.05		1.19	
Minimum Charge	Job		365		365	
3' High w / Top Rail						
Demolish	L.F.		2.14	.66	2.80	Includes material and labor to install
Install	L.F.	3.94	1.45	.29	5.68	residential galvanized fence 3' high
Demolish and Install	L.F.	3.94	3.59	.95	8.48	with 1-3/8" steel post 10' O.C. set
Reinstall	L.F.		1.16	.24	1.40	without concrete.
Clean	L.F.	.09	.87		.96	
Minimum Charge	Job		365		365	
4' High w / Top Rail						
Demolish	L.F.		2.14	.66	2.80	Includes material and labor to install
Install	L.F.	4.47	1.81	.37	6.65	residential galvanized fence 4' high
Demolish and Install	L.F.	4.47	3.95	1.03	9.45	with 1-5/8" steel post 10' O.C. set
Reinstall	L.F.		1.45	.29	1.74	without concrete and 1-5/8" top rail
Clean	L.F.	.12	.96		1.08	with sleeves.
Minimum Charge	Job		365		365	

Site Work

Fencing

Fencing	Unit	Material	Labor	Equip.	Total	Specification
5' High w / Top Rail						Includes material and labor to install
Demolish	L.F.		2.14	.66	2.80	residential galvanized fence 5' high
Install	L.F.	5.05	2.42	.49	7.96	with 1-5/8" steel post 10' O.C. set
Demolish and Install	L.F.	5.05	4.56	1.15	10.76	without concrete and 1-5/8" top rail
Reinstall	L.F.		1.93	.39	2.32	with sleeves.
Clean	L.F.	.14	1.07		1.21	
Minimum Charge	Job		365		365	
6' High w / Top Rail						Includes material and labor to install
Demolish	L.F.		2.14	.66	2.80	residential galvanized fence 6' high
Install	L.F.	5.40	3.62	.74	9.76	with 2" steel post 10' O.C. set without
Demolish and Install	L.F.	5.40	5.76	1.40	12.56	concrete and 1-3/8" top rail with
Reinstall	L.F.		2.90	.59	3.49	sleeves.
Clean	L.F.	.18	1.21		1.39	
Minimum Charge	Job		365		365	
Barbed Wire						Cost includes material and labor to
Demolish	L.F.		1.22		1.22	install 3 strand galvanized barbed
Install	L.F.	.33	.47	.13	.93	wire fence.
Demolish and Install	L.F.	.33	1.69	.13	2.15	
Reinstall	L.F.		.38	.10	.48	
Minimum Charge	Job		365	294	659	
6' Redwood						Cost includes material and labor to
Demolish	L.F.		8.15		8.15	install 6' high redwood fencing.
Install	L.F.	10.35	5.80	1.18	17.33	
Demolish and Install	L.F.	10.35	13.95	1.18	25.48	
Reinstall	L.F.		4.64	.94	5.58	
Clean	L.F.	.18	1.10		1.28	
Paint	L.F.	.63	3.01		3.64	
Minimum Charge	Job		365		365	
6' Red Cedar						Cost includes material and labor to
Demolish	L.F.		8.15		8.15	install 6' cedar fence.
Install	L.F.	15.80	5.80	1.18	22.78	
Demolish and Install	L.F.	15.80	13.95	1.18	30.93	
Reinstall	L.F.		4.64	.94	5.58	
Clean	L.F.	.18	1.10		1.28	
Paint	L.F.	.63	3.01		3.64	
Minimum Charge	Job		365		365	
6' Shadowbox						Cost includes material and labor to
Demolish	L.F.		8.15		8.15	install 6' pressure treated pine fence.
Install	L.F.	12.05	4.83	.98	17.86	
Demolish and Install	L.F.	12.05	12.98	.98	26.01	
Reinstall	L.F.		3.87	.78	4.65	
Clean	L.F.	.18	1.10		1.28	
Paint	L.F.	.63	3.01		3.64	
Minimum Charge	Job		365		365	
3' Picket						Includes material, labor and
Demolish	L.F.		8.15		8.15	equipment to install pressure treated
Install	L.F.	4.58	5.20	1.05	10.83	pine picket fence.
Demolish and Install	L.F.	4.58	13.35	1.05	18.98	
Reinstall	L.F.		4.14	.84	4.98	
Clean	L.F.	.09	.94		1.03	
Paint	L.F.	.32	1.63		1.95	
Minimum Charge	Job		365		365	

Fencing	Unit	Material	Labor	Equip.	Total	Specification
Wrought Iron						
Demolish	L.F.		17.45		17.45	Cost includes material and labor to
Install	L.F.	22.50	9.80		32.30	install 5' high wrought iron fence with
Demolish and Install	L.F.	22.50	27.25		49.75	pickets at 4-1/2" O.C.
Reinstall	L.F.		7.83		7.83	
Clean	L.F.		.94		.94	
Paint	L.F.	.58	3.36		3.94	
Minimum Charge	Job		365		365	
Post						
Dirt Based Wood						
Demolish	Ea.		20.50		20.50	Includes material and labor to install
Install	Ea.	11.15	91.50		102.65	wood post set in earth.
Demolish and Install	Ea.	11.15	112		123.15	
Reinstall	Ea.		73.39		73.39	
Clean	Ea.		2.95		2.95	
Paint	Ea.	.69	1.14		1.83	
Minimum Charge	Job		365		365	
Concrete Based Wood						
Demolish	Ea.		28.50	8.80	37.30	Includes material and labor to install
Install	Ea.	11.15	122		133.15	wood post set in concrete.
Demolish and Install	Ea.	11.15	150.50	8.80	170.45	
Clean	Ea.		2.95		2.95	
Paint	Ea.	.69	1.14		1.83	
Minimum Charge	Job		365		365	
Block Pilaster						
Demolish	Ea.		112		112	Includes material and labor to install
Install	Ea.	61	116		177	block pilaster for fence.
Demolish and Install	Ea.	61	228		289	
Reinstall	Ea.		115.94		115.94	
Paint	Ea.	3.88	12.70		16.58	
Minimum Charge	Job		164		164	
Gate						
Chain Link						
Demolish	Ea.		27		27	Cost includes material and labor to
Install	Ea.	102	41		143	install 3' wide chain link pass gate
Demolish and Install	Ea.	102	68		170	and hardware.
Reinstall	Ea.		32.62		32.62	
Clean	Ea.		3.27		3.27	
Minimum Charge	Job		365		365	
Redwood						
Demolish	Ea.		23.50		23.50	Includes material, labor and
Install	Ea.	132	41		173	equipment to install redwood pass
Demolish and Install	Ea.	132	64.50		196.50	gate and hardware.
Reinstall	Ea.		32.62		32.62	
Clean	Ea.		3.27		3.27	
Paint	Ea.	1.03	5.70		6.73	
Minimum Charge	Job		365		365	
Red Cedar						
Demolish	Ea.		23.50		23.50	Cost includes material and labor to
Install	Ea.	70.50	41		111.50	install 3' wide red cedar pass gate
Demolish and Install	Ea.	70.50	64.50		135	and hardware.
Reinstall	Ea.		32.62		32.62	
Clean	Ea.		3.27		3.27	
Paint	Ea.	1.03	5.70		6.73	
Minimum Charge	Job		365		365	

Fencing	Unit	Material	Labor	Equip.	Total	Specification
Wrought Iron						
Demolish	Ea.		17.45		17.45	Includes material, labor and
Install	Ea.	118	91.50		209.50	equipment to install wrought iron gate
Demolish and Install	Ea.	118	108.95		226.95	with hardware.
Reinstall	Ea.		73.39		73.39	
Clean	Ea.		3.27		3.27	
Paint	Ea.	1.17	5.70		6.87	
Minimum Charge	Job		365		365	

Fire-extinguishing System	Space Heater	Hot-air Furnace

Plumbing	Unit	Material	Labor	Equip.	Total	Specification
Components						
3/4" Supply						
Demolish	L.F.		1.97		1.97	Includes material and labor to install
Install	L.F.	2.42	4.66		7.08	type L copper tubing including
Demolish and Install	L.F.	2.42	6.63		9.05	couplings and hangers at 10' O.C.
Reinstall	L.F.		4.66		4.66	
Minimum Charge	Job		177		177	
1/2" Fixture Supply						
Install	Ea.	27	118		145	Includes labor and materials to install supply pipe only to a plumbing fixture.
2" Single Vent Stack						
Demolish	Ea.		33		33	Includes material and labor to install
Install	Ea.	13.80	177		190.80	PVC vent stack pipe with couplings at
Demolish and Install	Ea.	13.80	210		223.80	10' O.C.
Reinstall	Ea.		177.20		177.20	
Minimum Charge	Job		177		177	
4" Drain						
Demolish	L.F.		2.63		2.63	Includes material and labor to install
Install	L.F.	7.45	5.80		13.25	DVW type PVC pipe with couplings at
Demolish and Install	L.F.	7.45	8.43		15.88	10' O.C. and 3 strap hangers per 10'
Reinstall	L.F.		5.80		5.80	run.
Minimum Charge	Job		177		177	
1-1/2" Drain						
Demolish	L.F.		1.97		1.97	Includes material and labor to install
Install	L.F.	.73	3.77		4.50	DVW type PVC pipe with couplings at
Demolish and Install	L.F.	.73	5.74		6.47	10' O.C. and 3 strap hangers per 10'
Reinstall	L.F.		3.77		3.77	run.
Minimum Charge	Job		177		177	
Sewer / Septic						
Install	L.F.	2.48	3.28	1.01	6.77	Includes labor and material to install sewerage and drainage piping, 6" diameter.
Maintenance						
Flush Out / Unclog						
Clean	Job		177		177	Minimum labor and equipment charge
Minimum Charge	Job		177		177	to install water cooler.

Rough Mechanical

Plumbing	Unit	Material	Labor	Equip.	Total	Specification
Vent Cap						
Large						
Demolish	Ea.		8.95		8.95	Includes labor and material to install a
Install	Ea.	34.50	17.70		52.20	PVC plumbing pipe vent cap.
Demolish and Install	Ea.	34.50	26.65		61.15	
Clean	Ea.	.41	5.35		5.76	
Paint	Ea.	.15	6.50		6.65	
Minimum Charge	Job		88.50		88.50	
Medium						
Demolish	Ea.		8.95		8.95	Includes material and labor to install a
Install	Ea.	28.50	16.10		44.60	PVC plumbing pipe vent cap.
Demolish and Install	Ea.	28.50	25.05		53.55	
Clean	Ea.	.41	5.35		5.76	
Paint	Ea.	.15	6.50		6.65	
Minimum Charge	Job		88.50		88.50	
Small						
Demolish	Ea.		8.95		8.95	Includes material and labor to install a
Install	Ea.	22	14.75		36.75	PVC plumbing pipe vent cap.
Demolish and Install	Ea.	22	23.70		45.70	
Clean	Ea.	.41	5.35		5.76	
Paint	Ea.	.15	6.50		6.65	
Minimum Charge	Job		88.50		88.50	
Sump Pump						
1/4 HP						
Demolish	Ea.		35.50		35.50	Includes material and labor to install a
Install	Ea.	116	59		175	submersible sump pump.
Demolish and Install	Ea.	116	94.50		210.50	
Reinstall	Ea.		47.25		47.25	
Clean	Ea.	1.02	14.75		15.77	
Minimum Charge	Job		177		177	
1/3 HP						
Demolish	Ea.		35.50		35.50	Includes material and labor to install a
Install	Ea.	140	71		211	submersible sump pump.
Demolish and Install	Ea.	140	106.50		246.50	
Reinstall	Ea.		56.70		56.70	
Clean	Ea.	1.02	14.75		15.77	
Minimum Charge	Job		177		177	
1/2 HP						
Demolish	Ea.		35.50		35.50	Includes material and labor to install a
Install	Ea.	168	71		239	submersible sump pump.
Demolish and Install	Ea.	168	106.50		274.50	
Reinstall	Ea.		56.70		56.70	
Clean	Ea.	1.02	14.75		15.77	
Minimum Charge	Job		177		177	
Clean and Service						
Clean	Ea.		22		22	Includes labor to clean and service a
Minimum Charge	Job		177		177	sump pump.

Rough Mechanical

Water Heater

	Unit	Material	Labor	Equip.	Total	Specification
30 Gallon						
Demolish	Ea.		111		111	Includes material and labor to install a
Install	Ea.	310	161		471	30 gallon electric water heater.
Demolish and Install	Ea.	310	272		582	
Reinstall	Ea.		128.87		128.87	
Clean	Ea.	1.02	14.75		15.77	
Minimum Charge	Job		177		177	
40 Gallon						
Demolish	Ea.		111		111	Includes material and labor to install a
Install	Ea.	330	177		507	40 gallon electric water heater.
Demolish and Install	Ea.	330	288		618	
Reinstall	Ea.		141.76		141.76	
Clean	Ea.	1.02	14.75		15.77	
Minimum Charge	Job		177		177	
52 Gallon						
Demolish	Ea.		111		111	Includes material and labor to install a
Install	Ea.	390	177		567	52 gallon electric water heater.
Demolish and Install	Ea.	390	288		678	
Reinstall	Ea.		141.76		141.76	
Clean	Ea.	1.02	14.75		15.77	
Minimum Charge	Job		177		177	
82 Gallon						
Demolish	Ea.		111		111	Includes material and labor to install
Install	Ea.	590	222		812	an 82 gallon electric water heater.
Demolish and Install	Ea.	590	333		923	
Reinstall	Ea.		177.20		177.20	
Clean	Ea.	1.02	14.75		15.77	
Minimum Charge	Job		177		177	
Gas-fired						
30 Gallon Gas-fired						
Demolish	Ea.		136		136	Includes material and labor to install a
Install	Ea.	425	177		602	30 gallon gas water heater.
Demolish and Install	Ea.	425	313		738	
Reinstall	Ea.		141.76		141.76	
Clean	Ea.	1.02	14.75		15.77	
Minimum Charge	Job		177		177	
30 Gallon Gas-fired, Flue						
Install	Ea.	35	50.50		85.50	Includes labor and materials to install
Minimum Charge	Job		177		177	a flue for a water heater (up to 50 gallon size).
40 Gallon Gas-fired						
Demolish	Ea.		136		136	Includes material and labor to install a
Install	Ea.	445	187		632	40 gallon gas water heater.
Demolish and Install	Ea.	445	323		768	
Reinstall	Ea.		149.22		149.22	
Clean	Ea.	1.02	14.75		15.77	
Minimum Charge	Job		177		177	
40 Gallon Gas-fired, Flue						
Install	Ea.	35	50.50		85.50	Includes labor and materials to install
Minimum Charge	Job		177		177	a flue for a water heater (up to 50 gallon size).
50 Gallon Gas-fired						
Demolish	Ea.		136		136	Includes material and labor to install a
Install	Ea.	590	197		787	50 gallon gas water heater.
Demolish and Install	Ea.	590	333		923	
Reinstall	Ea.		157.51		157.51	
Clean	Ea.	1.02	14.75		15.77	
Minimum Charge	Job		177		177	

Water Heater	Unit	Material	Labor	Equip.	Total	Specification
50 Gallon Gas-fired, Flue						
Install	Ea.	35	50.50		85.50	Includes labor and materials to install
Minimum Charge	Job		177		177	a flue for a water heater (up to 50 gallon size).
Solar Hot Water System						
Water Heater						
Demolish	Ea.		88.50		88.50	Includes material and labor to install a
Install	Ea.	865	236		1101	solar hot water heat exchanger
Demolish and Install	Ea.	865	324.50		1189.50	storage tank, 100 gallon.
Reinstall	Ea.		189.01		189.01	
Minimum Charge	Job		178		178	
Solar Panel						
Demolish	S.F.		1.30		1.30	Includes material and labor to install
Install	S.F.	18.75	2.33		21.08	solar panel.
Demolish and Install	S.F.	18.75	3.63		22.38	
Reinstall	S.F.		1.87		1.87	
Minimum Charge	Job		178		178	
Repair Plumbing						
Install	Job		178		178	Includes 4 hour minimum labor charge
Minimum Charge	Job		178		178	for a steamfitter or pipefitter.
Insulating Wrap						
Demolish	Ea.		4.10		4.10	Includes material and labor to install
Install	Ea.	71.50	11.10		82.60	insulating wrap around a 100 gallon
Demolish and Install	Ea.	71.50	15.20		86.70	solar hot water heat exchanger
Reinstall	Ea.		8.86		8.86	storage tank.
Minimum Charge	Job		177		177	
Water Softener						
Demolish	Ea.		65.50		65.50	Includes material and labor to install a
Install	Ea.	875	159		1034	14 GPM water softener.
Demolish and Install	Ea.	875	224.50		1099.50	
Reinstall	Ea.		127.52		127.52	
Clean	Ea.	1.02	14.75		15.77	
Minimum Charge	Ea.		177		177	

Ductwork	Unit	Material	Labor	Equip.	Total	Specification
Galvanized, Square or Rectangular						
6"						
Demolish	L.F.		1.65		1.65	Includes material and labor to install
Install	L.F.	2.64	14.85		17.49	rectangular ductwork with a greatest
Demolish and Install	L.F.	2.64	16.50		19.14	dimension of 6".
Reinstall	L.F.		11.88		11.88	
Clean	L.F.		1.90		1.90	
Minimum Charge	Job		178		178	
8"						
Demolish	L.F.		1.65		1.65	Includes material and labor to install
Install	L.F.	3.52	14.85		18.37	rectangular ductwork with a greatest
Demolish and Install	L.F.	3.52	16.50		20.02	dimension of 8".
Reinstall	L.F.		11.88		11.88	
Clean	L.F.		2.53		2.53	
Minimum Charge	Job		178		178	

Ductwork		Unit	Material	Labor	Equip.	Total	Specification
10″							
	Demolish	L.F.		1.65		1.65	Includes material and labor to install
	Install	L.F.	4.40	14.85		19.25	rectangular ductwork with a greatest
	Demolish and Install	L.F.	4.40	16.50		20.90	dimension of 10″.
	Reinstall	L.F.		11.88		11.88	
	Clean	L.F.		3.17		3.17	
	Minimum Charge	Job		178		178	
Galvanized, Spiral							
4″							
	Demolish	L.F.		1.98		1.98	Cost includes material and labor to
	Install	L.F.	.94	1.78		2.72	install 4″ diameter spiral pre-formed
	Demolish and Install	L.F.	.94	3.76		4.70	steel galvanized ductwork.
	Reinstall	L.F.		1.43		1.43	
	Clean	L.F.		1		1	
	Minimum Charge	Job		178		178	
6″							
	Demolish	L.F.		1.98		1.98	Cost includes material and labor to
	Install	L.F.	1.35	2.29		3.64	install 6″ diameter spiral pre-formed
	Demolish and Install	L.F.	1.35	4.27		5.62	steel galvanized ductwork.
	Reinstall	L.F.		1.83		1.83	
	Clean	L.F.		1.49		1.49	
	Minimum Charge	Job		178		178	
8″							
	Demolish	L.F.		1.98		1.98	Cost includes material and labor to
	Install	L.F.	1.78	3.21		4.99	install 8″ diameter spiral pre-formed
	Demolish and Install	L.F.	1.78	5.19		6.97	steel galvanized ductwork.
	Reinstall	L.F.		2.57		2.57	
	Clean	L.F.		1.99		1.99	
	Minimum Charge	Job		178		178	
10″							
	Demolish	L.F.		2.48		2.48	Cost includes material and labor to
	Install	L.F.	2.20	4.01		6.21	install 10″ diameter spiral pre-formed
	Demolish and Install	L.F.	2.20	6.49		8.69	steel galvanized ductwork.
	Reinstall	L.F.		3.21		3.21	
	Clean	L.F.		2.49		2.49	
	Minimum Charge	Job		178		178	
12″							
	Demolish	L.F.		3.30		3.30	Cost includes material and labor to
	Install	L.F.	2.71	5.35		8.06	install 12″ diameter spiral pre-formed
	Demolish and Install	L.F.	2.71	8.65		11.36	steel galvanized ductwork.
	Reinstall	L.F.		4.28		4.28	
	Clean	L.F.		2.99		2.99	
	Minimum Charge	Job		178		178	
16″							
	Demolish	L.F.		6.60		6.60	Cost includes material and labor to
	Install	L.F.	4.66	10.70		15.36	install 16″ diameter spiral pre-formed
	Demolish and Install	L.F.	4.66	17.30		21.96	steel galvanized ductwork.
	Reinstall	L.F.		8.55		8.55	
	Clean	L.F.		3.98		3.98	
	Minimum Charge	Job		178		178	
Flexible Fiberglass							
6″ Non-insulated							
	Demolish	L.F.		1.52		1.52	Cost includes material and labor to
	Install	L.F.	1.67	2.29		3.96	install 6″ diameter non-insulated
	Demolish and Install	L.F.	1.67	3.81		5.48	flexible fiberglass fabric on corrosion
	Reinstall	L.F.		1.83		1.83	resistant metal ductwork.
	Clean	L.F.		1.49		1.49	
	Minimum Charge	Job		178		178	

Ductwork	Unit	Material	Labor	Equip.	Total	Specification
8″ Non-insulated						
Demolish	L.F.		2.20		2.20	Cost includes material and labor to
Install	L.F.	2.30	3.21		5.51	install 8″ diameter non-insulated
Demolish and Install	L.F.	2.30	5.41		7.71	flexible fiberglass fabric on corrosion
Reinstall	L.F.		2.57		2.57	resistant metal ductwork.
Clean	L.F.		1.99		1.99	
Minimum Charge	Job		178		178	
10″ Non-insulated						
Demolish	L.F.		2.83		2.83	Cost includes material and labor to
Install	L.F.	2.89	4.01		6.90	install 10″ diameter non-insulated
Demolish and Install	L.F.	2.89	6.84		9.73	flexible fiberglass fabric on corrosion
Reinstall	L.F.		3.21		3.21	resistant metal ductwork.
Clean	L.F.		2.49		2.49	
Minimum Charge	Job		178		178	
12″ Non-insulated						
Demolish	L.F.		3.96		3.96	Cost includes material and labor to
Install	L.F.	3.50	5.35		8.85	install 12″ diameter non-insulated
Demolish and Install	L.F.	3.50	9.31		12.81	flexible fiberglass fabric on corrosion
Reinstall	L.F.		4.28		4.28	resistant metal ductwork.
Clean	L.F.		2.99		2.99	
Minimum Charge	Job		178		178	
4″ Insulated						
Demolish	L.F.		1.16		1.16	Cost includes material and labor to
Install	L.F.	1.35	1.89		3.24	install 4″ diameter insulated flexible
Demolish and Install	L.F.	1.35	3.05		4.40	fiberglass fabric on corrosion resistant
Reinstall	L.F.		1.51		1.51	metal ductwork.
Clean	L.F.		1		1	
Minimum Charge	Job		178		178	
6″ Insulated						
Demolish	L.F.		1.52		1.52	Cost includes material and labor to
Install	L.F.	1.72	2.47		4.19	install 6″ diameter insulated flexible
Demolish and Install	L.F.	1.72	3.99		5.71	fiberglass fabric on corrosion resistant
Reinstall	L.F.		1.97		1.97	metal ductwork.
Clean	L.F.		1.49		1.49	
Minimum Charge	Job		178		178	
8″ Insulated						
Demolish	L.F.		2.20		2.20	Cost includes material and labor to
Install	L.F.	2.16	3.56		5.72	install 8″ diameter insulated flexible
Demolish and Install	L.F.	2.16	5.76		7.92	fiberglass fabric on corrosion resistant
Reinstall	L.F.		2.85		2.85	metal ductwork.
Clean	L.F.		1.99		1.99	
Minimum Charge	Job		178		178	
10″ Insulated						
Demolish	L.F.		2.83		2.83	Cost includes material and labor to
Install	L.F.	2.64	4.58		7.22	install 10″ diameter insulated flexible
Demolish and Install	L.F.	2.64	7.41		10.05	fiberglass fabric on corrosion resistant
Reinstall	L.F.		3.67		3.67	metal ductwork.
Clean	L.F.		2.49		2.49	
Minimum Charge	Job		178		178	
12″ Insulated						
Demolish	L.F.		3.96		3.96	Cost includes material and labor to
Install	L.F.	3.25	6.40		9.65	install 12″ diameter insulated flexible
Demolish and Install	L.F.	3.25	10.36		13.61	fiberglass fabric on corrosion resistant
Reinstall	L.F.		5.13		5.13	metal ductwork.
Clean	L.F.		2.99		2.99	
Minimum Charge	Job		178		178	

Heating	Unit	Material	Labor	Equip.	Total	Specification
Hot Water Baseboard						
Demolish	L.F.		6.10		6.10	Includes material and labor to install
Install	L.F.	5.90	10.65		16.55	baseboard radiation.
Demolish and Install	L.F.	5.90	16.75		22.65	
Reinstall	L.F.		8.53		8.53	
Clean	L.F.	1.02	3.68		4.70	
Minimum Charge	Job		177		177	
3'						
Demolish	Ea.		19.70		19.70	Includes material and labor to install
Install	Ea.	18.70	35.50		54.20	fin tube baseboard radiation in 3'
Demolish and Install	Ea.	18.70	55.20		73.90	sections.
Reinstall	Ea.		28.44		28.44	
Clean	Ea.	.30	11.20		11.50	
Minimum Charge	Job		177		177	
4'						
Demolish	Ea.		26.50		26.50	Includes material and labor to install
Install	Ea.	25	40		65	fin tube baseboard radiation in 4'
Demolish and Install	Ea.	25	66.50		91.50	sections.
Reinstall	Ea.		32		32	
Clean	Ea.	.58	14.75		15.33	
Minimum Charge	Job		177		177	
5'						
Demolish	Ea.		33		33	Includes material and labor to install
Install	Ea.	31.50	45.50		77	fin tube baseboard radiation in 5'
Demolish and Install	Ea.	31.50	78.50		110	sections.
Reinstall	Ea.		36.57		36.57	
Clean	Ea.	.58	18.10		18.68	
Minimum Charge	Job		177		177	
6'						
Demolish	Ea.		39.50		39.50	Includes material and labor to install
Install	Ea.	37.50	53.50		91	fin tube baseboard radiation in 6'
Demolish and Install	Ea.	37.50	93		130.50	sections.
Reinstall	Ea.		42.67		42.67	
Clean	Ea.	.58	21.50		22.08	
Minimum Charge	Job		177		177	
7'						
Demolish	Ea.		44		44	Includes material and labor to install
Install	Ea.	43.50	71		114.50	fin tube baseboard radiation in 7'
Demolish and Install	Ea.	43.50	115		158.50	sections.
Reinstall	Ea.		56.70		56.70	
Clean	Ea.	.73	26		26.73	
Minimum Charge	Job		177		177	
9'						
Demolish	Ea.		65.50		65.50	Includes material and labor to install
Install	Ea.	56	101		157	fin tube baseboard radiation in 9'
Demolish and Install	Ea.	56	166.50		222.50	sections.
Reinstall	Ea.		81.01		81.01	
Clean	Ea.	1.17	33.50		34.67	
Minimum Charge	Job		177		177	
50 MBH						
Demolish	Ea.		32		32	Includes material and labor to install a
Install	Ea.	610	91.50		701.50	cabinet space heater with an output of
Demolish and Install	Ea.	610	123.50		733.50	50 MBH.
Reinstall	Ea.		73.14		73.14	
Clean	Ea.	.73	29.50		30.23	
Minimum Charge	Job		178		178	

Rough Mechanical

Heating		Unit	Material	Labor	Equip.	Total	Specification
75 MBH							
	Demolish	Ea.		32		32	Includes material and labor to install a
	Install	Ea.	700	107		807	cabinet space heater with an output of
	Demolish and Install	Ea.	700	139		839	75 MBH.
	Reinstall	Ea.		85.33		85.33	
	Clean	Ea.	.73	29.50		30.23	
	Minimum Charge	Job		178		178	
125 MBH							
	Demolish	Ea.		38.50		38.50	Includes material and labor to install a
	Install	Ea.	910	128		1038	cabinet space heater with an output of
	Demolish and Install	Ea.	910	166.50		1076.50	125 MBH.
	Reinstall	Ea.		102.40		102.40	
	Clean	Ea.	.73	29.50		30.23	
	Minimum Charge	Job		178		178	
175 MBH							
	Demolish	Ea.		48		48	Includes material and labor to install a
	Install	Ea.	.1125	183		1308	cabinet space heater with an output of
	Demolish and Install	Ea.	1125	231		1356	175 MBH.
	Reinstall	Ea.		146.29		146.29	
	Clean	Ea.	.73	29.50		30.23	
	Minimum Charge	Job		178		178	

Furnace Unit		Unit	Material	Labor	Equip.	Total	Specification
Wall (gas fired)							
25 MBH w / Thermostat							
	Demolish	Ea.		80		80	Includes material and labor to install a
	Install	Ea.	405	128		533	gas fired hot air furnace rated at 25
	Demolish and Install	Ea.	405	208		613	MBH.
	Reinstall	Ea.		102.66		102.66	
	Clean	Ea.	1.46	65.50		66.96	
	Minimum Charge	Job		178		178	
35 MBH w / Thermostat							
	Demolish	Ea.		80		80	Includes material and labor to install a
	Install	Ea.	415	143		558	gas fired hot air furnace rated at 35
	Demolish and Install	Ea.	415	223		638	MBH.
	Reinstall	Ea.		114.06		114.06	
	Clean	Ea.	1.46	65.50		66.96	
	Minimum Charge	Job		178		178	
50 MBH w / Thermostat							
	Demolish	Ea.		80		80	Includes material and labor to install a
	Install	Ea.	785	160		945	gas fired hot air furnace rated at 50
	Demolish and Install	Ea.	785	240		1025	MBH.
	Reinstall	Ea.		128.32		128.32	
	Clean	Ea.	1.46	65.50		66.96	
	Minimum Charge	Job		178		178	
60 MBH w / Thermostat							
	Demolish	Ea.		80		80	Includes material and labor to install a
	Install	Ea.	870	178		1048	gas fired hot air furnace rated at 60
	Demolish and Install	Ea.	870	258		1128	MBH.
	Reinstall	Ea.		142.58		142.58	
	Clean	Ea.	1.46	65.50		66.96	
	Minimum Charge	Job		178		178	

Rough Mechanical

Furnace Unit	Unit	Material	Labor	Equip.	Total	Specification
Floor (gas fired)						
30 MBH w / Thermostat						
Demolish	Ea.		80		80	Includes material and labor to install a
Install	Ea.	670	128		798	gas fired hot air floor furnace rated at
Demolish and Install	Ea.	670	208		878	30 MBH.
Reinstall	Ea.		102.66		102.66	
Clean	Ea.	1.46	65.50		66.96	
Minimum Charge	Job		178		178	
50 MBH w / Thermostat						
Demolish	Ea.		80		80	Includes material and labor to install a
Install	Ea.	735	160		895	gas fired hot air floor furnace rated at
Demolish and Install	Ea.	735	240		975	50 MBH.
Reinstall	Ea.		128.32		128.32	
Clean	Ea.	1.46	65.50		66.96	
Minimum Charge	Job		178		178	
Gas Forced Air						
50 MBH						
Demolish	Ea.		80		80	Includes material and labor to install
Install	Ea.	755	160		915	an up flow gas fired hot air furnace
Demolish and Install	Ea.	755	240		995	rated at 50 MBH.
Reinstall	Ea.		128.32		128.32	
Clean	Ea.	1.46	65.50		66.96	
Minimum Charge	Job		178		178	
80 MBH						
Demolish	Ea.		80		80	Includes material and labor to install
Install	Ea.	830	178		1008	an up flow gas fired hot air furnace
Demolish and Install	Ea.	830	258		1088	rated at 80 MBH.
Reinstall	Ea.		142.58		142.58	
Clean	Ea.	1.46	65.50		66.96	
Minimum Charge	Job		178		178	
100 MBH						
Demolish	Ea.		80		80	Includes material and labor to install
Install	Ea.	950	201		1151	an up flow gas fired hot air furnace
Demolish and Install	Ea.	950	281		1231	rated at 100 MBH.
Reinstall	Ea.		160.40		160.40	
Clean	Ea.	1.46	65.50		66.96	
Minimum Charge	Job		178		178	
120 MBH						
Demolish	Ea.		80		80	Includes material and labor to install
Install	Ea.	1100	214		1314	an up flow gas fired hot air furnace
Demolish and Install	Ea.	1100	294		1394	rated at 120 MBH.
Reinstall	Ea.		171.09		171.09	
Clean	Ea.	1.46	65.50		66.96	
Minimum Charge	Job		178		178	
Electric Forced Air						
31.4 MBH						
Demolish	Ea.		80		80	Includes material and labor to install
Install	Ea.	840	182		1022	an electric hot air furnace rated at
Demolish and Install	Ea.	840	262		1102	31.4 MBH.
Reinstall	Ea.		145.67		145.67	
Clean	Ea.	1.46	65.50		66.96	
Minimum Charge	Job		178		178	

Rough Mechanical

Furnace Unit

Furnace Unit	Unit	Material	Labor	Equip.	Total	Specification
47.1 MBH						
Demolish	Ea.		80		80	Includes material and labor to install
Install	Ea.	935	191		1126	an electric hot air furnace rated at
Demolish and Install	Ea.	935	271		1206	47.1 MBH.
Reinstall	Ea.		152.45		152.45	
Clean	Ea.	1.46	65.50		66.96	
Minimum Charge	Job		178		178	
62.9 MBH						
Demolish	Ea.		80		80	Includes material and labor to install
Install	Ea.	1075	205		1280	an electric hot air furnace rated at
Demolish and Install	Ea.	1075	285		1360	62.9 MBH.
Reinstall	Ea.		163.88		163.88	
Clean	Ea.	1.46	65.50		66.96	
Minimum Charge	Job		178		178	
78.5 MBH						
Demolish	Ea.		80		80	Includes material and labor to install
Install	Ea.	1200	210		1410	an electric hot air furnace rated at
Demolish and Install	Ea.	1200	290		1490	78.5 MBH.
Reinstall	Ea.		168.08		168.08	
Clean	Ea.	1.46	65.50		66.96	
Minimum Charge	Job		178		178	
110 MBH						
Demolish	Ea.		80		80	Includes material and labor to install
Install	Ea.	1700	221		1921	an electric hot air furnace rated at
Demolish and Install	Ea.	1700	301		2001	110 MBH.
Reinstall	Ea.		177.17		177.17	
Clean	Ea.	1.46	65.50		66.96	
Minimum Charge	Job		178		178	
A/C Junction Box						
Install	Ea.	49.50	89		138.50	Includes labor and materials to install
Minimum Charge	Job		178		178	a heating system emergency shut-off switch. Cost includes up to 40 feet of non-metallic cable.

Boiler

Boiler	Unit	Material	Labor	Equip.	Total	Specification
Gas Fired						
60 MBH						
Demolish	Ea.		645		645	Includes material and labor to install a
Install	Ea.	2075	1175		3250	gas fired cast iron boiler rated at 60
Demolish and Install	Ea.	2075	1820		3895	MBH.
Reinstall	Ea.		930.91		930.91	
Clean	Ea.	2.92	39.50		42.42	
Minimum Charge	Job		178		178	
Gas Fired						
125 MBH						
Demolish	Ea.		1775		1775	Includes material and labor to install a
Install	Ea.	2900	1425		4325	gas fired cast iron boiler rated at 125
Demolish and Install	Ea.	2900	3200		6100	MBH.
Reinstall	Ea.		1137.78		1137.78	
Clean	Ea.	2.92	39.50		42.42	
Minimum Charge	Job		178		178	

Rough Mechanical

Boiler

Boiler	Unit	Material	Labor	Equip.	Total	Specification
Gas Fired						
400 MBH						
Demolish	Ea.		1775		1775	Includes material and labor to install a
Install	Ea.	3750	2000		5750	gas fired cast iron boiler rated at 400
Demolish and Install	Ea.	3750	3775		7525	MBH.
Reinstall	Ea.		2000		2000	
Clean	Ea.	2.92	39.50		42.42	
Minimum Charge	Job		178		178	

Heat Pump

Heat Pump	Unit	Material	Labor	Equip.	Total	Specification
2 Ton						
Demolish	Ea.		84		84	Includes material and labor to install a
Install	Ea.	1575	535		2110	2 ton heat pump.
Demolish and Install	Ea.	1575	619		2194	
Reinstall	Ea.		426.67		426.67	
Clean	Ea.	2.92	29.50		32.42	
Minimum Charge	Job		178		178	
3 Ton						
Demolish	Ea.		159		159	Includes material and labor to install a
Install	Ea.	1900	800		2700	3 ton heat pump.
Demolish and Install	Ea.	1900	959		2859	
Reinstall	Ea.		640		640	
Clean	Ea.	2.92	29.50		32.42	
Minimum Charge	Job		178		178	
4 Ton						
Demolish	Ea.		289		289	Includes material and labor to install a
Install	Ea.	2375	1075		3450	4 ton heat pump.
Demolish and Install	Ea.	2375	1364		3739	
Reinstall	Ea.		853.33		853.33	
Clean	Ea.	2.92	29.50		32.42	
Minimum Charge	Job		178		178	
5 Ton						
Demolish	Ea.		425		425	Includes material and labor to install a
Install	Ea.	2700	1275		3975	5 ton heat pump.
Demolish and Install	Ea.	2700	1700		4400	
Reinstall	Ea.		1024		1024	
Clean	Ea.	2.92	29.50		32.42	
Minimum Charge	Job		178		178	
Thermostat						
Demolish	Ea.		9.90		9.90	Includes material and labor to install
Install	Ea.	196	44.50		240.50	an electric thermostat, 2 set back.
Demolish and Install	Ea.	196	54.40		250.40	
Reinstall	Ea.		35.64		35.64	
Clean	Ea.	.06	7.35		7.41	
Minimum Charge	Job		178		178	
A/C Junction Box						
Install	Ea.	425	274		699	Includes labor and materials to install
Minimum Charge	Job		178		178	a heat pump circuit including up to 40 feet of non-metallic cable.

Heating	Unit	Material	Labor	Equip.	Total	Specification
Electric Baseboard						
Demolish	L.F.		4.94		4.94	Includes material and labor to install
Install	L.F.	10.75	11.85		22.60	electric baseboard heating elements.
Demolish and Install	L.F.	10.75	16.79		27.54	
Reinstall	L.F.		9.48		9.48	
Clean	L.F.	1.02	3.68		4.70	
Minimum Charge	Job		178		178	
2' 6"						
Demolish	Ea.		16.45		16.45	Includes material and labor to install a
Install	Ea.	32	44.50		76.50	2' 6" long section of electric
Demolish and Install	Ea.	32	60.95		92.95	baseboard heating elements.
Reinstall	Ea.		35.56		35.56	
Clean	Ea.	.30	9.05		9.35	
Minimum Charge	Job		178		178	
3'						
Demolish	Ea.		17.95		17.95	Includes material and labor to install a
Install	Ea.	38	44.50		82.50	3' long section of electric baseboard
Demolish and Install	Ea.	38	62.45		100.45	heating elements.
Reinstall	Ea.		35.56		35.56	
Clean	Ea.	.30	11.20		11.50	
Minimum Charge	Job		178		178	
4'						
Demolish	Ea.		19.75		19.75	Includes material and labor to install a
Install	Ea.	45	53		98	4' long section of electric baseboard
Demolish and Install	Ea.	45	72.75		117.75	heating elements.
Reinstall	Ea.		42.46		42.46	
Clean	Ea.	.12	2.45		2.57	
Minimum Charge	Job		178		178	
5'						
Demolish	Ea.		22		22	Includes material and labor to install a
Install	Ea.	53.50	62.50		116	5' long section of electric baseboard
Demolish and Install	Ea.	53.50	84.50		138	heating elements.
Reinstall	Ea.		49.91		49.91	
Clean	Ea.	.58	18.10		18.68	
Minimum Charge	Job		178		178	
6'						
Demolish	Ea.		24.50		24.50	Includes material and labor to install a
Install	Ea.	59.50	71		130.50	6' long section of electric baseboard
Demolish and Install	Ea.	59.50	95.50		155	heating elements.
Reinstall	Ea.		56.90		56.90	
Clean	Ea.	.12	2.95		3.07	
Minimum Charge	Job		178		178	
8'						
Demolish	Ea.		33		33	Includes material and labor to install
Install	Ea.	75	89		164	an 8' long section of electric
Demolish and Install	Ea.	75	122		197	baseboard heating elements.
Reinstall	Ea.		71.12		71.12	
Clean	Ea.	.12	3.68		3.80	
Minimum Charge	Job		178		178	
10'						
Demolish	Ea.		39.50		39.50	Includes material and labor to install a
Install	Ea.	93	108		201	10' long section of electric baseboard
Demolish and Install	Ea.	93	147.50		240.50	heating elements.
Reinstall	Ea.		86.21		86.21	
Clean	Ea.	1.17	37		38.17	
Minimum Charge	Job		178		178	

Rough Mechanical

Heating	Unit	Material	Labor	Equip.	Total	Specification
Electric Radiant Wall Panel						Includes material and labor to install an 18" x 23" electric radiant heat wall panel.
Demolish	Ea.		26.50		26.50	
Install	Ea.	405	71		476	
Demolish and Install	Ea.	405	97.50		502.50	
Reinstall	Ea.		56.90		56.90	
Clean	Ea.	.14	7.35		7.49	
Minimum Charge	Job		178		178	

Heating Control	Unit	Material	Labor	Equip.	Total	Specification
Thermostat						Includes material and labor to install an electric thermostat, 2 set back.
Demolish	Ea.		9.90		9.90	
Install	Ea.	196	44.50		240.50	
Demolish and Install	Ea.	196	54.40		250.40	
Reinstall	Ea.		35.64		35.64	
Clean	Ea.	.06	7.35		7.41	
Minimum Charge	Job		178		178	

Air Conditioning	Unit	Material	Labor	Equip.	Total	Specification
2 Ton						Includes material and labor to install a rooftop air conditioning unit, 2 ton. No hoisting is included.
Demolish	Ea.		330		330	
Install	Ea.	1400	550		1950	
Demolish and Install	Ea.	1400	880		2280	
Reinstall	Ea.		552.20		552.20	
Clean	Ea.	2.92	29.50		32.42	
Minimum Charge	Job		178		178	
3 Ton						Includes material and labor to install a rooftop air conditioning unit, 3 ton. No hoisting is included.
Demolish	Ea.		330		330	
Install	Ea.	1525	730		2255	
Demolish and Install	Ea.	1525	1060		2585	
Reinstall	Ea.		728.10		728.10	
Clean	Ea.	2.92	29.50		32.42	
Minimum Charge	Job		178		178	
4 Ton						Includes material and labor to install a rooftop air conditioning unit, 4 ton. No hoisting is included.
Demolish	Ea.		330		330	
Install	Ea.	1900	845		2745	
Demolish and Install	Ea.	1900	1175		3075	
Reinstall	Ea.		844.33		844.33	
Clean	Ea.	2.92	29.50		32.42	
Minimum Charge	Job		178		178	
5 Ton						Includes material and labor to install a rooftop air conditioning unit, 5 ton. No hoisting is included.
Demolish	Ea.		330		330	
Install	Ea.	2150	910		3060	
Demolish and Install	Ea.	2150	1240		3390	
Reinstall	Ea.		911.68		911.68	
Clean	Ea.	2.92	29.50		32.42	
Minimum Charge	Job		178		178	
Compressor						
2 Ton						Includes material and labor to install a refrigeration compressor.
Demolish	Ea.		296		296	
Install	Ea.	535	178		713	
Demolish and Install	Ea.	535	474		1009	
Reinstall	Ea.		142.24		142.24	
Clean	Ea.	2.92	29.50		32.42	
Minimum Charge	Job		178		178	

Rough Mechanical

Air Conditioning	Unit	Material	Labor	Equip.	Total	Specification
2.5 Ton						
Demolish	Ea.		296		296	Includes material and labor to install a
Install	Ea.	620	193		813	refrigeration compressor.
Demolish and Install	Ea.	620	489		1109	
Reinstall	Ea.		154.11		154.11	
Clean	Ea.	2.92	29.50		32.42	
Minimum Charge	Job		178		178	
3 Ton						
Demolish	Ea.		296		296	Includes material and labor to install a
Install	Ea.	650	201		851	refrigeration compressor.
Demolish and Install	Ea.	650	497		1147	
Reinstall	Ea.		160.81		160.81	
Clean	Ea.	2.92	29.50		32.42	
Minimum Charge	Job		178		178	
4 Ton						
Demolish	Ea.		296		296	Includes material and labor to install a
Install	Ea.	690	162		852	refrigeration compressor.
Demolish and Install	Ea.	690	458		1148	
Reinstall	Ea.		129.31		129.31	
Clean	Ea.	2.92	29.50		32.42	
Minimum Charge	Job		178		178	
5 Ton						
Demolish	Ea.		296		296	Includes material and labor to install a
Install	Ea.	855	220		1075	refrigeration compressor.
Demolish and Install	Ea.	855	516		1371	
Reinstall	Ea.		176.15		176.15	
Clean	Ea.	2.92	29.50		32.42	
Minimum Charge	Job		178		178	
Condensing Unit						
2 Ton Air-cooled						
Demolish	Ea.		84.50		84.50	Includes material and labor to install a
Install	Ea.	645	305		950	2 ton condensing unit.
Demolish and Install	Ea.	645	389.50		1034.50	
Reinstall	Ea.		243.81		243.81	
Clean	Ea.	2.92	29.50		32.42	
Minimum Charge	Job		178		178	
2.5 Ton Air-cooled						
Demolish	Ea.		99		99	Includes material and labor to install a
Install	Ea.	775	375		1150	2.5 ton condensing unit.
Demolish and Install	Ea.	775	474		1249	
Reinstall	Ea.		301.18		301.18	
Clean	Ea.	2.92	29.50		32.42	
Minimum Charge	Job		178		178	
3 Ton Air-cooled						
Demolish	Ea.		119		119	Includes material and labor to install a
Install	Ea.	915	490		1405	3 ton condensing unit.
Demolish and Install	Ea.	915	609		1524	
Reinstall	Ea.		393.85		393.85	
Clean	Ea.	2.92	29.50		32.42	
Minimum Charge	Job		178		178	
4 Ton Air-cooled						
Demolish	Ea.		178		178	Includes material and labor to install a
Install	Ea.	1175	710		1885	4 ton condensing unit.
Demolish and Install	Ea.	1175	888		2063	
Reinstall	Ea.		568.89		568.89	
Clean	Ea.	2.92	29.50		32.42	
Minimum Charge	Job		178		178	

Rough Mechanical

Air Conditioning		Unit	Material	Labor	Equip.	Total	Specification
5 Ton Air-cooled							
	Demolish	Ea.		267		267	Includes material and labor to install a
	Install	Ea.	1425	1075		2500	5 ton condensing unit.
	Demolish and Install	Ea.	1425	1342		2767	
	Reinstall	Ea.		853.33		853.33	
	Clean	Ea.	2.92	29.50		32.42	
	Minimum Charge	Job		178		178	
Air Handler							
1200 CFM							
	Demolish	Ea.		214		214	Includes material and labor to install a
	Install	Ea.	2075	490		2565	1200 CFM capacity air handling unit.
	Demolish and Install	Ea.	2075	704		2779	
	Reinstall	Ea.		393.85		393.85	
	Clean	Ea.	2.92	29.50		32.42	
	Minimum Charge	Job		178		178	
2000 CFM							
	Demolish	Ea.		214		214	Includes material and labor to install a
	Install	Ea.	3450	580		4030	2000 CFM capacity air handling unit.
	Demolish and Install	Ea.	3450	794		4244	
	Reinstall	Ea.		465.45		465.45	
	Clean	Ea.	2.92	29.50		32.42	
	Minimum Charge	Job		178		178	
3000 CFM							
	Demolish	Ea.		237		237	Includes material and labor to install a
	Install	Ea.	4300	640		4940	3000 CFM capacity air handling unit.
	Demolish and Install	Ea.	4300	877		5177	
	Reinstall	Ea.		512		512	
	Clean	Ea.	2.92	29.50		32.42	
	Minimum Charge	Job		178		178	
Evaporative Cooler							
4,300 CFM, 1/3 HP							
	Demolish	Ea.		87.50		87.50	Includes material and labor to install
	Install	Ea.	655	201		856	an evaporative cooler.
	Demolish and Install	Ea.	655	288.50		943.50	
	Reinstall	Ea.		160.40		160.40	
	Clean	Ea.	1.17	29.50		30.67	
	Minimum Charge	Job		178		178	
4,700 CFM, 1/2 HP							
	Demolish	Ea.		87.50		87.50	Includes material and labor to install
	Install	Ea.	620	207		827	an evaporative cooler.
	Demolish and Install	Ea.	620	294.50		914.50	
	Reinstall	Ea.		165.57		165.57	
	Clean	Ea.	1.17	29.50		30.67	
	Minimum Charge	Job		178		178	
5,600 CFM, 3/4 HP							
	Demolish	Ea.		87.50		87.50	Includes material and labor to install
	Install	Ea.	1075	214		1289	an evaporative cooler.
	Demolish and Install	Ea.	1075	301.50		1376.50	
	Reinstall	Ea.		171.09		171.09	
	Clean	Ea.	1.17	29.50		30.67	
	Minimum Charge	Job		178		178	
Master Cool Brand							
	Demolish	Ea.		87.50		87.50	Includes material and labor to install
	Install	Ea.	655	221		876	an evaporative cooler.
	Demolish and Install	Ea.	655	308.50		963.50	
	Reinstall	Ea.		176.99		176.99	
	Clean	Ea.	1.17	29.50		30.67	
	Minimum Charge	Job		178		178	

Rough Mechanical

Air Conditioning		Unit	Material	Labor	Equip.	Total	Specification
Motor							
	Demolish	Ea.		49		49	Includes material and labor to install
	Install	Ea.	222	156		378	an evaporative cooler motor.
	Demolish and Install	Ea.	222	205		427	
	Reinstall	Ea.		124.72		124.72	
	Clean	Ea.	1.17	29.50		30.67	
	Minimum Charge	Job		178		178	
Water Pump							
	Demolish	Ea.		49		49	Includes material and labor to install
	Install	Ea.	42	89		131	an evaporative cooler water pump.
	Demolish and Install	Ea.	42	138		180	
	Reinstall	Ea.		71.28		71.28	
	Clean	Ea.	1.17	29.50		30.67	
	Minimum Charge	Job		178		178	
Fan Belt							
	Demolish	Ea.		9.85		9.85	Includes material and labor to install
	Install	Ea.	27.50	17.80		45.30	an evaporative cooler fan belt.
	Demolish and Install	Ea.	27.50	27.65		55.15	
	Reinstall	Ea.		14.26		14.26	
	Minimum Charge	Job		178		178	
Pads (4 ea.)							
	Demolish	Ea.		9.85		9.85	Includes material and labor to install
	Install	Ea.	31	44.50		75.50	pre-fabricated roof pads for an
	Demolish and Install	Ea.	31	54.35		85.35	evaporative cooler.
	Reinstall	Ea.		35.64		35.64	
	Clean	Ea.	1.17	29.50		30.67	
	Minimum Charge	Job		178		178	
Service call							
	Install	Job		178		178	Includes 4 hour minimum labor charge for an sheetmetal worker.
Window Mounted							
Low Output BTU							
	Install	Ea.	224	41		265	Includes material and labor to install a
	Reinstall	Ea.		32.60		32.60	window air conditioning unit.
	Clean	Ea.	.44	7.35		7.79	
	Minimum Charge	Job		178		178	
Mid-range Output BTU							
	Demolish	Ea.		28.50		28.50	Includes material and labor to install a
	Install	Ea.	320	54.50		374.50	window air conditioning unit.
	Demolish and Install	Ea.	320	83		403	
	Reinstall	Ea.		43.47		43.47	
	Clean	Ea.	.44	7.35		7.79	
	Minimum Charge	Job		178		178	
High Output BTU							
	Demolish	Ea.		36		36	Includes material and labor to install a
	Install	Ea.	430	54.50		484.50	window air conditioning unit.
	Demolish and Install	Ea.	430	90.50		520.50	
	Reinstall	Ea.		43.47		43.47	
	Clean	Ea.	.44	7.35		7.79	
	Minimum Charge	Job		178		178	
Hotel-type Hot / Cold							
	Demolish	Ea.		60.50		60.50	Includes material and labor to install
	Install	Ea.	1425	213		1638	packaged air conditioning unit,
	Demolish and Install	Ea.	1425	273.50		1698.50	15000 BTU cooling, 13900 BTU
	Reinstall	Ea.		170.67		170.67	heating.
	Clean	Ea.	.58	14.75		15.33	
	Minimum Charge	Job		178		178	

Rough Mechanical

Air Conditioning

Air Conditioning		Unit	Material	Labor	Equip.	Total	Specification
Thermostat							
	Demolish	Ea.		9.90		9.90	Includes material and labor to install
	Install	Ea.	196	44.50		240.50	an electric thermostat, 2 set back.
	Demolish and Install	Ea.	196	54.40		250.40	
	Reinstall	Ea.		35.64		35.64	
	Clean	Ea.	.06	7.35		7.41	
	Minimum Charge	Job		178		178	
Concrete Pad							
	Demolish	S.F.		1.64		1.64	Includes material and labor to install
	Install	S.F.	1.10	3.71		4.81	air conditioner pad.
	Demolish and Install	S.F.	1.10	5.35		6.45	
	Reinstall	S.F.		2.97		2.97	
	Clean	S.F.	.06	.47		.53	
	Paint	S.F.	.12	.29		.41	
	Minimum Charge	Job		178		178	
A/C Junction Box							
	Install	Ea.	174	102		276	Includes labor and materials to install
	Minimum Charge	Job		178		178	an air conditioning circuit including up to 40 feet of non-metallic cable.

Sprinkler System

Sprinkler System		Unit	Material	Labor	Equip.	Total	Specification
Exposed							
Wet Type							
	Demolish	SF Flr.		.16		.16	Includes material and labor to install
	Install	SF Flr.	.86	1.56		2.42	an exposed wet pipe fire sprinkler
	Demolish and Install	SF Flr.	.86	1.72		2.58	system.
	Reinstall	SF Flr.		1.56		1.56	
	Minimum Charge	Job		177		177	
Dry Type							
	Demolish	SF Flr.		.16		.16	Includes material and labor to install
	Install	SF Flr.	.98	1.56		2.54	an exposed dry pipe fire sprinkler
	Demolish and Install	SF Flr.	.98	1.72		2.70	system.
	Reinstall	SF Flr.		1.56		1.56	
	Minimum Charge	Job		177		177	
Concealed							
Wet Type							
	Demolish	SF Flr.		.31		.31	Includes material and labor to install a
	Install	SF Flr.	.86	2.13		2.99	concealed wet pipe fire sprinkler
	Demolish and Install	SF Flr.	.86	2.44		3.30	system.
	Reinstall	SF Flr.		2.13		2.13	
	Minimum Charge	Job		177		177	
Dry Type							
	Demolish	SF Flr.		.31		.31	Includes material and labor to install a
	Install	SF Flr.	.98	2.13		3.11	concealed dry pipe fire sprinkler
	Demolish and Install	SF Flr.	.98	2.44		3.42	system.
	Reinstall	SF Flr.		2.13		2.13	
	Minimum Charge	Job		177		177	

| Junction Boxes | | Cable | | | Switch Box | |

Electrical Per S.F.	Unit	Material	Labor	Equip.	Total	Specification
Residential						
Wiring, Lighting and Fixtures						
Install	S.F.	1.30	1.48		2.78	Includes labor and material to replace
Minimum Charge	Job		178		178	house wiring, lighting fixtures, and
						built-in appliances. Apply S.F. cost to
						floor area including garages but not
						basements.
Wiring Only						
Install	S.F.	.15	.58		.73	Includes labor and material to replace
						house wiring only. Apply S.F. cost to
						floor area including garages but not
						basements.

Main Service	Unit	Material	Labor	Equip.	Total	Specification
100 Amp.						
Demolish	Ea.		146		146	Cost includes material and labor to
Install	Ea.	490	299		789	install 100 Amp. residential service
Demolish and Install	Ea.	490	445		935	panel including 12 breaker interior
Reinstall	Ea.		239.06		239.06	panel, 10 single pole breakers and
Clean	Ea.	.30	59		59.30	hook-up. Wiring is not included.
Minimum Charge	Job		178		178	
150 Amp.						
Demolish	Ea.		146		146	Cost includes material and labor to
Install	Ea.	760	345		1105	install 150 Amp. residential service,
Demolish and Install	Ea.	760	491		1251	incl. 20 breaker exterior panel, main
Reinstall	Ea.		276.19		276.19	switch, 1 GFI and 14 single pole
Clean	Ea.	.30	59		59.30	breakers, meter socket including
Minimum Charge	Job		178		178	hookup. Wiring is not included.
200 Amp.						
Demolish	Ea.		146		146	Cost includes material and labor to
Install	Ea.	1000	395		1395	install 200 Amp. residential service,
Demolish and Install	Ea.	1000	541		1541	incl. 20 breaker exterior panel, main
Reinstall	Ea.		316.09		316.09	switch, 1 GFI and 18 single pole
Clean	Ea.	.30	59		59.30	breakers, meter socket including
Minimum Charge	Job		178		178	hookup. Wiring is not included.

Rough Electrical

Main Service		Unit	Material	Labor	Equip.	Total	Specification
Masthead							
	Demolish	Ea.		24.50		24.50	Includes material and labor to install
	Install	Ea.	97	148		245	galvanized steel conduit including
	Demolish and Install	Ea.	97	172.50		269.50	masthead.
	Reinstall	Ea.		118.53		118.53	
	Minimum Charge	Job		178		178	
Allowance to Repair							
	Install	Job		178		178	Minimum labor charge for residential
	Minimum Charge	Job		178		178	wiring.

Insulation

Wall Insulation

Floor Insulation		Unit	Material	Labor	Equip.	Total	Specification
Kraft Faced Batt							
3-1/2″ (R-11)							
	Demolish	S.F.		.11		.11	Cost includes material and labor to
	Install	S.F.	.35	.47		.82	install 3-1/2″ (R-11) kraft paper or foil
	Demolish and Install	S.F.	.35	.58		.93	faced roll and batt fiberglass
	Minimum Charge	Job		163		163	insulation, joists 16″ O.C.
6″ (R-19)							
	Demolish	S.F.		.11		.11	Cost includes material and labor to
	Install	S.F.	.44	.54		.98	install 6″ (R-19) kraft paper or foil
	Demolish and Install	S.F.	.44	.65		1.09	faced roll and batt fiberglass
	Minimum Charge	Job		163		163	insulation, joists 16″ O.C.
10″ (R-30)							
	Demolish	S.F.		.11		.11	Cost includes material and labor to
	Install	S.F.	.75	.65		1.40	install 10″ (R-30) kraft paper-faced roll
	Demolish and Install	S.F.	.75	.76		1.51	and batt fiberglass insulation, joists
	Minimum Charge	Job		163		163	16″ O.C.
12″ (R-38)							
	Demolish	S.F.		.11		.11	Cost includes material and labor to
	Install	S.F.	.80	.69		1.49	install 12″ (R-38) kraft paper-faced roll
	Demolish and Install	S.F.	.80	.80		1.60	and batt fiberglass insulation, joists
	Minimum Charge	Job		163		163	16″ O.C.
Foil Faced Batt							
3-1/2″ (R-11)							
	Demolish	S.F.		.11		.11	Cost includes material and labor to
	Install	S.F.	.35	.47		.82	install 3-1/2″ (R-11) kraft paper or foil
	Demolish and Install	S.F.	.35	.58		.93	faced roll and batt fiberglass
	Minimum Charge	Job		163		163	insulation, joists 16″ O.C.
6″ (R-19)							
	Demolish	S.F.		.11		.11	Cost includes material and labor to
	Install	S.F.	.44	.54		.98	install 6″ (R-19) kraft paper or foil
	Demolish and Install	S.F.	.44	.65		1.09	faced roll and batt fiberglass
	Minimum Charge	Job		163		163	insulation, joists 16″ O.C.
Unfaced Batt							
3-1/2″ (R-11)							
	Demolish	S.F.		.11		.11	Cost includes material and labor to
	Install	S.F.	.26	.54		.80	install 3-1/2″ (R-11) unfaced roll and
	Demolish and Install	S.F.	.26	.65		.91	batt fiberglass insulation, joists 16″
	Minimum Charge	Job		163		163	O.C.

Insulation

Floor Insulation

Floor Insulation	Unit	Material	Labor	Equip.	Total	Specification
6″ (R-19)						
Demolish	S.F.		.11		.11	Cost includes material and labor to
Install	S.F.	.51	.65		1.16	install 6″ (R-19) unfaced roll and batt
Demolish and Install	S.F.	.51	.76		1.27	fiberglass insulation, joists 16″ O.C.
Minimum Charge	Job		163		163	
10″ (R-30)						
Demolish	S.F.		.11		.11	Cost includes material and labor to
Install	S.F.	.75	.72		1.47	install 10″ (R-30) unfaced roll and batt
Demolish and Install	S.F.	.75	.83		1.58	fiberglass insulation, joists 16″ O.C.
Minimum Charge	Job		163		163	
12″ (R-38)						
Demolish	S.F.		.11		.11	Cost includes material and labor to
Install	S.F.	.84	.77		1.61	install 12″ (R-38) unfaced roll and batt
Demolish and Install	S.F.	.84	.88		1.72	fiberglass insulation, joists 16″ O.C.
Minimum Charge	Job		163		163	
Rigid Board						
1/2″						
Demolish	S.F.		.11		.11	Includes material and labor to install
Install	S.F.	.32	.41		.73	rigid foam insulation board foil faced
Demolish and Install	S.F.	.32	.52		.84	two sides.
Minimum Charge	Job		163		163	
3/4″						
Demolish	S.F.		.11		.11	Includes material and labor to install
Install	S.F.	.26	.41		.67	rigid foam insulation board foil faced
Demolish and Install	S.F.	.26	.52		.78	two sides.
Minimum Charge	Job		163		163	
1-1/2″						
Demolish	S.F.		.11		.11	Includes material and labor to install
Install	S.F.	.39	.43		.82	rigid foam insulation board foil faced
Demolish and Install	S.F.	.39	.54		.93	two sides.
Minimum Charge	Job		163		163	
Support Wire						
Demolish	S.F.		.04		.04	Includes material and labor to install
Install	S.F.	.01	.09		.10	friction fit support wire.
Demolish and Install	S.F.	.01	.13		.14	
Minimum Charge	Job		163		163	
Vapor Barrier						
Demolish	S.F.		.06		.06	Cost includes material and labor to
Install	S.F.	.02	.09		.11	install 4 mil polyethylene vapor barrier
Demolish and Install	S.F.	.02	.15		.17	10′ wide sheets with 6″ overlaps.
Minimum Charge	Job		163		163	

Wall Insulation

Wall Insulation	Unit	Material	Labor	Equip.	Total	Specification
Kraft Faced Batt						
3-1/2″ (R-11)						
Demolish	S.F.		.10		.10	Cost includes material and labor to
Install	S.F.	.30	.20		.50	install 3-1/2″ (R-11) kraft paper-faced
Demolish and Install	S.F.	.30	.30		.60	roll and batt fiberglass insulation, joist
Minimum Charge	Job		163		163	or studs 16″ O.C.
6″ (R-19)						
Demolish	S.F.		.10		.10	Cost includes material and labor to
Install	S.F.	.39	.24		.63	install 6″ (R-19) kraft paper-faced roll
Demolish and Install	S.F.	.39	.34		.73	and batt fiberglass insulation, joist or
Minimum Charge	Job		163		163	studs 16″ O.C.

Insulation

Wall Insulation		Unit	Material	Labor	Equip.	Total	Specification
10" (R-30)							
	Demolish	S.F.		.10		.10	Cost includes material and labor to
	Install	S.F.	.66	.28		.94	install 10" (R-30) kraft paper-faced roll
	Demolish and Install	S.F.	.66	.38		1.04	and batt fiberglass insulation, joist or
	Minimum Charge	Job		163		163	studs 16" O.C.
12" (R-38)							
	Demolish	S.F.		.10		.10	Cost includes material and labor to
	Install	S.F.	.84	.33		1.17	install 12" (R-38) kraft paper-faced roll
	Demolish and Install	S.F.	.84	.43		1.27	and batt fiberglass insulation, joist or
	Minimum Charge	Job		163		163	studs 16" O.C.
Foil Faced Batt							
3-1/2" (R-11)							
	Demolish	S.F.		.10		.10	Cost includes material and labor to
	Install	S.F.	.44	.20		.64	install 3-1/2" (R-11) foil faced roll and
	Demolish and Install	S.F.	.44	.30		.74	batt fiberglass insulation, joist or studs
	Minimum Charge	Job		163		163	16" O.C.
6" (R-19)							
	Demolish	S.F.		.10		.10	Cost includes material and labor to
	Install	S.F.	.50	.24		.74	install 6" (R-19) foil faced roll and batt
	Demolish and Install	S.F.	.50	.34		.84	fiberglass insulation, joist or studs 16"
	Minimum Charge	Job		163		163	O.C.
Unfaced Batt							
3-1/2" (R-11)							
	Demolish	S.F.		.10		.10	Cost includes material and labor to
	Install	S.F.	.26	.24		.50	install 3-1/2" (R-11) unfaced roll and
	Demolish and Install	S.F.	.26	.34		.60	batt fiberglass insulation, joist or studs
	Minimum Charge	Job		163		163	16" O.C.
6" (R-19)							
	Demolish	S.F.		.10		.10	Cost includes material and labor to
	Install	S.F.	.51	.28		.79	install 6" (R-19) unfaced roll and batt
	Demolish and Install	S.F.	.51	.38		.89	fiberglass insulation, joist or studs 16"
	Minimum Charge	Job		163		163	O.C.
10" (R-30)							
	Demolish	S.F.		.10		.10	Cost includes material and labor to
	Install	S.F.	.66	.33		.99	install 10" (R-30) unfaced roll and batt
	Demolish and Install	S.F.	.66	.43		1.09	fiberglass insulation, joist or studs 16"
	Minimum Charge	Job		163		163	O.C.
12" (R-38)							
	Demolish	S.F.		.10		.10	Cost includes material and labor to
	Install	S.F.	.84	.33		1.17	install 12" (R-38) unfaced roll and batt
	Demolish and Install	S.F.	.84	.43		1.27	fiberglass insulation, joist or studs 16"
	Minimum Charge	Job		163		163	O.C.
Blown							
R-11							
	Demolish	S.F.		.10		.10	Includes material and labor to install
	Install	S.F.	.24	1.06	.34	1.64	fiberglass blown-in insulation in wall
	Demolish and Install	S.F.	.24	1.16	.34	1.74	with wood siding.
	Minimum Charge	Job		365	119	484	
R-19							
	Demolish	S.F.		.10		.10	Includes material and labor to install
	Install	S.F.	.41	1.22	.40	2.03	fiberglass blown-in insulation in wall
	Demolish and Install	S.F.	.41	1.32	.40	2.13	with wood siding.
	Minimum Charge	Job		365	119	484	

167

Insulation

Wall Insulation

Wall Insulation	Unit	Material	Labor	Equip.	Total	Specification
Rigid Board						
1/2"						
Demolish	S.F.		.13		.13	Includes material and labor to install
Install	S.F.	.32	.41		.73	rigid foam insulation board, foil faced,
Demolish and Install	S.F.	.32	.54		.86	two sides.
Minimum Charge	Job		163		163	
3/4"						
Demolish	S.F.		.13		.13	Includes material and labor to install
Install	S.F.	.26	.41		.67	rigid foam insulation board, foil faced,
Demolish and Install	S.F.	.26	.54		.80	two sides.
Minimum Charge	Job		163		163	
1-1/2"						
Demolish	S.F.		.13		.13	Includes material and labor to install
Install	S.F.	.39	.45		.84	rigid foam insulation board, foil faced,
Demolish and Install	S.F.	.39	.58		.97	two sides.
Minimum Charge	Job		163		163	
Vapor Barrier						
Demolish	S.F.		.04		.04	Cost includes material and labor to
Install	S.F.	.02	.09		.11	install 4 mil polyethylene vapor barrier
Demolish and Install	S.F.	.02	.13		.15	10' wide sheets with 6" overlaps.
Minimum Charge	Job		163		163	

Ceiling Insulation

Ceiling Insulation	Unit	Material	Labor	Equip.	Total	Specification
Kraft Faced Batt						
3-1/2" (R-11)						
Demolish	S.F.		.11		.11	Cost includes material and labor to
Install	S.F.	.30	.20		.50	install 3-1/2" (R-11) kraft paper-faced
Demolish and Install	S.F.	.30	.31		.61	roll and batt fiberglass insulation, joist
Minimum Charge	Job		163		163	or studs 16" O.C.
6" (R-19)						
Demolish	S.F.		.11		.11	Cost includes material and labor to
Install	S.F.	.39	.24		.63	install 6" (R-19) kraft paper-faced roll
Demolish and Install	S.F.	.39	.35		.74	and batt fiberglass insulation, joist or
Minimum Charge	Job		163		163	studs 16" O.C.
10" (R-30)						
Demolish	S.F.		.11		.11	Cost includes material and labor to
Install	S.F.	.66	.28		.94	install 10" (R-30) kraft paper-faced roll
Demolish and Install	S.F.	.66	.39		1.05	and batt fiberglass insulation, joist or
Minimum Charge	Job		163		163	studs 16" O.C.
12" (R-38)						
Demolish	S.F.		.11		.11	Cost includes material and labor to
Install	S.F.	.84	.33		1.17	install 12" (R-38) kraft paper-faced roll
Demolish and Install	S.F.	.84	.44		1.28	and batt fiberglass insulation, joist or
Minimum Charge	Job		163		163	studs 16" O.C.
Foil Faced Batt						
3-1/2" (R-11)						
Demolish	S.F.		.11		.11	Cost includes material and labor to
Install	S.F.	.44	.20		.64	install 3-1/2" (R-11) foil faced roll and
Demolish and Install	S.F.	.44	.31		.75	batt fiberglass insulation, joist or studs
Minimum Charge	Job		163		163	16" O.C.

Insulation

Ceiling Insulation

Ceiling Insulation		Unit	Material	Labor	Equip.	Total	Specification
6" (R-19)							
	Demolish	S.F.		.11		.11	Cost includes material and labor to
	Install	S.F.	.50	.24		.74	install 6" (R-19) foil faced roll and batt
	Demolish and Install	S.F.	.50	.35		.85	fiberglass insulation, joist or studs 16"
	Minimum Charge	Job		163		163	O.C.
Unfaced Batt							
3-1/2" (R-11)							
	Demolish	S.F.		.11		.11	Cost includes material and labor to
	Install	S.F.	.26	.24		.50	install 3-1/2" (R-11) unfaced roll and
	Demolish and Install	S.F.	.26	.35		.61	batt fiberglass insulation, joist or studs
	Minimum Charge	Job		163		163	16" O.C.
6" (R-19)							
	Demolish	S.F.		.11		.11	Cost includes material and labor to
	Install	S.F.	.51	.28		.79	install 6" (R-19) unfaced roll and batt
	Demolish and Install	S.F.	.51	.39		.90	fiberglass insulation, joist or studs 16"
	Minimum Charge	Job		163		163	O.C.
10" (R-30)							
	Demolish	S.F.		.11		.11	Cost includes material and labor to
	Install	S.F.	.66	.33		.99	install 10" (R-30) unfaced roll and batt
	Demolish and Install	S.F.	.66	.44		1.10	fiberglass insulation, joist or studs 16"
	Minimum Charge	Job		163		163	O.C.
12" (R-38)							
	Demolish	S.F.		.11		.11	Cost includes material and labor to
	Install	S.F.	.84	.33		1.17	install 12" (R-38) unfaced roll and batt
	Demolish and Install	S.F.	.84	.44		1.28	fiberglass insulation, joist or studs 16"
	Minimum Charge	Job		163		163	O.C.
Blown							
5"							
	Demolish	S.F.		.33		.33	Includes material and labor to install
	Install	S.F.	.18	.19	.06	.43	fiberglass blown-in insulation.
	Demolish and Install	S.F.	.18	.52	.06	.76	
	Minimum Charge	Job		365	119	484	
6"							
	Demolish	S.F.		.33		.33	Includes material and labor to install
	Install	S.F.	.24	.24	.08	.56	fiberglass blown-in insulation.
	Demolish and Install	S.F.	.24	.57	.08	.89	
	Minimum Charge	Job		365	119	484	
9"							
	Demolish	S.F.		.33		.33	Includes material and labor to install
	Install	S.F.	.34	.33	.11	.78	fiberglass blown-in insulation.
	Demolish and Install	S.F.	.34	.66	.11	1.11	
	Minimum Charge	Job		365	119	484	
12"							
	Demolish	S.F.		.33		.33	Includes material and labor to install
	Install	S.F.	.47	.49	.16	1.12	fiberglass blown-in insulation.
	Demolish and Install	S.F.	.47	.82	.16	1.45	
	Minimum Charge	Job		365	119	484	
Rigid Board							
1/2"							
	Demolish	S.F.		.18		.18	Includes material and labor to install
	Install	S.F.	.32	.41		.73	rigid foam insulation board, foil faced,
	Demolish and Install	S.F.	.32	.59		.91	two sides.
	Minimum Charge	Job		163		163	

Insulation

Ceiling Insulation

	Unit	Material	Labor	Equip.	Total	Specification
3/4"						
Demolish	S.F.		.18		.18	Includes material and labor to install
Install	S.F.	.26	.41		.67	rigid foam insulation board, foil faced,
Demolish and Install	S.F.	.26	.59		.85	two sides.
Minimum Charge	Job		163		163	
1-1/2"						
Demolish	S.F.		.18		.18	Includes material and labor to install
Install	S.F.	.39	.45		.84	rigid foam insulation board, foil faced,
Demolish and Install	S.F.	.39	.63		1.02	two sides.
Minimum Charge	Job		163		163	

Exterior Insulation

	Unit	Material	Labor	Equip.	Total	Specification
Rigid Board						
1/2"						
Demolish	S.F.		.10		.10	Includes material and labor to install
Install	S.F.	.32	.41		.73	rigid foam insulation board foil faced
Demolish and Install	S.F.	.32	.51		.83	two sides.
Minimum Charge	Job		163		163	
3/4"						
Demolish	S.F.		.10		.10	Includes material and labor to install
Install	S.F.	.26	.41		.67	rigid foam insulation board foil faced
Demolish and Install	S.F.	.26	.51		.77	two sides.
Minimum Charge	Job		163		163	
1-1/2"						
Demolish	S.F.		.10		.10	Includes material and labor to install
Install	S.F.	.39	.43		.82	rigid foam insulation board foil faced
Demolish and Install	S.F.	.39	.53		.92	two sides.
Minimum Charge	Job		163		163	
Vapor Barrier						
Demolish	S.F.		.06		.06	Cost includes material and labor to
Install	S.F.	.02	.09		.11	install 4 mil polyethylene vapor barrier
Demolish and Install	S.F.	.02	.15		.17	10' wide sheets with 6" overlaps.
Minimum Charge	Job		163		163	

Roof Insulation Board

	Unit	Material	Labor	Equip.	Total	Specification
Fiberglass						
3/4" Fiberglass						
Demolish	Sq.		34.50		34.50	Includes material and labor to install
Install	Sq.	50.50	24		74.50	fiberglass board insulation 3/4" thick,
Demolish and Install	Sq.	50.50	58.50		109	R-2.80 and C-0.36 values.
Minimum Charge	Job		163		163	
1" Fiberglass						
Demolish	Sq.		34.50		34.50	Includes material and labor to install
Install	Sq.	72	27		99	fiberglass board insulation 1" thick,
Demolish and Install	Sq.	72	61.50		133.50	R-4.20 and C-0.24 values.
Minimum Charge	Job		163		163	
1-3/8" Fiberglass						
Demolish	Sq.		34.50		34.50	Includes material and labor to install
Install	Sq.	116	29.50		145.50	fiberglass board insulation 1-3/8"
Demolish and Install	Sq.	116	64		180	thick, R-5.30 and C-0.19 values.
Minimum Charge	Job		163		163	

Insulation

Roof Insulation Board

Roof Insulation Board		Unit	Material	Labor	Equip.	Total	Specification
1-5/8" Fiberglass							
	Demolish	Sq.		34.50		34.50	Includes material and labor to install
	Install	Sq.	118	31		149	fiberglass board insulation 1-5/8"
	Demolish and Install	Sq.	118	65.50		183.50	thick, R-6.70 and C-0.15 values.
	Minimum Charge	Job		163		163	
2-1/4" Fiberglass							
	Demolish	Sq.		34.50		34.50	Includes material and labor to install
	Install	Sq.	141	34.50		175.50	fiberglass board insulation 2-1/4"
	Demolish and Install	Sq.	141	69		210	thick, R-8.30 and C-0.12 values.
	Minimum Charge	Job		163		163	
Perlite							
1" Perlite							
	Demolish	Sq.		34.50		34.50	Includes material and labor to install
	Install	Sq.	34	27		61	perlite board insulation 1" thick,
	Demolish and Install	Sq.	34	61.50		95.50	R-2.80 and C-0.36 values.
	Minimum Charge	Job		163		163	
4" Perlite							
	Demolish	Sq.		34.50		34.50	Includes material and labor to install
	Install	Sq.	136	52		188	perlite board insulation 4" thick,
	Demolish and Install	Sq.	136	86.50		222.50	R-10.0 and C-0.10 values.
	Minimum Charge	Job		163		163	
1-1/2" Perlite							
	Demolish	Sq.		34.50		34.50	Includes material and labor to install
	Install	Sq.	43	29.50		72.50	perlite board insulation 1-1/2" thick,
	Demolish and Install	Sq.	43	64		107	R-4.20 and C-0.24 values.
	Minimum Charge	Job		163		163	
2" Perlite							
	Demolish	Sq.		34.50		34.50	Includes material and labor to install
	Install	Sq.	68	36		104	perlite board insulation 2" thick,
	Demolish and Install	Sq.	68	70.50		138.50	R-5.30 and C-0.19 values.
	Minimum Charge	Job		163		163	
2-1/2" Perlite							
	Demolish	Sq.		34.50		34.50	Includes material and labor to install
	Install	Sq.	84.50	41		125.50	perlite board insulation 2-1/2" thick,
	Demolish and Install	Sq.	84.50	75.50		160	R-6.70 and C-0.15 values.
	Minimum Charge	Job		163		163	
3" Perlite							
	Demolish	Sq.		34.50		34.50	Includes material and labor to install
	Install	Sq.	102	43.50		145.50	perlite board insulation 3" thick,
	Demolish and Install	Sq.	102	78		180	R-8.30 and C-0.12 values.
	Minimum Charge	Job		163		163	
Polystyrene							
2" Polystyrene							
	Demolish	Sq.		34.50		34.50	Includes material and labor to install
	Install	Sq.	29.50	34.50		64	extruded polystyrene board, R5.0/ 2"
	Demolish and Install	Sq.	29.50	69		98.50	in thickness @25PSI compressive
	Minimum Charge	Job		163		163	strength.
1" Polystyrene							
	Demolish	Sq.		34.50		34.50	Includes material and labor to install
	Install	Sq.	28.50	23.50		52	extruded polystyrene board, R5.0/ 1"
	Demolish and Install	Sq.	28.50	58		86.50	in thickness @25PSI compressive
	Minimum Charge	Job		163		163	strength.
1-1/2" Polystyrene							
	Demolish	Sq.		34.50		34.50	Includes material and labor to install
	Install	Sq.	29	27		56	extruded polystyrene board, R5.0/
	Demolish and Install	Sq.	29	61.50		90.50	1-1/2" in thickness @25PSI
	Minimum Charge	Job		163		163	compressive strength.

171

Insulation

Roof Insulation Board	Unit	Material	Labor	Equip.	Total	Specification
Urethane						
3/4″ Urethane						
Demolish	Sq.		34.50		34.50	Includes material and labor to install
Install	Sq.	35	20.50		55.50	urethane board insulation 3/4″ thick,
Demolish and Install	Sq.	35	55		90	R-5.30 and C-0.19 values.
Minimum Charge	Job		163		163	
1″ Urethane						
Demolish	Sq.		34.50		34.50	Includes material and labor to install
Install	Sq.	38.50	21.50		60	urethane board insulation 1″ thick,
Demolish and Install	Sq.	38.50	56		94.50	R-6.70 and C-0.15 values.
Minimum Charge	Job		163		163	
1-1/2″ Urethane						
Demolish	Sq.		34.50		34.50	Includes material and labor to install
Install	Sq.	40.50	25		65.50	urethane board insulation 1-1/2″
Demolish and Install	Sq.	40.50	59.50		100	thick, R-10.1 and C-0.09 values.
Minimum Charge	Job		163		163	
2″ Urethane						
Demolish	Sq.		34.50		34.50	Includes material and labor to install
Install	Sq.	51	29.50		80.50	urethane board insulation 2″ thick,
Demolish and Install	Sq.	51	64		115	R-13.4 and C-0.06 values.
Minimum Charge	Job		163		163	

Gypsum Wallboard	Tile Ceiling	Gypsum Wallboard Ceiling

Wall Plaster	Unit	Material	Labor	Equip.	Total	Specification
Interior						
Cement						
Demolish	S.Y.		4.36		4.36	Includes material and labor to install
Install	S.Y.	13.75	18.65	1.53	33.93	cement plaster on interior wood
Demolish and Install	S.Y.	13.75	23.01	1.53	38.29	framing with metal lath.
Clean	S.Y.		2.14		2.14	
Paint	S.Y.	1.64	3.81		5.45	
Minimum Charge	Job		146		146	
Acoustic						
Demolish	S.Y.		4.36		4.36	Includes material and labor to install
Install	S.Y.	9.35	22.50	1.53	33.38	acoustical cement plaster on interior
Demolish and Install	S.Y.	9.35	26.86	1.53	37.74	wood framing with metal lath.
Clean	S.Y.		2.14		2.14	
Paint	S.Y.	1.64	3.81		5.45	
Minimum Charge	Job		146		146	
Gypsum						
Demolish	S.Y.		4.36		4.36	Includes material and labor to install
Install	S.Y.	8.05	19.15	1.30	28.50	gypsum wall or ceiling plaster on
Demolish and Install	S.Y.	8.05	23.51	1.30	32.86	interior wood framing with metal lath.
Clean	S.Y.		2.14		2.14	
Paint	S.Y.	1.64	3.81		5.45	
Minimum Charge	Job		146		146	
Stucco						
Demolish	S.Y.		10.45		10.45	Includes material and labor to install
Install	S.Y.	5.90	26.50	1.80	34.20	stucco on wire mesh over wood
Demolish and Install	S.Y.	5.90	36.95	1.80	44.65	framing.
Clean	S.Y.		2.14		2.14	
Paint	S.Y.	1.64	3.81		5.45	
Minimum Charge	Job		146		146	
Thin Coat						
Veneer						
Demolish	S.Y.		4.36		4.36	Includes materials and labor to install
Install	S.Y.	.64	3.45	.28	4.37	thin coat veneer plaster.
Demolish and Install	S.Y.	.64	7.81	.28	8.73	
Clean	S.Y.		2.14		2.14	
Paint	S.Y.	1.64	3.81		5.45	
Minimum Charge	Job		146		146	

Walls / Ceilings

Wall Plaster	Unit	Material	Labor	Equip.	Total	Specification
Lath						
Metal						
Demolish	S.Y.		1.03		1.03	Includes material and labor to install
Install	S.Y.	4	3.81		7.81	wire lath on steel framed walls.
Demolish and Install	S.Y.	4	4.84		8.84	
Clean	S.Y.		2.14		2.14	
Minimum Charge	Job		67		67	
Gypsum						
Demolish	S.Y.		1.03		1.03	Includes material and labor to install
Install	S.Y.	4.46	3.57		8.03	gypsum lath on walls.
Demolish and Install	S.Y.	4.46	4.60		9.06	
Clean	S.Y.		2.14		2.14	
Minimum Charge	Job		146		146	
Repairs						
Fill in Small Cracks						
Install	L.F.	.04	2.34		2.38	Includes labor and material to fill-in
Minimum Charge	Job		146		146	hairline cracks up to 1/8" with filler, sand and prep for paint.
Repair up to 1" Crack						
Install	L.F.	.19	2.92		3.11	Includes labor and material to fill-in
Minimum Charge	Job		146		146	hairline cracks up to 1" with filler, sand and prep for paint.
Patch Small Hole (2")						
Install	Ea.	.04	3.07		3.11	Includes labor and material to patch
Minimum Charge	Job		146		146	up to a 2" diameter hole in plaster.
Patch Large Hole						
Install	Ea.	5.65	48.50		54.15	Includes labor and material to patch
Minimum Charge	Job		146		146	up to a 20" diameter hole in plaster.
Patch Plaster Section						
Install	Ea.	1.94	2.92		4.86	Includes labor and material to repair
Minimum Charge	Job		146		146	section of plaster wall, area less than 20 S.F.
Repair Inside Corner						
Install	L.F.	2.26	7.30		9.56	Includes material and labor to patch
Minimum Charge	Job		146		146	plaster corners.

Ceiling Plaster	Unit	Material	Labor	Equip.	Total	Specification
Interior						
Cement						
Demolish	S.Y.		4.36		4.36	Includes material and labor to install
Install	S.Y.	13.75	18.65	1.53	33.93	cement plaster on interior wood
Demolish and Install	S.Y.	13.75	23.01	1.53	38.29	framing with metal lath.
Clean	S.Y.		2.24		2.24	
Paint	S.Y.	1.64	3.81		5.45	
Minimum Charge	Job		146		146	
Acoustic						
Demolish	S.Y.		4.36		4.36	Includes material and labor to install
Install	S.Y.	9.35	22.50	1.53	33.38	acoustical cement plaster on interior
Demolish and Install	S.Y.	9.35	26.86	1.53	37.74	wood framing with metal lath.
Clean	S.Y.		2.24		2.24	
Paint	S.Y.	1.64	3.81		5.45	
Minimum Charge	Job		146		146	

Walls / Ceilings

Ceiling Plaster

	Unit	Material	Labor	Equip.	Total	Specification
Gypsum						
Demolish	S.Y.		4.36		4.36	Includes material and labor to install
Install	S.Y.	8.05	19.15	1.30	28.50	gypsum wall or ceiling plaster on
Demolish and Install	S.Y.	8.05	23.51	1.30	32.86	interior wood framing with metal lath.
Clean	S.Y.		2.24		2.24	
Paint	S.Y.	1.64	3.81		5.45	
Minimum Charge	Job		146		146	
Stucco						
Demolish	S.Y.		10.45		10.45	Includes material and labor to install
Install	S.Y.	5.90	26.50	1.80	34.20	stucco on wire mesh over wood
Demolish and Install	S.Y.	5.90	36.95	1.80	44.65	framing.
Clean	S.Y.		2.24		2.24	
Paint	S.Y.	1.64	3.81		5.45	
Minimum Charge	Job		146		146	

Thin Coat

	Unit	Material	Labor	Equip.	Total	Specification
Veneer						
Demolish	S.Y.		4.36		4.36	Includes materials and labor to install
Install	S.Y.	.64	3.45	.28	4.37	thin coat veneer plaster.
Demolish and Install	S.Y.	.64	7.81	.28	8.73	
Clean	S.Y.		2.24		2.24	
Paint	S.Y.	1.64	3.81		5.45	
Minimum Charge	Job		146		146	

Lath

	Unit	Material	Labor	Equip.	Total	Specification
Metal						
Demolish	S.Y.		1.03		1.03	Includes material and labor to install
Install	S.Y.	4	4.76		8.76	wire lath on steel framed ceilings.
Demolish and Install	S.Y.	4	5.79		9.79	
Minimum Charge	Job		146		146	
Gypsum						
Demolish	S.Y.		1.03		1.03	Includes material and labor to install
Install	S.Y.	4.46	2.97		7.43	gypsum lath on ceilings.
Demolish and Install	S.Y.	4.46	4		8.46	
Minimum Charge	Job		146		146	

Repairs

	Unit	Material	Labor	Equip.	Total	Specification
Fill in Small Cracks						
Install	S.F.	.04	.53		.57	Includes labor and material to fill in
Minimum Charge	Job		292		292	hairline cracks up to 1/8" with filler,
						sand and prep for paint.

Gypsum Wallboard

	Unit	Material	Labor	Equip.	Total	Specification
Finished						
1/4"						
Demolish	S.F.		.26		.26	Cost includes material and labor to
Install	S.F.	.19	.98		1.17	install 1/4" drywall, including joint
Demolish and Install	S.F.	.19	1.24		1.43	taping, premixed compound and
Clean	S.F.		.21		.21	screws.
Paint	S.F.	.18	.25		.43	
Minimum Charge	Job		163		163	
3/8"						
Demolish	S.F.		.26		.26	Cost includes material and labor to
Install	S.F.	.19	.99		1.18	install 3/8" drywall including joint
Demolish and Install	S.F.	.19	1.25		1.44	tape, premixed compound and
Clean	S.F.		.21		.21	screws.
Paint	S.F.	.18	.25		.43	
Minimum Charge	Job		163		163	

Walls / Ceilings

Gypsum Wallboard		Unit	Material	Labor	Equip.	Total	Specification
1/2"							
	Demolish	S.F.		.26		.26	Cost includes material and labor to
	Install	S.F.	.21	1		1.21	install 1/2" drywall including joint
	Demolish and Install	S.F.	.21	1.26		1.47	tape, premixed compound and
	Clean	S.F.		.21		.21	screws.
	Paint	S.F.	.18	.25		.43	
	Minimum Charge	Job		163		163	
1/2" Water Resistant							
	Demolish	S.F.		.26		.26	Cost includes material and labor to
	Install	S.F.	.17	1		1.17	install 1/2" WR drywall including joint
	Demolish and Install	S.F.	.17	1.26		1.43	tape, premixed compound and
	Clean	S.F.		.21		.21	screws.
	Paint	S.F.	.18	.25		.43	
	Minimum Charge	Job		163		163	
1/2" Fire-rated							
	Demolish	S.F.		.26		.26	Cost includes material and labor to
	Install	S.F.	.21	1		1.21	install 1/2" FR drywall including
	Demolish and Install	S.F.	.21	1.26		1.47	screws. Taping and finishing not
	Clean	S.F.		.21		.21	included.
	Paint	S.F.	.18	.25		.43	
	Minimum Charge	Job		163		163	
5/8"							
	Demolish	S.F.		.26		.26	Cost includes material and labor to
	Install	S.F.	.25	1.14		1.39	install 5/8" drywall including joint
	Demolish and Install	S.F.	.25	1.40		1.65	tape, premixed compound and
	Clean	S.F.		.21		.21	screws.
	Paint	S.F.	.18	.25		.43	
	Minimum Charge	Job		163		163	
5/8" Fire-rated							
	Demolish	S.F.		.26		.26	Cost includes material and labor to
	Install	S.F.	.26	1.14		1.40	install 5/8" FR drywall, including joint
	Demolish and Install	S.F.	.26	1.40		1.66	taping, premixed compound and
	Clean	S.F.		.21		.21	screws.
	Paint	S.F.	.18	.25		.43	
	Minimum Charge	Job		163		163	
2-hour fire-rated							
	Demolish	S.F.		.65		.65	Cost includes material and labor to
	Install	S.F.	.56	2.26		2.82	install 2 layers of 5/8" type X drywall
	Demolish and Install	S.F.	.56	2.91		3.47	including joint taping, premixed
	Clean	S.F.		.21		.21	compound and screws.
	Paint	S.F.	.18	.25		.43	
	Minimum Charge	Job		163		163	
Unfinished							
1/4"							
	Demolish	S.F.		.26		.26	Cost includes material and labor to
	Install	S.F.	.17	.34		.51	install 1/4" drywall including screws.
	Demolish and Install	S.F.	.17	.60		.77	Taping and finishing not included.
	Clean	S.F.		.21		.21	
	Minimum Charge	Job		163		163	
3/8"							
	Demolish	S.F.		.26		.26	Cost includes material and labor to
	Install	S.F.	.17	.34		.51	install 3/8" drywall including screws.
	Demolish and Install	S.F.	.17	.60		.77	Tape, finish or texture is not included.
	Clean	S.F.		.21		.21	
	Minimum Charge	Job		163		163	

Gypsum Wallboard		Unit	Material	Labor	Equip.	Total	Specification
1/2″							
	Demolish	S.F.		.26		.26	Cost includes material and labor to
	Install	S.F.	.18	.34		.52	install 1/2″ drywall including screws.
	Demolish and Install	S.F.	.18	.60		.78	Taping and finishing not included.
	Clean	S.F.		.21		.21	
	Minimum Charge	Job		163		163	
1/2″ Water Resistant							
	Demolish	S.F.		.26		.26	Cost includes material and labor to
	Install	S.F.	.15	.34		.49	install 1/2″ WR drywall including
	Demolish and Install	S.F.	.15	.60		.75	screws. Taping and finishing is not
	Clean	S.F.		.21		.21	included.
	Minimum Charge	Job		163		163	
1/2″ Fire-rated							
	Demolish	S.F.		.26		.26	Cost includes labor and material to
	Install	S.F.	.18	.24		.42	install 1/2″ FR drywall including
	Demolish and Install	S.F.	.18	.50		.68	screws. Tape, finish or texture is not
	Clean	S.F.		.21		.21	included.
	Minimum Charge	Job		163		163	
5/8″							
	Demolish	S.F.		.26		.26	Cost includes material and labor to
	Install	S.F.	.22	.39		.61	install 5/8″ drywall including screws.
	Demolish and Install	S.F.	.22	.65		.87	Taping and finishing not included.
	Clean	S.F.		.21		.21	
	Minimum Charge	Job		163		163	
5/8″ Fire-rated							
	Demolish	S.F.		.26		.26	Cost includes material and labor to
	Install	S.F.	.23	.39		.62	install 5/8″ FR drywall including
	Demolish and Install	S.F.	.23	.65		.88	screws. Taping and finishing not
	Clean	S.F.		.21		.21	included.
	Minimum Charge	Job		163		163	
2-hour Fire-rated							
	Demolish	S.F.		.65		.65	Cost includes material and labor to
	Install	S.F.	.56	2.26		2.82	install 2 layers of 5/8″ type X drywall
	Demolish and Install	S.F.	.56	2.91		3.47	including joint taping, premixed
	Clean	S.F.		.21		.21	compound and screws.
	Minimum Charge	Job		163		163	
Accessories							
Tape & Finish							
	Install	S.F.	.03	.33		.36	Includes labor and material for taping,
	Minimum Charge	Job		163		163	sanding and finish.
Texture							
Sprayed							
	Install	S.F.	.04	.36		.40	Includes labor and material to install
	Minimum Charge	Job		163		163	by spray, texture finish.
Trowel							
	Install	S.F.		.19		.19	Includes labor and material to apply
	Minimum Charge	Job		163		163	hand troweled texture.
Patch Hole in Gypsum Wallboard							
	Install	S.F.	.34	43.50		43.84	Includes labor and material to patch a
	Minimum Charge	Job		163		163	one square foot area of damaged
							drywall.
Furring Strips							
	Demolish	S.F.		.07		.07	Cost includes material and labor to
	Install	S.F.	.15	.72		.87	install 1″ x 2″ furring strips, 16″ O.C.
	Demolish and Install	S.F.	.15	.79		.94	per S.F. of surface area to be covered.
	Minimum Charge	Job		81.50		81.50	

Walls / Ceilings

Ceramic Wall Tile

	Unit	Material	Labor	Equip.	Total	Specification
Thin Set						
2″ x 2″						
Demolish	S.F.		.87		.87	Cost includes material and labor to
Install	S.F.	6.15	4.91		11.06	install 2″ x 2″ ceramic tile in organic
Demolish and Install	S.F.	6.15	5.78		11.93	adhesive including cutting and grout.
Clean	S.F.		.45		.45	
Minimum Charge	Job		147		147	
4-1/4″ x 4-1/4″						
Demolish	S.F.		.87		.87	Cost includes material and labor to
Install	S.F.	2.30	2.76		5.06	install 4-1/4″ x 4-1/4″ ceramic tile in
Demolish and Install	S.F.	2.30	3.63		5.93	organic adhesive including grout.
Clean	S.F.		.45		.45	
Minimum Charge	Job		147		147	
6″ x 6″						
Demolish	S.F.		.87		.87	Cost includes material and labor to
Install	S.F.	2.94	2.62		5.56	install 6″ x 6″ ceramic tile in organic
Demolish and Install	S.F.	2.94	3.49		6.43	adhesive including and grout.
Clean	S.F.		.45		.45	
Minimum Charge	Job		147		147	
8″ x 8″						
Demolish	S.F.		.87		.87	Cost includes material and labor to
Install	S.F.	3.51	2.33		5.84	install 8″ x 8″ ceramic tile in organic
Demolish and Install	S.F.	3.51	3.20		6.71	adhesive including cutting and grout.
Clean	S.F.		.45		.45	
Minimum Charge	Job		147		147	
Thick Set						
2″ x 2″						
Demolish	S.F.		1.05		1.05	Cost includes material and labor to
Install	S.F.	4.92	5.90		10.82	install 2″ x 2″ ceramic tile in mortar
Demolish and Install	S.F.	4.92	6.95		11.87	bed including cutting and grout.
Clean	S.F.		.45		.45	
Minimum Charge	Job		147		147	
4-1/4″ x 4-1/4″						
Demolish	S.F.		1.05		1.05	Cost includes material and labor to
Install	S.F.	6.80	5.90		12.70	install 4-1/4″ x 4-1/4″ ceramic tile in
Demolish and Install	S.F.	6.80	6.95		13.75	mortar bed including cutting and
Clean	S.F.		.45		.45	grout.
Minimum Charge	Job		147		147	
6″ x 6″						
Demolish	S.F.		1.05		1.05	Cost includes material and labor to
Install	S.F.	7.45	5.35		12.80	install 6″ x 6″ ceramic tile in mortar
Demolish and Install	S.F.	7.45	6.40		13.85	bed including cutting and grout.
Clean	S.F.		.45		.45	
Minimum Charge	Job		147		147	
8″ x 8″						
Demolish	S.F.		1.05		1.05	Cost includes material and labor to
Install	S.F.	9.75	4.91		14.66	install 8″ x 8″ ceramic tile in mortar
Demolish and Install	S.F.	9.75	5.96		15.71	bed including cutting and grout.
Clean	S.F.		.45		.45	
Minimum Charge	Job		147		147	
Re-grout						
Install	S.F.	.14	2.95		3.09	Includes material and labor to regrout
Minimum Charge	Job		147		147	ceramic tile walls.

Walls / Ceilings

Wall Tile		Unit	Material	Labor	Equip.	Total	Specification
Mirror							
Custom Cut Wall							
	Demolish	S.F.		.50		.50	Includes material and labor to install
	Install	S.F.	1.75	3.18		4.93	distortion-free float glass mirror, 1/8"
	Demolish and Install	S.F.	1.75	3.68		5.43	to 3/16" thick with cut and polished
	Clean	S.F.	.02	.14		.16	edges.
	Minimum Charge	Job		155		155	
Beveled Wall							
	Demolish	S.F.		.50		.50	Includes material and labor to install
	Install	S.F.	4.50	5.65		10.15	distortion-free float glass mirror, 1/8"
	Demolish and Install	S.F.	4.50	6.15		10.65	to 3/16" thick with beveled edges.
	Clean	S.F.	.02	.14		.16	
	Minimum Charge	Job		155		155	
Smoked Wall							
	Demolish	S.F.		.50		.50	Includes material and labor to install
	Install	S.F.	7.15	5.65		12.80	distortion-free smoked float glass
	Demolish and Install	S.F.	7.15	6.15		13.30	mirror, 1/8" to 3/16" thick with cut
	Clean	S.F.	.02	.14		.16	and polished edges.
	Minimum Charge	Job		155		155	
Marble							
Good Grade							
	Demolish	S.F.		1.64		1.64	Cost includes material and labor to
	Install	S.F.	7.60	12.80		20.40	install 3/8" x 12" x 12" marble tile in
	Demolish and Install	S.F.	7.60	14.44		22.04	mortar bed with grout.
	Clean	S.F.		.45		.45	
	Minimum Charge	Job		147		147	
Better Grade							
	Demolish	S.F.		1.64		1.64	Cost includes material and labor to
	Install	S.F.	9.80	12.80		22.60	install 3/8" x 12" x 12" marble, in
	Demolish and Install	S.F.	9.80	14.44		24.24	mortar bed with grout.
	Clean	S.F.		.45		.45	
	Minimum Charge	Job		147		147	
Premium Grade							
	Demolish	S.F.		1.64		1.64	Cost includes material and labor to
	Install	S.F.	18.90	12.80		31.70	install 3/8" x 12" x 12" marble, in
	Demolish and Install	S.F.	18.90	14.44		33.34	mortar bed with grout.
	Clean	S.F.		.45		.45	
	Minimum Charge	Job		147		147	
Glass Block							
	Demolish	S.F.		1.31		1.31	Cost includes material and labor to
	Install	S.F.	16.80	12.95		29.75	install 3-7/8" thick, 6" x 6" smooth
	Demolish and Install	S.F.	16.80	14.26		31.06	face glass block including mortar and
	Clean	S.F.		.45		.45	cleaning.
	Minimum Charge	Job		290		290	
Plastic							
	Demolish	S.F.		.40		.40	Cost includes material and labor to
	Install	S.F.	.95	1.68		2.63	install 4" plastic tile.
	Demolish and Install	S.F.	.95	2.08		3.03	
	Clean	S.F.		.45		.45	
	Minimum Charge	Job		147		147	
Cork							
	Demolish	S.F.		.20		.20	Cost includes material and labor to
	Install	S.F.	2.81	1.18		3.99	install 3/16" thick cork tile and
	Demolish and Install	S.F.	2.81	1.38		4.19	installation.
	Clean	S.F.		.45		.45	
	Minimum Charge	Job		141		141	

Walls / Ceilings

Wall Tile		Unit	Material	Labor	Equip.	Total	Specification
Re-grout							
	Install	S.F.	.14	2.95		3.09	Includes material and labor to regrout
	Minimum Charge	Job		147		147	ceramic tile walls.

Wall Paneling		Unit	Material	Labor	Equip.	Total	Specification
Plywood							
Good Grade							
	Demolish	S.F.		.26		.26	Cost includes material and labor to
	Install	S.F.	.87	1.30		2.17	install 4' x 8' x 1/4" paneling,
	Demolish and Install	S.F.	.87	1.56		2.43	including nails and adhesive.
	Clean	S.F.		.21		.21	
	Paint	S.F.	.17	.38		.55	
	Minimum Charge	Job		163		163	
Better Grade							
	Demolish	S.F.		.26		.26	Cost includes material and labor to
	Install	S.F.	1.33	1.55		2.88	install 4' x 8' x 1/4" plywood wall
	Demolish and Install	S.F.	1.33	1.81		3.14	panel, including nails and adhesive.
	Clean	S.F.		.21		.21	
	Paint	S.F.	.17	.38		.55	
	Minimum Charge	Job		163		163	
Premium Grade							
	Demolish	S.F.		.26		.26	Cost includes material and labor to
	Install	S.F.	1.94	1.86		3.80	install 4' x 8' x 1/4" plywood wall
	Demolish and Install	S.F.	1.94	2.12		4.06	panel, including nails and adhesive.
	Clean	S.F.		.21		.21	
	Paint	S.F.	.17	.38		.55	
	Minimum Charge	Job		163		163	
Hardboard							
Embossed Paneling							
	Demolish	S.F.		.26		.26	Includes material and labor to install
	Install	S.F.	.61	1.30		1.91	hardboard embossed paneling
	Demolish and Install	S.F.	.61	1.56		2.17	including adhesives.
	Clean	S.F.		.21		.21	
	Minimum Charge	Job		163		163	
Plastic Molding							
	Demolish	L.F.		.26		.26	Includes material and labor to install
	Install	L.F.	.54	.82		1.36	plastic trim for paneling.
	Demolish and Install	L.F.	.54	1.08		1.62	
	Clean	L.F.	.02	.14		.16	
	Minimum Charge	Job		163		163	
Hardboard							
Embossed Brick							
	Demolish	S.F.		.26		.26	Cost includes material and labor to
	Install	S.F.	1.98	1.45		3.43	install 4' x 8' non-ceramic mineral
	Demolish and Install	S.F.	1.98	1.71		3.69	brick-like veneer wall panels including
	Clean	S.F.		.21		.21	adhesive.
	Paint	S.F.	.18	.43		.61	
	Minimum Charge	Job		163		163	
Embossed Ceramic Tile							
	Demolish	S.F.		.26		.26	Includes material and labor to install
	Install	S.F.	.61	1.30		1.91	hardboard embossed paneling
	Demolish and Install	S.F.	.61	1.56		2.17	including adhesives.
	Clean	S.F.		.21		.21	
	Paint	S.F.	.18	.43		.61	
	Minimum Charge	Job		163		163	

Walls / Ceilings

Wall Paneling		Unit	Material	Labor	Equip.	Total	Specification
Pegboard							Cost includes material and labor to
	Demolish	S.F.		.26		.26	install 4' x 8' x 1/4" pegboard panel,
	Install	S.F.	.41	1.30		1.71	including adhesive.
	Demolish and Install	S.F.	.41	1.56		1.97	
	Clean	S.F.		.21		.21	
	Paint	S.F.	.18	.43		.61	
	Minimum Charge	Job		163		163	
Furring Strips							Cost includes material and labor to
	Demolish	S.F.		.07		.07	install 1" x 2" furring strips, 16" O.C.
	Install	S.F.	.15	.72		.87	per S.F. of surface area to be covered.
	Demolish and Install	S.F.	.15	.79		.94	
	Minimum Charge	Job		163		163	
Molding							
Unfinished Base							Cost includes material and labor to
	Demolish	L.F.		.44		.44	install 1/2" x 2-1/2" pine or birch
	Install	L.F.	.49	1		1.49	base molding.
	Demolish and Install	L.F.	.49	1.44		1.93	
	Reinstall	L.F.		.80		.80	
	Clean	L.F.	.02	.14		.16	
	Paint	L.F.	.03	.31		.34	
	Minimum Charge	Job		163		163	
Unfinished Corner							Includes material and labor to install
	Demolish	L.F.		.44		.44	unfinished pine corner molding.
	Install	L.F.	1.13	1.36		2.49	
	Demolish and Install	L.F.	1.13	1.80		2.93	
	Reinstall	L.F.		1.09		1.09	
	Clean	Ea.	.73	21		21.73	
	Paint	L.F.	.12	.46		.58	
	Minimum Charge	Job		163		163	
Unfinished Cove							Cost includes material and labor to
	Demolish	L.F.		.44		.44	install 3/4" x 3-1/2" unfinished pine
	Install	L.F.	1.56	1.21		2.77	cove molding.
	Demolish and Install	L.F.	1.56	1.65		3.21	
	Reinstall	L.F.		.97		.97	
	Clean	Ea.	.73	21		21.73	
	Paint	L.F.	.12	.46		.58	
	Minimum Charge	Job		163		163	
Unfinished Crown							Includes material and labor to install
	Demolish	L.F.		.52		.52	unfinished pine crown molding.
	Install	L.F.	1.84	1.30		3.14	
	Demolish and Install	L.F.	1.84	1.82		3.66	
	Reinstall	L.F.		1.04		1.04	
	Clean	L.F.	.03	.18		.21	
	Paint	L.F.	.04	.31		.35	
	Minimum Charge	Job		163		163	
Prefinished Base							Cost includes material and labor to
	Demolish	L.F.		.44		.44	install 9/16" x 3-5/16" prefinished
	Install	L.F.	1.69	1.04		2.73	base molding.
	Demolish and Install	L.F.	1.69	1.48		3.17	
	Reinstall	L.F.		.84		.84	
	Clean	L.F.	.02	.14		.16	
	Paint	L.F.	.03	.31		.34	
	Minimum Charge	Job		163		163	

Walls / Ceilings

Wall Paneling

Wall Paneling	Unit	Material	Labor	Equip.	Total	Specification
Prefinished Corner						
Demolish	L.F.		.44		.44	Includes material and labor to install
Install	L.F.	1.98	1.36		3.34	inside or outside prefinished corner
Demolish and Install	L.F.	1.98	1.80		3.78	molding.
Reinstall	L.F.		1.09		1.09	
Clean	L.F.		.09		.09	
Paint	L.F.	.12	.46		.58	
Minimum Charge	Job		163		163	
Prefinished Crown						
Demolish	L.F.		.52		.52	Cost includes material and labor to
Install	L.F.	2.24	1.30		3.54	install 3/4" x 3-13/16" prefinished
Demolish and Install	L.F.	2.24	1.82		4.06	crown molding.
Reinstall	L.F.		1.04		1.04	
Clean	L.F.	.03	.18		.21	
Paint	L.F.	.04	.31		.35	
Minimum Charge	Job		163		163	

Wallcovering

Wallcovering	Unit	Material	Labor	Equip.	Total	Specification
Surface Prep						
Patch Walls						
Install	S.F.	.10	2.92		3.02	Includes labor and material to repair
Minimum Charge	Job		141		141	section of plaster wall, area less than
						20 S.F.
Sizing						
Install	S.F.	.07	.14		.21	Includes labor and materials to apply
Minimum Charge	Job		141		141	wall sizing.
Good Grade						
Walls						
Demolish	S.F.		.52		.52	Includes material and labor to install
Install	S.F.	.80	.59		1.39	pre-pasted, vinyl-coated wallcovering.
Demolish and Install	S.F.	.80	1.11		1.91	
Clean	S.F.	.01	.05		.06	
Minimum Charge	Job		141		141	
Ceiling						
Demolish	S.F.		.52		.52	Includes material and labor to install
Install	S.F.	.80	.59		1.39	pre-pasted, vinyl-coated wallcovering.
Demolish and Install	S.F.	.80	1.11		1.91	
Clean	S.F.	.02	.14		.16	
Minimum Charge	Job		141		141	
Better Grade						
Walls						
Demolish	S.F.		.52		.52	Includes material and labor to install
Install	S.F.	.90	.59		1.49	pre-pasted, vinyl-coated wallcovering.
Demolish and Install	S.F.	.90	1.11		2.01	
Clean	S.F.	.01	.05		.06	
Minimum Charge	Job		141		141	
Ceiling						
Demolish	S.F.		.52		.52	Includes material and labor to install
Install	S.F.	.90	.59		1.49	pre-pasted, vinyl-coated wallcovering.
Demolish and Install	S.F.	.90	1.11		2.01	
Clean	S.F.	.02	.14		.16	
Minimum Charge	Job		141		141	

Walls / Ceilings

Wallcovering		Unit	Material	Labor	Equip.	Total	Specification
Premium Grade							
Walls							
	Demolish	S.F.		.65		.65	Includes material and labor to install
	Install	S.F.	1.14	.59		1.73	pre-pasted, vinyl-coated wallcovering.
	Demolish and Install	S.F.	1.14	1.24		2.38	
	Clean	S.F.	.01	.05		.06	
	Minimum Charge	Job		141		141	
Ceiling							
	Demolish	S.F.		.65		.65	Includes material and labor to install
	Install	S.F.	1.14	.59		1.73	pre-pasted, vinyl-coated wallcovering.
	Demolish and Install	S.F.	1.14	1.24		2.38	
	Clean	S.F.	.02	.14		.16	
	Minimum Charge	Job		141		141	
Mural							
Pre-printed							
	Demolish	Ea.		82		82	Includes material and labor to install
	Install	Ea.	213	213		426	wallcovering with a pre-printed mural.
	Demolish and Install	Ea.	213	295		508	
	Clean	Ea.		5.90		5.90	
	Minimum Charge	Job		141		141	
Hand-painted							
	Demolish	Ea.		82		82	Includes material and labor to install
	Install	Ea.	430	430		860	wallcovering with a hand-painted
	Demolish and Install	Ea.	430	512		942	mural.
	Clean	Ea.		5.90		5.90	
	Minimum Charge	Job		141		141	
Vinyl-coated							
Good Grade							
	Demolish	Ea.		14.55		14.55	Includes material and labor to install
	Install	Ea.	17.75	19.80		37.55	pre-pasted, vinyl-coated wallcovering.
	Demolish and Install	Ea.	17.75	34.35		52.10	Based upon 30 S.F. rolls.
	Clean	Ea.		5.90		5.90	
	Minimum Charge	Job		141		141	
Better Grade							
	Demolish	Ea.		14.55		14.55	Includes material and labor to install
	Install	Ea.	23.50	19.80		43.30	pre-pasted, vinyl-coated wallcovering.
	Demolish and Install	Ea.	23.50	34.35		57.85	Based upon 30 S.F. rolls.
	Clean	Ea.		5.90		5.90	
	Minimum Charge	Job		141		141	
Premium Grade							
	Demolish	Ea.		14.55		14.55	Includes material and labor to install
	Install	Ea.	41.50	19.80		61.30	pre-pasted, vinyl-coated wallcovering.
	Demolish and Install	Ea.	41.50	34.35		75.85	Based upon 30 S.F. rolls.
	Clean	Ea.		5.90		5.90	
	Minimum Charge	Job		141		141	
Blank Underliner							
	Demolish	Ea.		14.55		14.55	Includes material and labor to install
	Install	Ea.	10.50	20		30.50	blank stock liner paper or bridging
	Demolish and Install	Ea.	10.50	34.55		45.05	paper. Based upon 30 S.F. rolls.
	Clean	Ea.		5.90		5.90	
	Minimum Charge	Job		141		141	
Solid Vinyl							
Good Grade							
	Demolish	Roll		17.45		17.45	Includes material and labor to install
	Install	Ea.	23.50	19.80		43.30	pre-pasted, vinyl wallcovering. Based
	Demolish and Install	Ea.	23.50	37.25		60.75	upon 30 S.F. rolls.
	Clean	Ea.		5.90		5.90	
	Minimum Charge	Job		141		141	

Walls / Ceilings

Wallcovering		Unit	Material	Labor	Equip.	Total	Specification
Better Grade							
	Demolish	Roll		17.45		17.45	Includes material and labor to install
	Install	Ea.	33	19.80		52.80	pre-pasted, vinyl wallcovering. Based
	Demolish and Install	Ea.	33	37.25		70.25	upon 30 S.F. rolls.
	Clean	Ea.		5.90		5.90	
	Minimum Charge	Job		141		141	
Premium Grade							
	Demolish	Roll		17.45		17.45	Includes material and labor to install
	Install	Ea.	42.50	19.80		62.30	pre-pasted, vinyl wallcovering. Based
	Demolish and Install	Ea.	42.50	37.25		79.75	upon 30 S.F. rolls.
	Clean	Ea.		5.90		5.90	
	Minimum Charge	Job		141		141	
Blank Underliner							
	Demolish	Ea.		14.55		14.55	Includes material and labor to install
	Install	Ea.	10.50	20		30.50	blank stock liner paper or bridging
	Demolish and Install	Ea.	10.50	34.55		45.05	paper. Based upon 30 S.F. rolls.
	Clean	Ea.		5.90		5.90	
	Minimum Charge	Job		141		141	
Grasscloth / String							
Good Grade							
	Demolish	Ea.		14.55		14.55	Includes material and labor to install
	Install	Ea.	33.50	25.50		59	grasscloth or string wallcovering.
	Demolish and Install	Ea.	33.50	40.05		73.55	Based upon 30 S.F. rolls.
	Clean	Ea.		7.85		7.85	
	Minimum Charge	Job		141		141	
Better Grade							
	Demolish	Ea.		14.55		14.55	Includes material and labor to install
	Install	Ea.	43.50	25.50		69	grasscloth or string wallcovering.
	Demolish and Install	Ea.	43.50	40.05		83.55	Based upon 30 S.F. rolls.
	Clean	Ea.		7.85		7.85	
	Minimum Charge	Job		141		141	
Premium Grade							
	Demolish	Ea.		14.55		14.55	Includes material and labor to install
	Install	Ea.	55.50	25.50		81	grasscloth or string wallcovering.
	Demolish and Install	Ea.	55.50	40.05		95.55	Based upon 30 S.F. rolls.
	Clean	Ea.		7.85		7.85	
	Minimum Charge	Job		141		141	
Blank Underliner							
	Demolish	Ea.		14.55		14.55	Includes material and labor to install
	Install	Ea.	10.50	20		30.50	blank stock liner paper or bridging
	Demolish and Install	Ea.	10.50	34.55		45.05	paper. Based upon 30 S.F. rolls.
	Clean	Ea.		5.90		5.90	
	Minimum Charge	Job		141		141	
Designer							
Premium Grade							
	Demolish	S.F.		.65		.65	Includes material and labor to install
	Install	Ea.	103	30		133	paperback vinyl, untrimmed and
	Demolish and Install	Ea.	103	30.65		133.65	hand-painted. Based upon 30 S.F.
	Clean	Ea.		5.90		5.90	rolls.
	Minimum Charge	Job		141		141	
Blank Underliner							
	Demolish	Ea.		14.55		14.55	Includes material and labor to install
	Install	Ea.	10.50	20		30.50	blank stock liner paper or bridging
	Demolish and Install	Ea.	10.50	34.55		45.05	paper. Based upon 30 S.F. rolls.
	Clean	Ea.		5.90		5.90	
	Minimum Charge	Job		141		141	

Walls / Ceilings

Wallcovering

Wallcovering		Unit	Material	Labor	Equip.	Total	Specification
Foil							
	Demolish	Ea.		14.55		14.55	Includes material and labor to install
	Install	Ea.	36	25.50		61.50	foil coated wallcovering. Based upon
	Demolish and Install	Ea.	36	40.05		76.05	30 S.F. rolls.
	Clean	Ea.		5.90		5.90	
	Minimum Charge	Job		141		141	
Wallcovering Border							
Good Grade							
	Demolish	L.F.		.95		.95	Includes material and labor install
	Install	L.F.	1.77	1.09		2.86	vinyl-coated border 4" to 6" wide.
	Demolish and Install	L.F.	1.77	2.04		3.81	
	Clean	L.F.		.09		.09	
	Minimum Charge	Job		141		141	
Better Grade							
	Demolish	L.F.		.95		.95	Includes material and labor install
	Install	L.F.	2.37	1.09		3.46	vinyl-coated border 4" to 6" wide.
	Demolish and Install	L.F.	2.37	2.04		4.41	
	Clean	L.F.		.09		.09	
	Minimum Charge	Job		141		141	
Premium Grade							
	Demolish	L.F.		.95		.95	Includes material and labor install
	Install	L.F.	2.96	1.09		4.05	vinyl-coated border 4" to 6" wide.
	Demolish and Install	L.F.	2.96	2.04		5	
	Clean	L.F.		.09		.09	
	Minimum Charge	Job		141		141	

Acoustical Ceiling

Acoustical Ceiling		Unit	Material	Labor	Equip.	Total	Specification
Complete System							
2' x 2' Square Edge							
	Demolish	S.F.		.44		.44	Includes material and labor to install
	Install	S.F.	1.21	1		2.21	T-bar grid acoustical ceiling system,
	Demolish and Install	S.F.	1.21	1.44		2.65	random pinhole 5/8" thick square
	Reinstall	S.F.		.80		.80	edge tile.
	Clean	S.F.	.03	.18		.21	
	Minimum Charge	Job		163		163	
2' x 2' Reveal Edge							
	Demolish	S.F.		.44		.44	Includes material and labor to install
	Install	S.F.	1.97	1.30		3.27	T-bar grid acoustical ceiling system,
	Demolish and Install	S.F.	1.97	1.74		3.71	random pinhole 5/8" thick reveal
	Reinstall	S.F.		1.04		1.04	edge tile.
	Clean	S.F.	.03	.18		.21	
	Minimum Charge	Job		163		163	
2' x 2' Fire-rated							
	Demolish	S.F.		.44		.44	Includes material and labor to install
	Install	S.F.	2.34	1		3.34	T-bar grid FR acoustical ceiling system,
	Demolish and Install	S.F.	2.34	1.44		3.78	random pinhole 5/8" thick square
	Reinstall	S.F.		.80		.80	edge tile.
	Clean	S.F.	.03	.18		.21	
	Minimum Charge	Job		163		163	
2' x 4' Square Edge							
	Demolish	S.F.		.44		.44	Cost includes material and labor to
	Install	S.F.	1.01	.86		1.87	install 2' x 4' T-bar grid acoustical
	Demolish and Install	S.F.	1.01	1.30		2.31	ceiling system, random pinhole 5/8"
	Reinstall	S.F.		.69		.69	thick square edge tile.
	Clean	S.F.	.03	.18		.21	
	Minimum Charge	Job		163		163	

Acoustical Ceiling		Unit	Material	Labor	Equip.	Total	Specification
2' x 4' Reveal Edge							
	Demolish	S.F.		.44		.44	Includes material and labor to install
	Install	S.F.	1.97	1.19		3.16	T-bar grid acoustical ceiling system,
	Demolish and Install	S.F.	1.97	1.63		3.60	random pinhole 5/8" thick reveal
	Reinstall	S.F.		.95		.95	edge tile.
	Clean	S.F.	.03	.18		.21	
	Minimum Charge	Job		163		163	
2' x 4' Fire-rated							
	Demolish	S.F.		.44		.44	Includes material and labor to install
	Install	S.F.	2.34	1		3.34	T-bar grid FR acoustical ceiling system,
	Demolish and Install	S.F.	2.34	1.44		3.78	random pinhole 5/8" thick square
	Reinstall	S.F.		.80		.80	edge tile.
	Clean	S.F.	.03	.18		.21	
	Minimum Charge	Job		163		163	
T-bar Grid							
2' x 2'							
	Demolish	S.F.		.51		.51	Includes material and labor to install
	Install	S.F.	.67	.50		1.17	T-bar grid acoustical ceiling
	Demolish and Install	S.F.	.67	1.01		1.68	suspension system.
	Reinstall	S.F.		.40		.40	
	Clean	S.F.	.03	.18		.21	
	Minimum Charge	Job		146		146	
2' x 4'							
	Demolish	S.F.		.51		.51	Includes material and labor to install
	Install	S.F.	.54	.41		.95	T-bar grid acoustical ceiling
	Demolish and Install	S.F.	.54	.92		1.46	suspension system.
	Reinstall	S.F.		.33		.33	
	Clean	S.F.	.03	.18		.21	
	Minimum Charge	Job		146		146	
Tile Panels							
2' x 2' Square Edge							
	Demolish	S.F.		.14		.14	Includes material and labor to install
	Install	S.F.	.47	.52		.99	square edge panels 5/8" thick with
	Demolish and Install	S.F.	.47	.66		1.13	pinhole pattern. T-bar grid not
	Reinstall	S.F.		.42		.42	included.
	Clean	S.F.	.03	.18		.21	
	Minimum Charge	Job		163		163	
2' x 2' Reveal Edge							
	Demolish	S.F.		.14		.14	Includes material and labor to install
	Install	S.F.	1.14	.69		1.83	reveal edge panels 5/8" thick with
	Demolish and Install	S.F.	1.14	.83		1.97	pinhole pattern. T-bar grid not
	Reinstall	S.F.		.55		.55	included.
	Clean	S.F.	.03	.18		.21	
	Minimum Charge	Job		163		163	
2' x 2' Fire-rated							
	Demolish	S.F.		.14		.14	Includes material and labor to install
	Install	S.F.	.92	.48		1.40	FR square edge panels 5/8" thick with
	Demolish and Install	S.F.	.92	.62		1.54	pinhole pattern. T-bar grid not
	Reinstall	S.F.		.39		.39	included.
	Clean	S.F.	.03	.18		.21	
	Minimum Charge	Job		163		163	
2' x 4' Square Edge							
	Demolish	S.F.		.14		.14	Includes material and labor to install
	Install	S.F.	.47	.52		.99	square edge panels 5/8" thick with
	Demolish and Install	S.F.	.47	.66		1.13	pinhole pattern. T-bar grid not
	Reinstall	S.F.		.42		.42	included.
	Clean	S.F.	.03	.18		.21	
	Minimum Charge	Job		163		163	

Walls / Ceilings

Acoustical Ceiling

	Unit	Material	Labor	Equip.	Total	Specification
2' x 4' Reveal Edge						
Demolish	S.F.		.14		.14	Includes material and labor to install
Install	S.F.	1.14	.69		1.83	reveal edge panels 5/8" thick with
Demolish and Install	S.F.	1.14	.83		1.97	pinhole pattern. T-bar grid not
Reinstall	S.F.		.55		.55	included.
Clean	S.F.	.03	.18		.21	
Minimum Charge	Job		163		163	
2' x 4' Fire-rated						
Demolish	S.F.		.14		.14	Includes material and labor to install
Install	S.F.	.92	.48		1.40	FR square edge panels 5/8" thick with
Demolish and Install	S.F.	.92	.62		1.54	pinhole pattern. T-bar grid not
Reinstall	S.F.		.39		.39	included.
Clean	S.F.	.03	.18		.21	
Minimum Charge	Job		163		163	
Bleach Out Stains						
Install	S.F.	.09	.24		.33	Includes labor and material to apply
Minimum Charge	Job		163		163	laundry type chlorine bleach to
						remove light staining of acoustic tile
						material.
Adhesive Tile						
Good Grade						
Demolish	S.F.		.58		.58	Cost includes material and labor to
Install	S.F.	.85	1.09		1.94	install 12" x 12" tile with embossed
Demolish and Install	S.F.	.85	1.67		2.52	texture pattern. Furring strips not
Clean	S.F.	.02	.14		.16	included.
Minimum Charge	Job		163		163	
Better Grade						
Demolish	S.F.		.58		.58	Cost includes material and labor to
Install	S.F.	1.13	1.09		2.22	install 12" x 12" tile with embossed
Demolish and Install	S.F.	1.13	1.67		2.80	texture pattern. Furring strips not
Clean	S.F.	.02	.14		.16	included.
Minimum Charge	Job		163		163	
Premium Grade						
Demolish	S.F.		.58		.58	Cost includes material and labor to
Install	S.F.	1.75	1.09		2.84	install 12" x 12" tile with embossed
Demolish and Install	S.F.	1.75	1.67		3.42	texture pattern. Furring strips not
Clean	S.F.	.02	.14		.16	included.
Minimum Charge	Job		163		163	
Bleach Out Stains						
Install	S.F.	.09	.24		.33	Includes labor and material to apply
Minimum Charge	Job		163		163	laundry type chlorine bleach to
						remove light staining of acoustic tile
						material.
Stapled Tile						
Good Grade						
Demolish	S.F.		.35		.35	Cost includes material and labor to
Install	S.F.	.85	1.09		1.94	install 12" x 12" tile with embossed
Demolish and Install	S.F.	.85	1.44		2.29	texture pattern. Furring strips not
Clean	S.F.	.02	.14		.16	included.
Minimum Charge	Job		163		163	
Better Grade						
Demolish	S.F.		.35		.35	Cost includes material and labor to
Install	S.F.	1.13	1.09		2.22	install 12" x 12" tile with embossed
Demolish and Install	S.F.	1.13	1.44		2.57	texture pattern. Furring strips not
Clean	S.F.	.02	.14		.16	included.
Minimum Charge	Job		163		163	

Walls / Ceilings

Acoustical Ceiling		Unit	Material	Labor	Equip.	Total	Specification
Premium Grade							Cost includes material and labor to
	Demolish	S.F.		.35		.35	install 12" x 12" tile with embossed
	Install	S.F.	1.75	1.09		2.84	texture pattern. Furring strips not
	Demolish and Install	S.F.	1.75	1.44		3.19	included.
	Clean	S.F.	.02	.14		.16	
	Minimum Charge	Job		163		163	
Bleach Out Stains							Includes labor and material to apply
	Install	S.F.	.09	.24		.33	laundry type chlorine bleach to
	Minimum Charge	Job		163		163	remove light staining of acoustic tile
							material.
Blown							
Scrape Off Existing							Includes labor and material to remove
	Install	S.F.		.48		.48	existing blown acoustical ceiling
	Minimum Charge	Job		163		163	material.
Seal Ceiling Gypsum Wallboard							Includes labor and material to seal
	Install	S.F.	.06	.25		.31	existing drywall with shellac-based
	Minimum Charge	Job		143		143	material.
Reblow Existing							Includes labor and material to install
	Install	S.F.	.04	.36		.40	by spray, texture finish.
	Minimum Charge	Job		390	59.50	449.50	
Bleach Out Stains							Includes labor and material to apply
	Install	S.F.	.09	.24		.33	laundry type chlorine bleach to
	Minimum Charge	Job		163		163	remove light staining of acoustic tile
							material.
New Blown							Includes labor and material to install
	Install	S.F.	.04	.36		.40	by spray, texture finish.
	Minimum Charge	Job		143		143	
Apply Glitter To							Includes labor and material to apply
	Install	S.F.	.09	.13		.22	glitter material to existing or new
	Minimum Charge	Job		143		143	drywall.
Furring Strips							Cost includes material and labor to
	Demolish	S.F.		.07		.07	install 1" x 2" furring strips, 16" O.C.
	Install	S.F.	.15	.72		.87	per S.F. of surface area to be covered.
	Demolish and Install	S.F.	.15	.79		.94	
	Minimum Charge	Job		81.50		81.50	

Finished Stair

Bi-Fold Door

Raised Panel Door

Hollow Core Door		Unit	Material	Labor	Equip.	Total	Specification
Pre-hung Wood							
1' 6" x 6' 8"							
	Demolish	Ea.		30		30	Cost includes material and labor to
	Install	Ea.	146	15.50		161.50	install 1' 6" x 6' 8" pre-hung 1-3/8"
	Demolish and Install	Ea.	146	45.50		191.50	lauan door with split pine jamb
	Reinstall	Ea.		15.52		15.52	including casing, stops, and hinges.
	Clean	Ea.	2.92	9.80		12.72	
	Paint	Ea.	7.75	41		48.75	
	Minimum Charge	Job		163		163	
2' x 6' 8"							
	Demolish	Ea.		30		30	Cost includes material and labor to
	Install	Ea.	146	16.30		162.30	install 2' x 6' 8" pre-hung 1-3/8"
	Demolish and Install	Ea.	146	46.30		192.30	lauan door with split pine jamb
	Reinstall	Ea.		16.30		16.30	including casing, stops, and hinges.
	Clean	Ea.	2.92	9.80		12.72	
	Paint	Ea.	7.75	41		48.75	
	Minimum Charge	Job		163		163	
2' 4" x 6' 8"							
	Demolish	Ea.		30		30	Cost includes material and labor to
	Install	Ea.	157	16.30		173.30	install 2' 4" x 6' 8" pre-hung 1-3/8"
	Demolish and Install	Ea.	157	46.30		203.30	lauan door with split pine jamb
	Reinstall	Ea.		16.30		16.30	including casing, stops, and hinges.
	Clean	Ea.	2.92	9.80		12.72	
	Paint	Ea.	7.75	41		48.75	
	Minimum Charge	Job		163		163	
2' 6" x 6' 8"							
	Demolish	Ea.		30		30	Cost includes material and labor to
	Install	Ea.	157	16.30		173.30	install 2' 6" x 6' 8" pre-hung 1-3/8"
	Demolish and Install	Ea.	157	46.30		203.30	lauan door with split pine jamb
	Reinstall	Ea.		16.30		16.30	including casing, stops, and hinges.
	Clean	Ea.	2.92	9.80		12.72	
	Paint	Ea.	7.75	41		48.75	
	Minimum Charge	Job		163		163	
2' 8" x 6' 8"							
	Demolish	Ea.		30		30	Cost includes material and labor to
	Install	Ea.	157	16.30		173.30	install 2' 8" x 6' 8" pre-hung 1-3/8"
	Demolish and Install	Ea.	157	46.30		203.30	lauan door with split pine jamb,
	Reinstall	Ea.		16.30		16.30	including casing, stop and hinges.
	Clean	Ea.	2.92	9.80		12.72	
	Paint	Ea.	7.75	41		48.75	
	Minimum Charge	Job		163		163	

Finish Carpentry

Hollow Core Door		Unit	Material	Labor	Equip.	Total	Specification
3' x 6' 8"							
	Demolish	Ea.		30		30	Cost includes material and labor to
	Install	Ea.	163	17.15		180.15	install 3' x 6' 8" pre-hung 1-3/8"
	Demolish and Install	Ea.	163	47.15		210.15	lauan door with split pine jamb
	Reinstall	Ea.		17.16		17.16	including casing, stops, and hinges.
	Clean	Ea.	2.92	9.80		12.72	
	Paint	Ea.	7.75	41		48.75	
	Minimum Charge	Job		163		163	
Pocket Type							
	Demolish	Ea.		45.50		45.50	Cost includes material and labor to
	Install	Ea.	33	36		69	install 2' 8" x 6' 8" lauan door.
	Demolish and Install	Ea.	33	81.50		114.50	
	Reinstall	Ea.		36.22		36.22	
	Clean	Ea.	2.92	9.80		12.72	
	Paint	Ea.	7.75	41		48.75	
	Minimum Charge	Job		163		163	
Stain							
	Install	Ea.	4.05	24		28.05	Includes labor and material to stain
	Minimum Charge	Job		143		143	single door and trim on both sides.
Pre-hung Masonite							
1' 6" x 6' 8"							
	Demolish	Ea.		30		30	Cost includes material and labor to
	Install	Ea.	186	15.50		201.50	install 1' 6" x 6' 8" pre-hung 1-3/8"
	Demolish and Install	Ea.	186	45.50		231.50	masonite door with split pine jamb
	Reinstall	Ea.		15.52		15.52	including casing, stops, and hinges.
	Clean	Ea.	2.92	9.80		12.72	
	Paint	Ea.	7.75	41		48.75	
	Minimum Charge	Job		163		163	
2' x 6' 8"							
	Demolish	Ea.		30		30	Cost includes material and labor to
	Install	Ea.	194	16.30		210.30	install 2' x 6' 8" pre-hung 1-3/8"
	Demolish and Install	Ea.	194	46.30		240.30	masonite door with split pine jamb
	Reinstall	Ea.		16.30		16.30	including casing, stops, and hinges.
	Clean	Ea.	2.92	9.80		12.72	
	Paint	Ea.	7.75	41		48.75	
	Minimum Charge	Job		163		163	
2' 4" x 6' 8"							
	Demolish	Ea.		30		30	Cost includes material and labor to
	Install	Ea.	196	16.30		212.30	install 2' 4" x 6' 8" pre-hung 1-3/8"
	Demolish and Install	Ea.	196	46.30		242.30	masonite door with split pine jamb
	Reinstall	Ea.		16.30		16.30	including casing, stops, and hinges.
	Clean	Ea.	2.92	9.80		12.72	
	Paint	Ea.	7.75	41		48.75	
	Minimum Charge	Job		163		163	
2' 8" x 6' 8"							
	Demolish	Ea.		30		30	Cost includes material and labor to
	Install	Ea.	198	16.30		214.30	install 2' 8" x 6' 8" pre-hung 1-3/8"
	Demolish and Install	Ea.	198	46.30		244.30	masonite door with split pine jamb
	Reinstall	Ea.		16.30		16.30	including casing, stops, and hinges.
	Clean	Ea.	2.92	9.80		12.72	
	Paint	Ea.	7.75	41		48.75	
	Minimum Charge	Job		163		163	
3' x 6' 8"							
	Demolish	Ea.		30		30	Cost includes material and labor to
	Install	Ea.	202	17.15		219.15	install 3' x 6' 8" pre-hung 1-3/8"
	Demolish and Install	Ea.	202	47.15		249.15	masonite door with split pine jamb
	Reinstall	Ea.		17.16		17.16	including casing, stops, and hinges.
	Clean	Ea.	2.92	9.80		12.72	
	Paint	Ea.	7.75	41		48.75	
	Minimum Charge	Job		163		163	

Finish Carpentry

Hollow Core Door		Unit	Material	Labor	Equip.	Total	Specification
Pocket Type							
	Demolish	Ea.		45.50		45.50	Cost includes material and labor to
	Install	Ea.	33	36		69	install 2' 8" x 6' 8" lauan door.
	Demolish and Install	Ea.	33	81.50		114.50	
	Reinstall	Ea.		36.22		36.22	
	Clean	Ea.	2.92	9.80		12.72	
	Paint	Ea.	7.75	41		48.75	
	Minimum Charge	Job		163		163	
Stain							
	Install	Ea.	4.05	24		28.05	Includes labor and material to stain
	Minimum Charge	Job		143		143	single door and trim on both sides.
Hollow Core Door Only							
1' 6" x 6' 8'							
	Demolish	Ea.		6.45		6.45	Cost includes material and labor to
	Install	Ea.	36	17.15		53.15	install 1' 6" x 6' 8" lauan hollow core
	Demolish and Install	Ea.	36	23.60		59.60	door slab.
	Reinstall	Ea.		17.16		17.16	
	Clean	Ea.	1.21	7.35		8.56	
	Paint	Ea.	2.41	36		38.41	
	Minimum Charge	Job		163		163	
2' x 6' 8'							
	Demolish	Ea.		6.45		6.45	Cost includes material and labor to
	Install	Ea.	34.50	18.10		52.60	install 2' x 6' 8" lauan hollow core
	Demolish and Install	Ea.	34.50	24.55		59.05	door slab.
	Reinstall	Ea.		18.11		18.11	
	Clean	Ea.	1.21	7.35		8.56	
	Paint	Ea.	2.41	36		38.41	
	Minimum Charge	Job		163		163	
2' 4" x 6' 8'							
	Demolish	Ea.		6.45		6.45	Cost includes material and labor to
	Install	Ea.	38	18.10		56.10	install 2' 4" x 6' 8" lauan hollow core
	Demolish and Install	Ea.	38	24.55		62.55	door slab.
	Reinstall	Ea.		18.11		18.11	
	Clean	Ea.	1.21	7.35		8.56	
	Paint	Ea.	2.41	36		38.41	
	Minimum Charge	Job		163		163	
2' 8" x 6' 8'							
	Demolish	Ea.		6.45		6.45	Cost includes material and labor to
	Install	Ea.	39.50	18.10		57.60	install 2' 8" x 6' 8" lauan hollow core
	Demolish and Install	Ea.	39.50	24.55		64.05	door slab.
	Reinstall	Ea.		18.11		18.11	
	Clean	Ea.	1.21	7.35		8.56	
	Paint	Ea.	2.41	36		38.41	
	Minimum Charge	Job		163		163	
3' x 6' 8'							
	Demolish	Ea.		6.45		6.45	Cost includes material and labor to
	Install	Ea.	41.50	19.20		60.70	install 3' x 6' 8" lauan hollow core
	Demolish and Install	Ea.	41.50	25.65		67.15	door slab.
	Reinstall	Ea.		19.18		19.18	
	Clean	Ea.	1.21	7.35		8.56	
	Paint	Ea.	2.41	36		38.41	
	Minimum Charge	Job		163		163	
Pocket Type							
	Demolish	Ea.		6.45		6.45	Cost includes material and labor to
	Install	Ea.	33	36		69	install 2' 8" x 6' 8" lauan door.
	Demolish and Install	Ea.	33	42.45		75.45	
	Reinstall	Ea.		36.22		36.22	
	Clean	Ea.	1.21	7.35		8.56	
	Paint	Ea.	2.41	36		38.41	
	Minimum Charge	Job		163		163	

Finish Carpentry

Hollow Core Door		Unit	Material	Labor	Equip.	Total	Specification
Stain							
	Install	Ea.	2.23	32		34.23	Includes labor and materials to stain a
	Minimum Charge	Job		143		143	flush style door on both sides by brush.
Casing Trim							
Single Width							
	Demolish	Ea.		5.75		5.75	Cost includes material and labor to
	Install	Opng.	15.90	55.50		71.40	install 11/16" x 2-1/2" pine ranch
	Demolish and Install	Opng.	15.90	61.25		77.15	style casing for one side of a standard
	Reinstall	Opng.		44.20		44.20	door opening.
	Clean	Opng.	.42	2.36		2.78	
	Paint	Ea.	3.15	5.70		8.85	
	Minimum Charge	Job		163		163	
Double Width							
	Demolish	Ea.		6.85		6.85	Cost includes material and labor to
	Install	Opng.	18.70	65		83.70	install 11/16" x 2-1/2" pine ranch
	Demolish and Install	Opng.	18.70	71.85		90.55	style door casing for one side of a
	Reinstall	Opng.		52.16		52.16	double door opening.
	Clean	Opng.	.48	4.71		5.19	
	Paint	Ea.	3.15	6.50		9.65	
	Minimum Charge	Job		163		163	
Hardware							
Doorknob w / Lock							
	Demolish	Ea.		12.05		12.05	Includes material and labor to install
	Install	Ea.	21	20.50		41.50	privacy lockset.
	Demolish and Install	Ea.	21	32.55		53.55	
	Reinstall	Ea.		16.30		16.30	
	Clean	Ea.	.30	7.35		7.65	
	Minimum Charge	Job		163		163	
Doorknob							
	Demolish	Ea.		12.05		12.05	Includes material and labor to install
	Install	Ea.	18.20	20.50		38.70	residential passage lockset, keyless
	Demolish and Install	Ea.	18.20	32.55		50.75	bored type.
	Reinstall	Ea.		16.30		16.30	
	Clean	Ea.	.30	7.35		7.65	
	Minimum Charge	Job		163		163	
Deadbolt							
	Demolish	Ea.		11.30		11.30	Includes material and labor to install a
	Install	Ea.	38.50	23.50		62	deadbolt.
	Demolish and Install	Ea.	38.50	34.80		73.30	
	Reinstall	Ea.		18.63		18.63	
	Clean	Ea.	.30	7.35		7.65	
	Minimum Charge	Job		163		163	
Lever Handle							
	Demolish	Ea.		12.05		12.05	Includes material and labor to install
	Install	Ea.	27	32.50		59.50	residential passage lockset, keyless
	Demolish and Install	Ea.	27	44.55		71.55	bored type with lever handle.
	Reinstall	Ea.		26.08		26.08	
	Clean	Ea.	.30	7.35		7.65	
	Minimum Charge	Job		163		163	
Closer							
	Demolish	Ea.		6.80		6.80	Includes material and labor to install
	Install	Ea.	95.50	50		145.50	pneumatic light duty closer for interior
	Demolish and Install	Ea.	95.50	56.80		152.30	type doors.
	Reinstall	Ea.		40.12		40.12	
	Clean	Ea.	.14	9.80		9.94	
	Minimum Charge	Job		163		163	

Finish Carpentry

Hollow Core Door

	Unit	Material	Labor	Equip.	Total	Specification
Push Plate						
Demolish	Ea.		10.05		10.05	Includes material and labor to install
Install	Ea.	18.10	27		45.10	bronze push-pull plate.
Demolish and Install	Ea.	18.10	37.05		55.15	
Reinstall	Ea.		21.73		21.73	
Clean	Ea.	.30	7.35		7.65	
Minimum Charge	Job		163		163	
Kickplate						
Demolish	Ea.		10.05		10.05	Includes material and labor to install
Install	Ea.	20.50	21.50		42	aluminum kickplate, 10" x 28".
Demolish and Install	Ea.	20.50	31.55		52.05	
Reinstall	Ea.		17.39		17.39	
Clean	Ea.	.30	7.35		7.65	
Minimum Charge	Job		163		163	
Exit Sign						
Demolish	Ea.		17.95		17.95	Includes material and labor to install a
Install	Ea.	48.50	44.50		93	wall mounted interior electric exit sign.
Demolish and Install	Ea.	48.50	62.45		110.95	
Reinstall	Ea.		35.56		35.56	
Clean	Ea.	.02	5.30		5.32	
Minimum Charge	Job		163		163	
Jamb						
Demolish	Ea.		22.50		22.50	Includes material and labor to install
Install	Ea.	80	8.80		88.80	flat pine jamb with square cut heads
Demolish and Install	Ea.	80	31.30		111.30	and rabbeted sides for 6' 8" high and
Clean	Ea.	.09	5.90		5.99	3-9/16" door including trim sets for
Paint	Ea.	5.70	19.05		24.75	both sides.
Minimum Charge	Job		163		163	
Shave & Refit						
Install	Ea.		27		27	Includes labor to shave and rework
Minimum Charge	Job		163		163	door to fit opening at the job site.

Solid Core Door

	Unit	Material	Labor	Equip.	Total	Specification
Pre-hung Wood						
2' 4" x 6' 8"						
Demolish	Ea.		30		30	Cost includes material and labor to
Install	Ea.	186	16.30		202.30	install 2' 4" x 6' 8" pre-hung 1-3/4"
Demolish and Install	Ea.	186	46.30		232.30	door, 4-1/2" split jamb, including
Reinstall	Ea.		16.30		16.30	casing and stop, hinges, aluminum sill,
Clean	Ea.	2.92	9.80		12.72	weatherstripping.
Paint	Ea.	9.15	41		50.15	
Minimum Charge	Job		163		163	
2' 6" x 6' 8"						
Demolish	Ea.		30		30	Cost includes material and labor to
Install	Ea.	219	16.30		235.30	install 2' 6" x 6' 8" pre-hung 1-3/4"
Demolish and Install	Ea.	219	46.30		265.30	door, 4-1/2" split jamb, including
Reinstall	Ea.		16.30		16.30	casing and stop, hinges, aluminum sill,
Clean	Ea.	2.92	9.80		12.72	weatherstripping.
Paint	Ea.	9.15	41		50.15	
Minimum Charge	Job		163		163	

Finish Carpentry

Solid Core Door

Solid Core Door	Unit	Material	Labor	Equip.	Total	Specification
2' 8" x 6' 8"						
Demolish	Ea.		30		30	Cost includes material and labor to
Install	Ea.	315	16.30		331.30	install 2' 8" x 6' 8" pre-hung 1-3/4"
Demolish and Install	Ea.	315	46.30		361.30	door, 4-1/2" split jamb, including
Reinstall	Ea.		16.30		16.30	casing and stop, hinges, aluminum sill,
Clean	Ea.	2.92	9.80		12.72	weatherstripping.
Paint	Ea.	9.15	41		50.15	
Minimum Charge	Job		163		163	
3' x 6' 8"						
Demolish	Ea.		30		30	Cost includes material and labor to
Install	Ea.	320	17.15		337.15	install 3' x 6' 8" pre-hung 1-3/4"
Demolish and Install	Ea.	320	47.15		367.15	door, including casing and stop,
Reinstall	Ea.		17.16		17.16	hinges, jamb, aluminum sill,
Clean	Ea.	2.92	9.80		12.72	weatherstripped.
Paint	Ea.	9.15	41		50.15	
Minimum Charge	Job		163		163	
3' 6" x 6' 8"						
Demolish	Ea.		30		30	Cost includes material and labor to
Install	Ea.	440	18.10		458.10	install 3' 6" x 6' 8" pre-hung 1-3/4"
Demolish and Install	Ea.	440	48.10		488.10	door, 4-1/2" split jamb, including
Reinstall	Ea.		18.11		18.11	casing and stop, hinges, aluminum sill,
Clean	Ea.	2.92	9.80		12.72	weatherstripping.
Paint	Ea.	9.15	41		50.15	
Minimum Charge	Job		163		163	
Raised Panel						
Demolish	Ea.		30		30	Cost includes material and labor to
Install	Ea.	375	18.10		393.10	install 3' x 6' 8" pre-hung 1-3/4"
Demolish and Install	Ea.	375	48.10		423.10	door, jamb, including casing and
Reinstall	Ea.		18.11		18.11	stop, hinges, aluminum sill,
Clean	Ea.	2.92	9.80		12.72	weatherstripping.
Paint	Ea.	9.15	95.50		104.65	
Minimum Charge	Job		163		163	
Stain						
Install	Ea.	4.05	24		28.05	Includes labor and material to stain
Minimum Charge	Job		143		143	single door and trim on both sides.

Bifold Door

Bifold Door	Unit	Material	Labor	Equip.	Total	Specification
Wood, Single						
3'						
Demolish	Ea.		22.50		22.50	Cost includes material and labor to
Install	Ea.	54.50	50		104.50	install 3' wide solid wood bi-fold door
Demolish and Install	Ea.	54.50	72.50		127	not including jamb, trim, track, and
Reinstall	Ea.		50.15		50.15	hardware.
Clean	Ea.	2.92	9.80		12.72	
Paint	Ea.	7.75	41		48.75	
Minimum Charge	Job		163		163	
5'						
Demolish	Ea.		22.50		22.50	Includes material and labor to install
Install	Ea.	100	59.50		159.50	pair of 2' 6" wide solid wood bi-fold
Demolish and Install	Ea.	100	82		182	doors not including jamb, trim, track,
Reinstall	Ea.		59.27		59.27	and hardware.
Clean	Ea.	1.28	16.85		18.13	
Paint	Ea.	16.70	71.50		88.20	
Minimum Charge	Job		163		163	

Finish Carpentry

Bifold Door

Bifold Door	Unit	Material	Labor	Equip.	Total	Specification
6'						
Demolish	Ea.		22.50		22.50	Includes material and labor to install
Install	Ea.	108	65		173	pair of 3' wide solid wood bi-fold
Demolish and Install	Ea.	108	87.50		195.50	doors not including jamb, trim, track,
Reinstall	Ea.		65.20		65.20	and hardware.
Clean	Ea.	1.46	19.65		21.11	
Paint	Ea.	16.70	71.50		88.20	
Minimum Charge	Job		163		163	
8'						
Demolish	Ea.		22.50		22.50	Includes material and labor to install
Install	Ea.	190	36		226	pair of 4' wide solid wood bi-fold
Demolish and Install	Ea.	190	58.50		248.50	doors not including jamb, trim, track,
Reinstall	Ea.		36.22		36.22	and hardware.
Clean	Ea.	2	23.50		25.50	
Paint	Ea.	8.70	71.50		80.20	
Minimum Charge	Job		163		163	

Mirrored

	Unit	Material	Labor	Equip.	Total	Specification
5'						
Demolish	Ea.		22.50		22.50	Includes material and labor to install
Install	Ea.	375	29.50		404.50	pair of 2' 6" wide mirrored bi-fold
Demolish and Install	Ea.	375	52		427	doors including jamb, trim, track, and
Reinstall	Ea.		29.64		29.64	hardware.
Clean	Ea.	1.28	8.40		9.68	
Minimum Charge	Job		163		163	
6'						
Demolish	Ea.		22.50		22.50	Includes material and labor to install
Install	Ea.	415	36		451	pair of 3' wide mirrored bi-fold doors
Demolish and Install	Ea.	415	58.50		473.50	including jamb, trim, track, and
Reinstall	Ea.		36.22		36.22	hardware.
Clean	Ea.	1.46	9.80		11.26	
Minimum Charge	Job		163		163	
8'						
Demolish	Ea.		22.50		22.50	Includes material and labor to install
Install	Ea.	665	54.50		719.50	pair of 4' wide mirrored bi-fold doors
Demolish and Install	Ea.	665	77		742	including jamb, trim, track, and
Reinstall	Ea.		54.33		54.33	hardware.
Clean	Ea.	2	11.80		13.80	
Minimum Charge	Job		163		163	

Louvered

	Unit	Material	Labor	Equip.	Total	Specification
3'						
Demolish	Ea.		22.50		22.50	Cost includes material and labor to
Install	Ea.	162	50		212	install 3' wide louvered bi-fold door
Demolish and Install	Ea.	162	72.50		234.50	not including jamb, trim, track, and
Reinstall	Ea.		50.15		50.15	hardware.
Clean	Ea.	2.92	9.80		12.72	
Paint	Ea.	7.75	41		48.75	
Minimum Charge	Job		163		163	
5'						
Demolish	Ea.		22.50		22.50	Includes material and labor to install
Install	Ea.	204	59.50		263.50	pair of 2' 6" wide louvered wood
Demolish and Install	Ea.	204	82		286	bi-fold doors not including jamb, trim,
Reinstall	Ea.		59.27		59.27	track, and hardware.
Clean	Ea.	1.28	16.85		18.13	
Paint	Ea.	16.70	71.50		88.20	
Minimum Charge	Job		163		163	

Finish Carpentry

Bifold Door	Unit	Material	Labor	Equip.	Total	Specification
6'						
Demolish	Ea.		22.50		22.50	Includes material and labor to install
Install	Ea.	224	65		289	pair of 3' wide louvered wood bi-fold
Demolish and Install	Ea.	224	87.50		311.50	doors not including jamb, trim, track,
Reinstall	Ea.		65.20		65.20	and hardware.
Clean	Ea.	1.46	19.65		21.11	
Paint	Ea.	16.70	71.50		88.20	
Minimum Charge	Job		163		163	
8'						
Demolish	Ea.		22.50		22.50	Includes material and labor to install
Install	Ea.	395	32.50		427.50	pair of 4' wide solid wood bi-fold
Demolish and Install	Ea.	395	55		450	doors not including jamb, trim, track,
Reinstall	Ea.		32.60		32.60	and hardware.
Clean	Ea.	6.95	16.85		23.80	
Paint	Ea.	8.70	71.50		80.20	
Minimum Charge	Job		163		163	
Accordion Type						
Demolish	S.F.		.77		.77	Includes material and labor to install
Install	S.F.	1.82	1.63		3.45	vinyl folding accordion closet doors,
Demolish and Install	S.F.	1.82	2.40		4.22	including track and frame.
Reinstall	S.F.		1.63		1.63	
Clean	S.F.		.28		.28	
Minimum Charge	Job		163		163	
Casing Trim						
Single Width						
Demolish	Ea.		5.75		5.75	Cost includes material and labor to
Install	Opng.	15.90	55.50		71.40	install 11/16" x 2-1/2" pine ranch
Demolish and Install	Opng.	15.90	61.25		77.15	style casing for one side of a standard
Reinstall	Opng.		44.20		44.20	door opening.
Clean	Opng.	.42	2.36		2.78	
Paint	Ea.	3.15	5.70		8.85	
Minimum Charge	Job		163		163	
Double Width						
Demolish	Ea.		6.85		6.85	Cost includes material and labor to
Install	Opng.	18.70	65		83.70	install 11/16" x 2-1/2" pine ranch
Demolish and Install	Opng.	18.70	71.85		90.55	style door casing for one side of a
Reinstall	Opng.		52.16		52.16	double door opening.
Clean	Opng.	.48	4.71		5.19	
Paint	Ea.	3.15	6.50		9.65	
Minimum Charge	Job		163		163	
Jamb						
Demolish	Ea.		22.50		22.50	Includes material and labor to install
Install	Ea.	80	8.80		88.80	flat pine jamb with square cut heads
Demolish and Install	Ea.	80	31.30		111.30	and rabbeted sides for 6' 8" high and
Clean	Ea.	.09	5.90		5.99	3-9/16" door including trim sets for
Paint	Ea.	5.70	19.05		24.75	both sides.
Minimum Charge	Job		81.50		81.50	
Stain						
Install	Ea.	4.05	24		28.05	Includes labor and material to stain
Minimum Charge	Job		143		143	single door and trim on both sides.
Shave & Refit						
Install	Ea.		27		27	Includes labor to shave and rework
Minimum Charge	Job		81.50		81.50	door to fit opening at the job site.

Finish Carpentry

Louver Door

	Unit	Material	Labor	Equip.	Total	Specification
Full						
2' x 6' 8"						
Demolish	Ea.		6.45		6.45	Cost includes material and labor to
Install	Ea.	168	32.50		200.50	install 2' x 6' 8" pine full louver door,
Demolish and Install	Ea.	168	38.95		206.95	jamb, hinges.
Reinstall	Ea.		32.60		32.60	
Clean	Ea.	5.40	29.50		34.90	
Paint	Ea.	7.75	41		48.75	
Minimum Charge	Job		163		163	
2' 8" x 6' 8"						
Demolish	Ea.		6.45		6.45	Cost includes material and labor to
Install	Ea.	187	32.50		219.50	install 2' 8" x 6' 8" pine full louver
Demolish and Install	Ea.	187	38.95		225.95	door, jamb, hinges.
Reinstall	Ea.		32.60		32.60	
Clean	Ea.	5.40	29.50		34.90	
Paint	Ea.	7.75	41		48.75	
Minimum Charge	Job		163		163	
3' x 6' 8"						
Demolish	Ea.		6.45		6.45	Cost includes material and labor to
Install	Ea.	197	34.50		231.50	install 3' x 6' 8" pine full louver door,
Demolish and Install	Ea.	197	40.95		237.95	jamb, hinges.
Reinstall	Ea.		34.32		34.32	
Clean	Ea.	5.40	29.50		34.90	
Paint	Ea.	7.75	41		48.75	
Minimum Charge	Job		163		163	
Half						
2' x 6' 8"						
Demolish	Ea.		6.45		6.45	Cost includes material and labor to
Install	Ea.	282	32.50		314.50	install 2' x 6' 8" pine half louver door,
Demolish and Install	Ea.	282	38.95		320.95	jamb, hinges.
Reinstall	Ea.		32.60		32.60	
Clean	Ea.	5.40	29.50		34.90	
Paint	Ea.	7.75	41		48.75	
Minimum Charge	Job		163		163	
2' 8" x 6' 8"						
Demolish	Ea.		6.45		6.45	Cost includes material and labor to
Install	Ea.	310	32.50		342.50	install 2' 8" x 6' 8" pine half louver
Demolish and Install	Ea.	310	38.95		348.95	door, jamb, hinges.
Reinstall	Ea.		32.60		32.60	
Clean	Ea.	5.40	29.50		34.90	
Paint	Ea.	7.75	41		48.75	
Minimum Charge	Job		163		163	
3' x 6' 8"						
Demolish	Ea.		6.45		6.45	Cost includes material and labor to
Install	Ea.	325	34.50		359.50	install 3' x 6' 8" pine half louver door,
Demolish and Install	Ea.	325	40.95		365.95	jamb, hinges.
Reinstall	Ea.		34.32		34.32	
Clean	Ea.	5.40	29.50		34.90	
Paint	Ea.	7.75	41		48.75	
Minimum Charge	Job		163		163	
Jamb						
Demolish	Ea.		22.50		22.50	Includes material and labor to install
Install	Ea.	80	8.80		88.80	flat pine jamb with square cut heads
Demolish and Install	Ea.	80	31.30		111.30	and rabbeted sides for 6' 8" high and
Clean	Ea.	.09	5.90		5.99	3-9/16" door including trim sets for
Paint	Ea.	5.70	19.05		24.75	both sides.
Minimum Charge	Job		81.50		81.50	

Finish Carpentry

Louver Door

Louver Door		Unit	Material	Labor	Equip.	Total	Specification
Stain							
	Install	Ea.	4.05	24		28.05	Includes labor and material to stain
	Minimum Charge	Job		143		143	single door and trim on both sides.
Shave & Refit							
	Install	Ea.		27		27	Includes labor to shave and rework
	Minimum Charge	Job		143		143	door to fit opening at the job site.

Bypass Sliding Door

Bypass Sliding Door		Unit	Material	Labor	Equip.	Total	Specification
Wood							
5'							
	Demolish	Ea.		30		30	Includes material and labor to install
	Install	Opng.	188	59.50		247.50	pair of 2' 6" wide lauan hollow core
	Demolish and Install	Opng.	188	89.50		277.50	wood bi-pass doors with jamb, track
	Reinstall	Opng.		59.27		59.27	hardware.
	Clean	Ea.	1.28	16.85		18.13	
	Paint	Ea.	16.70	71.50		88.20	
	Minimum Charge	Job		163		163	
6'							
	Demolish	Ea.		30		30	Includes material and labor to install
	Install	Opng.	201	65		266	pair of 3' wide lauan hollow core
	Demolish and Install	Opng.	201	95		296	wood bi-pass doors with jamb, track
	Reinstall	Opng.		65.20		65.20	and hardware.
	Clean	Ea.	1.46	19.65		21.11	
	Paint	Ea.	16.70	71.50		88.20	
	Minimum Charge	Job		163		163	
8'							
	Demolish	Ea.		30		30	Includes material and labor to install
	Install	Opng.	335	81.50		416.50	pair of 4' wide lauan hollow core
	Demolish and Install	Opng.	335	111.50		446.50	wood bi-pass doors with jamb, track
	Reinstall	Opng.		81.50		81.50	and hardware.
	Clean	Ea.	2	23.50		25.50	
	Paint	Ea.	8.70	71.50		80.20	
	Minimum Charge	Job		163		163	
Mirrored							
5'							
	Demolish	Ea.		30		30	Includes material and labor to install
	Install	Opng.	223	65		288	pair of 2' 6 " wide mirrored bi-pass
	Demolish and Install	Opng.	223	95		318	doors with jamb, trim, track hardware.
	Reinstall	Opng.		65.20		65.20	
	Clean	Ea.	1.28	8.40		9.68	
	Minimum Charge	Job		163		163	
6'							
	Demolish	Ea.		30		30	Includes material and labor to install
	Install	Opng.	254	65		319	pair of 3' wide mirrored bi-pass doors
	Demolish and Install	Opng.	254	95		349	with jamb, trim, track hardware.
	Reinstall	Opng.		65.20		65.20	
	Clean	Ea.	1.46	9.80		11.26	
	Minimum Charge	Job		163		163	
8'							
	Demolish	Ea.		30		30	Includes material and labor to install
	Install	Opng.	530	72.50		602.50	pair of 4' wide mirrored bi-pass doors
	Demolish and Install	Opng.	530	102.50		632.50	with jamb, trim, track hardware.
	Reinstall	Opng.		72.44		72.44	
	Clean	Ea.	2	11.80		13.80	
	Minimum Charge	Job		163		163	

Finish Carpentry

Bypass Sliding Door

Bypass Sliding Door	Unit	Material	Labor	Equip.	Total	Specification
Casing Trim						
Single Width						
Demolish	Ea.		5.75		5.75	Cost includes material and labor to
Install	Opng.	15.90	55.50		71.40	install 11/16" x 2-1/2" pine ranch
Demolish and Install	Opng.	15.90	61.25		77.15	style casing for one side of a standard
Reinstall	Opng.		44.20		44.20	door opening.
Clean	Opng.	.42	2.36		2.78	
Paint	Ea.	3.15	5.70		8.85	
Minimum Charge	Job		163		163	
Double Width						
Demolish	Ea.		6.85		6.85	Cost includes material and labor to
Install	Opng.	18.70	65		83.70	install 11/16" x 2-1/2" pine ranch
Demolish and Install	Opng.	18.70	71.85		90.55	style door casing for one side of a
Reinstall	Opng.		52.16		52.16	double door opening.
Clean	Opng.	.48	4.71		5.19	
Paint	Ea.	3.15	6.50		9.65	
Minimum Charge	Job		163		163	
Jamb						
Demolish	Ea.		22.50		22.50	Includes material and labor to install
Install	Ea.	80	8.80		88.80	flat pine jamb with square cut heads
Demolish and Install	Ea.	80	31.30		111.30	and rabbeted sides for 6' 8" high and
Clean	Ea.	.09	5.90		5.99	3-9/16" door including trim sets for
Paint	Ea.	5.70	19.05		24.75	both sides.
Minimum Charge	Job		81.50		81.50	
Stain						
Install	Ea.	4.05	24		28.05	Includes labor and material to stain
Minimum Charge	Job		143		143	single door and trim on both sides.
Shave & Refit						
Install	Ea.		27		27	Includes labor to shave and rework
Minimum Charge	Job		81.50		81.50	door to fit opening at the job site.

Base Molding

Base Molding	Unit	Material	Labor	Equip.	Total	Specification
One Piece						
1-5/8"						
Demolish	L.F.		.44		.44	Includes material and labor to install
Install	L.F.	.27	.96		1.23	pine or birch base molding.
Demolish and Install	L.F.	.27	1.40		1.67	
Reinstall	L.F.		.77		.77	
Clean	L.F.	.02	.14		.16	
Paint	L.F.	.03	.31		.34	
Minimum Charge	Job		163		163	
2-1/2"						
Demolish	L.F.		.44		.44	Cost includes material and labor to
Install	L.F.	.49	1		1.49	install 1/2" x 2-1/2" pine or birch
Demolish and Install	L.F.	.49	1.44		1.93	base molding.
Reinstall	L.F.		.80		.80	
Clean	L.F.	.02	.14		.16	
Paint	L.F.	.03	.31		.34	
Minimum Charge	Job		163		163	

Base Molding

	Unit	Material	Labor	Equip.	Total	Specification
3-1/2"						
Demolish	L.F.		.44		.44	Cost includes material and labor to
Install	L.F.	1.18	1.04		2.22	install 9/16" x 3-1/2" pine or birch
Demolish and Install	L.F.	1.18	1.48		2.66	base molding.
Reinstall	L.F.		.84		.84	
Clean	L.F.	.02	.14		.16	
Paint	L.F.	.03	.34		.37	
Minimum Charge	Job		163		163	
6"						
Demolish	L.F.		.44		.44	Cost includes material and labor to
Install	L.F.	.77	1.14		1.91	install 1" x 6" pine or birch base
Demolish and Install	L.F.	.77	1.58		2.35	molding.
Reinstall	L.F.		.91		.91	
Clean	L.F.	.02	.20		.22	
Paint	L.F.	.06	.43		.49	
Minimum Charge	Job		163		163	
8"						
Demolish	L.F.		.52		.52	Cost includes material and labor to
Install	L.F.	1.20	1.25		2.45	install 1" x 8" pine or birch base
Demolish and Install	L.F.	1.20	1.77		2.97	molding.
Reinstall	L.F.		1		1	
Clean	L.F.	.03	.24		.27	
Paint	L.F.	.06	.48		.54	
Minimum Charge	Job		163		163	
Premium / Custom Grade						
Demolish	L.F.		.52		.52	Includes material and labor to install
Install	L.F.	3.83	3.13		6.96	custom 2 or 3 piece oak or other
Demolish and Install	L.F.	3.83	3.65		7.48	hardwood, stain grade trim.
Reinstall	L.F.		2.51		2.51	
Clean	L.F.	.03	.24		.27	
Paint	L.F.	.06	.48		.54	
Minimum Charge	Job		163		163	
One Piece Oak						
1-5/8"						
Demolish	L.F.		.44		.44	Cost includes material and labor to
Install	L.F.	.70	.96		1.66	install 1/2" x 1-5/8" oak base
Demolish and Install	L.F.	.70	1.40		2.10	molding.
Reinstall	L.F.		.77		.77	
Clean	L.F.	.02	.14		.16	
Paint	L.F.	.08	.42		.50	
Minimum Charge	Job		163		163	
2-1/2"						
Demolish	L.F.		.44		.44	Cost includes material and labor to
Install	L.F.	1.07	.98		2.05	install 1/2" x 2-1/2" oak base
Demolish and Install	L.F.	1.07	1.42		2.49	molding.
Reinstall	L.F.		.79		.79	
Clean	L.F.	.02	.14		.16	
Paint	L.F.	.08	.43		.51	
Minimum Charge	Job		163		163	
3-1/2"						
Demolish	L.F.		.44		.44	Cost includes material and labor to
Install	L.F.	2.40	1		3.40	install 1/2" x 3-1/2" oak base
Demolish and Install	L.F.	2.40	1.44		3.84	molding.
Reinstall	L.F.		.80		.80	
Clean	L.F.	.02	.14		.16	
Paint	L.F.	.08	.44		.52	
Minimum Charge	Job		163		163	

Finish Carpentry

Base Molding

Base Molding	Unit	Material	Labor	Equip.	Total	Specification
6"						
Demolish	L.F.		.44		.44	Cost includes material and labor to
Install	L.F.	5.55	1.14		6.69	install 1" x 6" oak base molding.
Demolish and Install	L.F.	5.55	1.58		7.13	
Reinstall	L.F.		.91		.91	
Clean	L.F.	.03	.24		.27	
Paint	L.F.	.08	.46		.54	
Minimum Charge	Job		163		163	
8"						
Demolish	L.F.		.52		.52	Cost includes material and labor to
Install	L.F.	6.45	1.25		7.70	install 1" x 8" oak base molding.
Demolish and Install	L.F.	6.45	1.77		8.22	
Reinstall	L.F.		1		1	
Clean	L.F.	.03	.24		.27	
Paint	L.F.	.09	.48		.57	
Minimum Charge	Job		163		163	

Vinyl Base Molding

Vinyl Base Molding	Unit	Material	Labor	Equip.	Total	Specification
2-1/2"						
Demolish	L.F.		.44		.44	Cost includes material and labor to
Install	L.F.	.51	.94		1.45	install 2-1/2" vinyl or rubber cove
Demolish and Install	L.F.	.51	1.38		1.89	base including adhesive.
Reinstall	L.F.		.94		.94	
Clean	L.F.		.21		.21	
Minimum Charge	Job		148		148	
4"						
Demolish	L.F.		.44		.44	Cost includes material and labor to
Install	L.F.	.56	.94		1.50	install 4" vinyl or rubber cove base
Demolish and Install	L.F.	.56	1.38		1.94	including adhesive.
Reinstall	L.F.		.94		.94	
Clean	L.F.		.21		.21	
Minimum Charge	Job		148		148	
6"						
Demolish	L.F.		.44		.44	Cost includes material and labor to
Install	L.F.	.92	.94		1.86	install 6" vinyl or rubber cove base
Demolish and Install	L.F.	.92	1.38		2.30	including adhesive.
Reinstall	L.F.		.94		.94	
Clean	L.F.		.21		.21	
Minimum Charge	Job		148		148	

Prefinished

Prefinished	Unit	Material	Labor	Equip.	Total	Specification
Demolish	L.F.		.44		.44	Cost includes material and labor to
Install	L.F.	1.69	1.04		2.73	install 9/16" x 3-5/16" prefinished
Demolish and Install	L.F.	1.69	1.48		3.17	base molding.
Reinstall	L.F.		.84		.84	
Clean	L.F.	.02	.14		.16	
Paint	L.F.	.03	.34		.37	
Minimum Charge	Job		163		163	

Shoe

Shoe	Unit	Material	Labor	Equip.	Total	Specification
Demolish	L.F.		.44		.44	Includes material and labor to install
Install	L.F.	.27	.96		1.23	pine or birch base molding.
Demolish and Install	L.F.	.27	1.40		1.67	
Reinstall	L.F.		.77		.77	
Clean	L.F.	.02	.14		.16	
Paint	L.F.	.12	.46		.58	
Minimum Charge	Job		163		163	

Finish Carpentry

Base Molding

Base Molding	Unit	Material	Labor	Equip.	Total	Specification
Carpeted						
Demolish	L.F.		.17		.17	Includes material and labor to install
Install	L.F.	1.25	1.79		3.04	carpet baseboard with adhesive.
Demolish and Install	L.F.	1.25	1.96		3.21	
Reinstall	L.F.		1.43		1.43	
Clean	L.F.		.09		.09	
Minimum Charge	Job		147		147	
Ceramic Tile						
Demolish	L.F.		.95		.95	Includes material and labor to install
Install	L.F.	3.45	4.09		7.54	thin set ceramic tile cove base
Demolish and Install	L.F.	3.45	5.04		8.49	including grout.
Clean	L.F.		.45		.45	
Minimum Charge	Job		147		147	
Quarry Tile						
Demolish	L.F.		.95		.95	Includes material and labor to install
Install	L.F.	4.27	4.76		9.03	quarry tile base including grout.
Demolish and Install	L.F.	4.27	5.71		9.98	
Clean	L.F.		.45		.45	
Minimum Charge	Job		147		147	
Caulk and Renail						
Install	L.F.	.01	.51		.52	Includes labor and material to renail
Minimum Charge	Job		163		163	base and caulk where necessary.

Molding

Molding	Unit	Material	Labor	Equip.	Total	Specification
Oak Base						
Demolish	L.F.		.44		.44	Cost includes material and labor to
Install	L.F.	1.07	.98		2.05	install 1/2" x 2-1/2" oak base
Demolish and Install	L.F.	1.07	1.42		2.49	molding.
Reinstall	L.F.		.79		.79	
Clean	L.F.	.02	.14		.16	
Paint	L.F.	.08	.43		.51	
Minimum Charge	Job		163		163	
Shoe						
Demolish	L.F.		.44		.44	Includes material and labor to install
Install	L.F.	.27	.96		1.23	pine or birch base molding.
Demolish and Install	L.F.	.27	1.40		1.67	
Reinstall	L.F.		.77		.77	
Clean	L.F.	.02	.14		.16	
Paint	L.F.	.12	.46		.58	
Minimum Charge	Job		163		163	
Quarter Round						
Demolish	L.F.		.44		.44	Includes material and labor to install
Install	L.F.	.40	1.28		1.68	pine base quarter round molding.
Demolish and Install	L.F.	.40	1.72		2.12	
Reinstall	L.F.		1.02		1.02	
Clean	L.F.	.02	.14		.16	
Paint	L.F.	.06	.72		.78	
Minimum Charge	Job		163		163	
Chair Rail						
Demolish	L.F.		.44		.44	Cost includes material and labor to
Install	L.F.	1.23	1.21		2.44	install 5/8" x 2-1/2" oak chair rail
Demolish and Install	L.F.	1.23	1.65		2.88	molding.
Reinstall	L.F.		.97		.97	
Clean	L.F.	.02	.14		.16	
Paint	L.F.	.06	.72		.78	
Minimum Charge	Job		163		163	

Finish Carpentry

Molding		Unit	Material	Labor	Equip.	Total	Specification
Crown							
	Demolish	L.F.		.44		.44	Includes material and labor to install
	Install	L.F.	1.84	1.30		3.14	unfinished pine crown molding.
	Demolish and Install	L.F.	1.84	1.74		3.58	
	Reinstall	L.F.		1.04		1.04	
	Clean	L.F.	.03	.18		.21	
	Paint	L.F.	.06	.72		.78	
	Minimum Charge	Job		163		163	
Cove							
	Demolish	L.F.		.44		.44	Cost includes material and labor to
	Install	L.F.	1.38	1.28		2.66	install 1/2" x 2-3/4" pine cove
	Demolish and Install	L.F.	1.38	1.72		3.10	molding.
	Reinstall	L.F.		1.02		1.02	
	Clean	L.F.	.02	.14		.16	
	Paint	L.F.	.06	.72		.78	
	Minimum Charge	Job		163		163	
Corner							
	Demolish	L.F.		.44		.44	Includes material and labor to install
	Install	L.F.	.56	.51		1.07	inside or outside pine corner molding.
	Demolish and Install	L.F.	.56	.95		1.51	
	Reinstall	L.F.		.41		.41	
	Clean	L.F.	.02	.14		.16	
	Paint	L.F.	.06	.72		.78	
	Minimum Charge	Job		163		163	
Picture							
	Demolish	L.F.		.44		.44	Cost includes material and labor to
	Install	L.F.	.94	1.36		2.30	install 9/16" x 2-1/2" pine casing.
	Demolish and Install	L.F.	.94	1.80		2.74	
	Reinstall	L.F.		1.09		1.09	
	Clean	L.F.	.02	.14		.16	
	Paint	L.F.	.06	.72		.78	
	Minimum Charge	Job		163		163	
Single Cased Opening							
	Demolish	Ea.		22.50		22.50	Includes material and labor to install
	Install	Ea.	13.20	20.50		33.70	flat pine jamb with square cut heads
	Demolish and Install	Ea.	13.20	43		56.20	and rabbeted sides for 6' 8" high
	Clean	Ea.	.20	4.91		5.11	opening including trim sets for both
	Paint	Ea.	1.02	7.15		8.17	sides.
	Minimum Charge	Job		163		163	
Double Cased Opening							
	Demolish	Ea.		22.50		22.50	Includes material and labor to install
	Install	Ea.	26.50	23.50		50	flat pine jamb with square cut heads
	Demolish and Install	Ea.	26.50	46		72.50	and rabbeted sides for 6' 8" high
	Clean	Ea.	.23	5.75		5.98	opening including trim sets for both
	Paint	Ea.	1.18	8.40		9.58	sides.
	Minimum Charge	Job		163		163	

Window Trim Set		Unit	Material	Labor	Equip.	Total	Specification
Single							
	Demolish	Ea.		5.80		5.80	Cost includes material and labor to
	Install	Opng.	26	32.50		58.50	install 11/16" x 2-1/2" pine ranch
	Demolish and Install	Opng.	26	38.30		64.30	style window casing.
	Reinstall	Opng.		26.08		26.08	
	Clean	Ea.	.30	3.37		3.67	
	Paint	Ea.	.96	6.80		7.76	
	Minimum Charge	Job		163		163	

Finish Carpentry

Window Trim Set

	Unit	Material	Labor	Equip.	Total	Specification
Double						
Demolish	Ea.		6.55		6.55	Cost includes material and labor to
Install	Ea.	35	41		76	install 11/16" x 2-1/2" pine ranch
Demolish and Install	Ea.	35	47.55		82.55	style trim for casing one side of a
Reinstall	Ea.		32.60		32.60	double window opening.
Clean	Ea.	.52	3.93		4.45	
Paint	Ea.	1.61	11.45		13.06	
Minimum Charge	Job		163		163	
Triple						
Demolish	Ea.		7.50		7.50	Cost includes material and labor to
Install	Opng.	64	54.50		118.50	install 11/16" x 2-1/2" pine ranch
Demolish and Install	Opng.	64	62		126	style trim for casing one side of a
Reinstall	Opng.		43.47		43.47	triple window opening.
Clean	Ea.	.73	4.71		5.44	
Paint	Ea.	2.24	15.90		18.14	
Minimum Charge	Job		163		163	
Window Casing Per L.F.						
Demolish	L.F.		.44		.44	Cost includes material and labor to
Install	L.F.	.94	1.36		2.30	install 11/16" x 2-1/2" pine ranch
Demolish and Install	L.F.	.94	1.80		2.74	style trim for casing.
Reinstall	L.F.		1.09		1.09	
Clean	L.F.	.02	.14		.16	
Paint	L.F.	.06	.72		.78	
Minimum Charge	Job		163		163	

Window Sill

	Unit	Material	Labor	Equip.	Total	Specification
Wood						
Demolish	L.F.		.44		.44	Includes material and labor to install
Install	L.F.	1.68	1.63		3.31	flat, wood window stool.
Demolish and Install	L.F.	1.68	2.07		3.75	
Reinstall	L.F.		1.30		1.30	
Clean	L.F.	.02	.14		.16	
Paint	L.F.	.06	.72		.78	
Minimum Charge	Job		163		163	
Marble						
Demolish	L.F.		.65		.65	Includes material and labor to install
Install	L.F.	8.05	6.80		14.85	marble window sills, 6" x 3/4" thick.
Demolish and Install	L.F.	8.05	7.45		15.50	
Reinstall	L.F.		5.46		5.46	
Clean	L.F.	.02	.14		.16	
Minimum Charge	Job		147		147	

Shelving

	Unit	Material	Labor	Equip.	Total	Specification
Pine						
18"						
Demolish	L.F.		.75		.75	Cost includes material and labor to
Install	L.F.	4.97	3.43		8.40	install 1" thick custom shelving board
Demolish and Install	L.F.	4.97	4.18		9.15	18" wide.
Reinstall	L.F.		2.75		2.75	
Clean	L.F.	.04	.39		.43	
Paint	L.F.	.74	.45		1.19	
Minimum Charge	Job		163		163	

Finish Carpentry

Shelving

	Unit	Material	Labor	Equip.	Total	Specification
24″						
Demolish	L.F.		.75		.75	Cost includes material and labor to
Install	L.F.	6.60	3.84		10.44	install 1″ thick custom shelving board.
Demolish and Install	L.F.	6.60	4.59		11.19	
Reinstall	L.F.		3.07		3.07	
Clean	L.F.	.06	.59		.65	
Paint	L.F.	1	.60		1.60	
Minimum Charge	Job		163		163	
Particle Board						
Demolish	L.F.		.69		.69	Cost includes material and labor to
Install	L.F.	1.14	1.72		2.86	install 3/4″ thick particle shelving
Demolish and Install	L.F.	1.14	2.41		3.55	board 18″ wide.
Reinstall	L.F.		1.37		1.37	
Clean	L.F.	.04	.39		.43	
Paint	S.F.	.11	.67		.78	
Minimum Charge	Job		163		163	
Plywood						
Demolish	L.F.		.69		.69	Cost includes material and labor to
Install	L.F.	1.03	5		6.03	install 3/4″ thick plywood shelving
Demolish and Install	L.F.	1.03	5.69		6.72	board.
Reinstall	L.F.		4.01		4.01	
Clean	L.F.	.04	.39		.43	
Paint	S.F.	.11	.67		.78	
Minimum Charge	Job		163		163	
Hardwood (oak)						
Demolish	L.F.		.69		.69	Includes material and labor to install
Install	L.F.	8.20	2.25		10.45	oak shelving, 1″ x 12″, incl. cleats and
Demolish and Install	L.F.	8.20	2.94		11.14	bracing.
Reinstall	L.F.		1.80		1.80	
Clean	L.F.	.04	.39		.43	
Paint	S.F.	.11	.67		.78	
Minimum Charge	Job		163		163	
Custom Bookcase						
Demolish	L.F.		8.75		8.75	Includes material and labor to install
Install	L.F.	22	32.50		54.50	custom modular bookcase unit with
Demolish and Install	L.F.	22	41.25		63.25	clear pine faced frames, shelves 12″
Reinstall	L.F.		26.08		26.08	O.C., 7′ high, 8″ deep.
Clean	L.F.	1.17	8.40		9.57	
Paint	L.F.	.96	22		22.96	
Minimum Charge	Job		163		163	
Glass						
Demolish	L.F.		.69		.69	Includes material and labor to install
Install	L.F.	24	2.25		26.25	glass shelving, 1″ x 12″, incl. cleats
Demolish and Install	L.F.	24	2.94		26.94	and bracing.
Reinstall	L.F.		1.80		1.80	
Clean	L.F.	.06	.36		.42	
Minimum Charge	Job		155		155	

Closet Shelving

	Unit	Material	Labor	Equip.	Total	Specification
Closet Shelf and Rod						
Demolish	L.F.		.69		.69	Cost includes material and labor to
Install	L.F.	2.35	5.45		7.80	install 1″ thick custom shelving board
Demolish and Install	L.F.	2.35	6.14		8.49	18″ wide and 1″ diameter clothes pole
Reinstall	L.F.		4.35		4.35	with brackets 3′ O.C.
Clean	L.F.	.03	.29		.32	
Paint	L.F.	.11	.72		.83	
Minimum Charge	Job		163		163	

Finish Carpentry

Closet Shelving

Clothing Rod	Unit	Material	Labor	Equip.	Total	Specification
Demolish	L.F.		.46		.46	Includes material and labor to install
Install	L.F.	1.40	1.63		3.03	fir closet pole, 1-5/8" diameter.
Demolish and Install	L.F.	1.40	2.09		3.49	
Reinstall	L.F.		1.30		1.30	
Clean	L.F.	.03	.41		.44	
Paint	L.F.	.12	.78		.90	
Minimum Charge	Job		163		163	

Stair Assemblies

Straight Hardwood	Unit	Material	Labor	Equip.	Total	Specification
Demolish	Ea.		315		315	Includes material and labor to install
Install	Flight	955	217		1172	factory cut and assembled straight
Demolish and Install	Flight	955	532		1487	closed box stairs with oak treads and
Clean	Flight	2.92	20.50		23.42	prefinished stair rail with balusters.
Paint	Flight	10.35	38		48.35	
Minimum Charge	Job		163		163	
Clear Oak Tread						
Demolish	Ea.		8.25		8.25	Includes material and labor to install
Install	Ea.	88.50	18.10		106.60	1-1/4" thick clear oak treads per riser.
Demolish and Install	Ea.	88.50	26.35		114.85	
Clean	Ea.	.09	2.03		2.12	
Paint	Ea.	.34	2.12		2.46	
Minimum Charge	Job		163		163	
Landing						
Demolish	S.F.		2.73		2.73	Includes material and labor to install
Install	S.F.	10.75	18.10		28.85	clear oak landing.
Demolish and Install	S.F.	10.75	20.83		31.58	
Clean	S.F.		.29		.29	
Paint	S.F.	.26	.71		.97	
Minimum Charge	Job		163		163	
Refinish / Stain						
Install	S.F.	1.14	1.81		2.95	Includes labor and material to sand
Minimum Charge	Job		143		143	and finish (3 passes) with two coats of
						urethane on a new floor.
Spiral Hardwood						
Demolish	Ea.		315		315	Cost includes material and labor to
Install	Flight	4850	435		5285	install 4' - 6' diameter spiral stairs with
Demolish and Install	Flight	4850	750		5600	oak treads, factory cut and assembled
Clean	Flight	2.92	20.50		23.42	with double handrails.
Paint	Flight	10.35	38		48.35	
Minimum Charge	Job		163		163	

Stairs

Disappearing	Unit	Material	Labor	Equip.	Total	Specification
Demolish	Ea.		16.35		16.35	Includes material and labor to install a
Install	Ea.	116	93		209	folding staircase.
Demolish and Install	Ea.	116	109.35		225.35	
Reinstall	Ea.		74.51		74.51	
Clean	Ea.	.44	14.75		15.19	
Minimum Charge	Job		163		163	

Finish Carpentry

Stair Components	Unit	Material	Labor	Equip.	Total	Specification
Handrail w / Balusters						
Demolish	L.F.		2.18		2.18	Includes material and labor to install
Install	L.F.	23	6.80		29.80	prefinished assembled stair rail with
Demolish and Install	L.F.	23	8.98		31.98	brackets, turned balusters and newel.
Clean	L.F.	.58	1.62		2.20	
Paint	L.F.	1.38	3.36		4.74	
Minimum Charge	Job		163		163	
Bannister						
Demolish	L.F.		2.18		2.18	Includes material and labor to install
Install	L.F.	33	5.45		38.45	built-up oak railings.
Demolish and Install	L.F.	33	7.63		40.63	
Clean	L.F.	.58	1.47		2.05	
Paint	L.F.	1.38	2.60		3.98	
Minimum Charge	Job		163		163	

Cabinets and Countertops

Wall and Base Cabinets

Countertop

Wall Cabinets	Unit	Material	Labor	Equip.	Total	Specification
Good Grade Laminated						
12″ Wide, 1 Door						
Demolish	Ea.		6.55		6.55	Includes material and labor to install
Install	Ea.	105	29.50		134.50	good modular unit with melamine
Demolish and Install	Ea.	105	36.05		141.05	laminated to particle board, in
Reinstall	Ea.		23.71		23.71	textured colors or wood grain print
Clean	Ea.	.44	4.53		4.97	finish, hinges and pulls.
Paint	Ea.	.41	10.20		10.61	
Minimum Charge	Job		163		163	
15″ Wide, 1 Door						
Demolish	Ea.		8.20		8.20	Includes material and labor to install
Install	Ea.	119	30.50		149.50	good modular unit with melamine
Demolish and Install	Ea.	119	38.70		157.70	laminated to particle board, in
Reinstall	Ea.		24.37		24.37	textured colors or wood grain print
Clean	Ea.	.55	5.60		6.15	finish, hinges and pulls.
Paint	Ea.	.52	13		13.52	
Minimum Charge	Job		163		163	
18″ Wide, 1 Door						
Demolish	Ea.		9.90		9.90	Includes material and labor to install
Install	Ea.	131	31		162	good modular unit with melamine
Demolish and Install	Ea.	131	40.90		171.90	laminated to particle board, in
Reinstall	Ea.		24.96		24.96	textured colors or wood grain print
Clean	Ea.	.66	6.75		7.41	finish, hinges and pulls.
Paint	Ea.	.62	15.05		15.67	
Minimum Charge	Job		163		163	
21″ Wide, 1 Door						
Demolish	Ea.		11.40		11.40	Includes material and labor to install
Install	Ea.	153	31.50		184.50	good modular unit with melamine
Demolish and Install	Ea.	153	42.90		195.90	laminated to particle board, in
Reinstall	Ea.		25.32		25.32	textured colors or wood grain print
Clean	Ea.	.77	7.85		8.62	finish, hinges and pulls.
Paint	Ea.	.73	17.90		18.63	
Minimum Charge	Job		163		163	
24″ w / Blind Corner						
Demolish	Ea.		13.10		13.10	Includes material and labor to install
Install	Ea.	147	32		179	good modular unit with melamine
Demolish and Install	Ea.	147	45.10		192.10	laminated to particle board, in
Reinstall	Ea.		25.69		25.69	textured colors or wood grain print
Clean	Ea.	.88	9.05		9.93	finish, hinges and pulls.
Paint	Ea.	.83	20.50		21.33	
Minimum Charge	Job		163		163	

Cabinets and Countertops

Wall Cabinets	Unit	Material	Labor	Equip.	Total	Specification
27″ Wide, 2 Door						
Demolish	Ea.		14.55		14.55	Includes material and labor to install
Install	Ea.	182	33		215	good modular unit with melamine
Demolish and Install	Ea.	182	47.55		229.55	laminated to particle board, in
Reinstall	Ea.		26.34		26.34	textured colors or wood grain print
Clean	Ea.	.98	10.25		11.23	finish, hinges and pulls.
Paint	Ea.	.94	24		24.94	
Minimum Charge	Job		163		163	
30″ Wide, 2 Door						
Demolish	Ea.		16.35		16.35	Includes material and labor to install
Install	Ea.	176	34		210	good modular unit with melamine
Demolish and Install	Ea.	176	50.35		226.35	laminated to particle board, in
Reinstall	Ea.		27.03		27.03	textured colors or wood grain print
Clean	Ea.	1.09	11.20		12.29	finish, hinges and pulls.
Paint	Ea.	1.03	26		27.03	
Minimum Charge	Job		163		163	
36″ Wide, 2 Door						
Demolish	Ea.		19.40		19.40	Includes material and labor to install
Install	Ea.	201	34.50		235.50	good modular unit with melamine
Demolish and Install	Ea.	201	53.90		254.90	laminated to particle board, in
Reinstall	Ea.		27.74		27.74	textured colors or wood grain print
Clean	Ea.	1.31	13.85		15.16	finish, hinges and pulls.
Paint	Ea.	1.24	32		33.24	
Minimum Charge	Job		163		163	
48″ Wide, 2 Door						
Demolish	Ea.		26		26	Includes material and labor to install
Install	Ea.	246	35.50		281.50	good modular unit with melamine
Demolish and Install	Ea.	246	61.50		307.50	laminated to particle board, in
Reinstall	Ea.		28.35		28.35	textured colors or wood grain print
Clean	Ea.	1.75	18.10		19.85	finish, hinges and pulls.
Paint	Ea.	1.65	41		42.65	
Minimum Charge	Job		163		163	
Above-Appliance						
Demolish	Ea.		16.35		16.35	Includes material and labor to install
Install	Ea.	117	26.50		143.50	good modular unit with melamine
Demolish and Install	Ea.	117	42.85		159.85	laminated to particle board, in
Reinstall	Ea.		21.03		21.03	textured colors or wood grain print
Clean	Ea.	1.09	11.20		12.29	finish, hinges and pulls.
Paint	Ea.	1.03	26		27.03	
Minimum Charge	Job		163		163	
Better Grade Wood						
12″ Wide, 1 Door						
Demolish	Ea.		6.55		6.55	Includes material and labor to install
Install	Ea.	141	29.50		170.50	custom modular unit with solid
Demolish and Install	Ea.	141	36.05		177.05	hardwood faced frames, hardwood
Reinstall	Ea.		23.71		23.71	door frames and drawer fronts,
Clean	Ea.	.44	4.53		4.97	hardwood veneer on raised door
Paint	Ea.	.41	11		11.41	panels, hinges and pulls.
Minimum Charge	Job		163		163	
15″ Wide, 1 Door						
Demolish	Ea.		8.20		8.20	Includes material and labor to install
Install	Ea.	160	30.50		190.50	custom modular unit with solid
Demolish and Install	Ea.	160	38.70		198.70	hardwood faced frames, hardwood
Reinstall	Ea.		24.37		24.37	door frames and drawer fronts,
Clean	Ea.	.55	5.60		6.15	hardwood veneer on raised door
Paint	Ea.	.52	13.60		14.12	panels, hinges and pulls.
Minimum Charge	Job		163		163	

Cabinets and Countertops

Wall Cabinets	Unit	Material	Labor	Equip.	Total	Specification
18" Wide, 1 Door						
Demolish	Ea.		9.90		9.90	Includes material and labor to install
Install	Ea.	174	31		205	custom modular unit with solid
Demolish and Install	Ea.	174	40.90		214.90	hardwood faced frames, hardwood
Reinstall	Ea.		24.96		24.96	door frames and drawer fronts,
Clean	Ea.	.66	6.75		7.41	hardwood veneer on raised door
Paint	Ea.	.62	16.80		17.42	panels, hinges and pulls.
Minimum Charge	Job		163		163	
21" Wide, 1 Door						
Demolish	Ea.		11.40		11.40	Includes material and labor to install
Install	Ea.	198	31.50		229.50	custom modular unit with solid
Demolish and Install	Ea.	198	42.90		240.90	hardwood faced frames, hardwood
Reinstall	Ea.		25.32		25.32	door frames and drawer fronts,
Clean	Ea.	.77	7.85		8.62	hardwood veneer on raised door
Paint	Ea.	.73	19.05		19.78	panels, hinges and pulls.
Minimum Charge	Job		163		163	
24" w / Blind Corner						
Demolish	Ea.		13.10		13.10	Includes material and labor to install
Install	Ea.	196	32		228	custom modular unit with solid
Demolish and Install	Ea.	196	45.10		241.10	hardwood faced frames, hardwood
Reinstall	Ea.		25.69		25.69	door frames and drawer fronts,
Clean	Ea.	.88	9.05		9.93	hardwood veneer on raised door
Paint	Ea.	.83	22		22.83	panels, hinges and pulls.
Minimum Charge	Job		163		163	
27" Wide, 2 Door						
Demolish	Ea.		14.55		14.55	Includes material and labor to install
Install	Ea.	221	33		254	custom modular unit with solid
Demolish and Install	Ea.	221	47.55		268.55	hardwood faced frames, hardwood
Reinstall	Ea.		26.34		26.34	door frames and drawer fronts,
Clean	Ea.	.98	10.25		11.23	hardwood veneer on raised door
Paint	Ea.	.94	24		24.94	panels, hinges and pulls.
Minimum Charge	Job		163		163	
30" Wide, 2 Door						
Demolish	Ea.		16.35		16.35	Includes material and labor to install
Install	Ea.	234	34		268	custom modular unit with solid
Demolish and Install	Ea.	234	50.35		284.35	hardwood faced frames, hardwood
Reinstall	Ea.		27.03		27.03	door frames and drawer fronts,
Clean	Ea.	1.09	11.20		12.29	hardwood veneer on raised door
Paint	Ea.	1.03	28.50		29.53	panels, hinges and pulls.
Minimum Charge	Job		163		163	
36" Wide, 2 Door						
Demolish	Ea.		19.40		19.40	Includes material and labor to install
Install	Ea.	268	34.50		302.50	custom modular unit with solid
Demolish and Install	Ea.	268	53.90		321.90	hardwood faced frames, hardwood
Reinstall	Ea.		27.74		27.74	door frames and drawer fronts,
Clean	Ea.	1.31	13.85		15.16	hardwood veneer on raised door
Paint	Ea.	1.24	32		33.24	panels, hinges and pulls.
Minimum Charge	Job		163		163	
48" Wide, 2 Door						
Demolish	Ea.		26		26	Includes material and labor to install
Install	Ea.	330	35.50		365.50	custom modular unit with solid
Demolish and Install	Ea.	330	61.50		391.50	hardwood faced frames, hardwood
Reinstall	Ea.		28.35		28.35	door frames and drawer fronts,
Clean	Ea.	1.75	18.10		19.85	hardwood veneer on raised door
Paint	Ea.	1.65	41		42.65	panels, hinges and pulls.
Minimum Charge	Job		163		163	

Cabinets and Countertops

Wall Cabinets	Unit	Material	Labor	Equip.	Total	Specification
Above-appliance						
Demolish	Ea.		16.35		16.35	Includes material and labor to install
Install	Ea.	160	26.50		186.50	custom modular unit with solid
Demolish and Install	Ea.	160	42.85		202.85	hardwood faced frames, hardwood
Reinstall	Ea.		21.03		21.03	door frames and drawer fronts,
Clean	Ea.	1.09	11.20		12.29	hardwood veneer on raised door
Paint	Ea.	1.03	28.50		29.53	panels, hinges and pulls.
Minimum Charge	Job		163		163	
Premium Grade Wood						
12″ Wide, 1 Door						
Demolish	Ea.		6.55		6.55	Includes material and labor to install
Install	Ea.	182	29.50		211.50	premium modular unit with solid
Demolish and Install	Ea.	182	36.05		218.05	hardwood faced frames, hardwood
Reinstall	Ea.		23.71		23.71	door frames and drawer fronts,
Clean	Ea.	.44	4.53		4.97	hardwood veneer on raised door
Paint	Ea.	.41	11.90		12.31	panels, hinges and pulls.
Minimum Charge	Job		163		163	
15″ Wide, 1 Door						
Demolish	Ea.		8.20		8.20	Includes material and labor to install
Install	Ea.	211	30.50		241.50	premium modular unit with solid
Demolish and Install	Ea.	211	38.70		249.70	hardwood faced frames, hardwood
Reinstall	Ea.		24.37		24.37	door frames and drawer fronts,
Clean	Ea.	.55	5.60		6.15	hardwood veneer on raised door
Paint	Ea.	.52	15.05		15.57	panels, hinges and pulls.
Minimum Charge	Job		163		163	
18″ Wide, 1 Door						
Demolish	Ea.		9.90		9.90	Includes material and labor to install
Install	Ea.	223	31		254	premium modular unit with solid
Demolish and Install	Ea.	223	40.90		263.90	hardwood faced frames, hardwood
Reinstall	Ea.		24.96		24.96	door frames and drawer fronts,
Clean	Ea.	.66	6.75		7.41	hardwood veneer on raised door
Paint	Ea.	.62	17.90		18.52	panels, hinges and pulls.
Minimum Charge	Job		163		163	
21″ Wide, 1 Door						
Demolish	Ea.		11.40		11.40	Includes material and labor to install
Install	Ea.	237	31.50		268.50	premium modular unit with solid
Demolish and Install	Ea.	237	42.90		279.90	hardwood faced frames, hardwood
Reinstall	Ea.		25.32		25.32	door frames and drawer fronts,
Clean	Ea.	.77	7.85		8.62	hardwood veneer on raised door
Paint	Ea.	.73	20.50		21.23	panels, hinges and pulls.
Minimum Charge	Job		163		163	
24″ w / Blind Corner						
Demolish	Ea.		13.10		13.10	Includes material and labor to install
Install	Ea.	260	32		292	premium modular unit with solid
Demolish and Install	Ea.	260	45.10		305.10	hardwood faced frames, hardwood
Reinstall	Ea.		25.69		25.69	door frames and drawer fronts,
Clean	Ea.	.88	9.05		9.93	hardwood veneer on raised door
Paint	Ea.	.83	24		24.83	panels, hinges and pulls.
Minimum Charge	Job		163		163	
27″ Wide, 2 Door						
Demolish	Ea.		14.55		14.55	Includes material and labor to install
Install	Ea.	243	33		276	premium modular unit with solid
Demolish and Install	Ea.	243	47.55		290.55	hardwood faced frames, hardwood
Reinstall	Ea.		26.34		26.34	door frames and drawer fronts,
Clean	Ea.	.98	10.25		11.23	hardwood veneer on raised door
Paint	Ea.	.94	26		26.94	panels, hinges and pulls.
Minimum Charge	Job		163		163	

Cabinets and Countertops

Wall Cabinets		Unit	Material	Labor	Equip.	Total	Specification
30″ Wide, 2 Door							
	Demolish	Ea.		16.35		16.35	Includes material and labor to install
	Install	Ea.	296	34		330	premium modular unit with solid
	Demolish and Install	Ea.	296	50.35		346.35	hardwood faced frames, hardwood
	Reinstall	Ea.		27.03		27.03	door frames and drawer fronts,
	Clean	Ea.	1.09	11.20		12.29	hardwood veneer on raised door
	Paint	Ea.	1.03	28.50		29.53	panels, hinges and pulls.
	Minimum Charge	Job		163		163	
36″ Wide, 2 Door							
	Demolish	Ea.		19.40		19.40	Includes material and labor to install
	Install	Ea.	335	34.50		369.50	premium modular unit with solid
	Demolish and Install	Ea.	335	53.90		388.90	hardwood faced frames, hardwood
	Reinstall	Ea.		27.74		27.74	door frames and drawer fronts,
	Clean	Ea.	1.31	13.85		15.16	hardwood veneer on raised door
	Paint	Ea.	1.24	36		37.24	panels, hinges and pulls.
	Minimum Charge	Job		163		163	
48″ Wide, 2 Door							
	Demolish	Ea.		26		26	Includes material and labor to install
	Install	Ea.	420	35.50		455.50	premium modular unit with solid
	Demolish and Install	Ea.	420	61.50		481.50	hardwood faced frames, hardwood
	Reinstall	Ea.		28.35		28.35	door frames and drawer fronts,
	Clean	Ea.	1.75	18.10		19.85	hardwood veneer on raised door
	Paint	Ea.	1.65	47.50		49.15	panels, hinges and pulls.
	Minimum Charge	Job		163		163	
Above-appliance							
	Demolish	Ea.		16.35		16.35	Includes material and labor to install
	Install	Ea.	198	26.50		224.50	premium modular unit with solid
	Demolish and Install	Ea.	198	42.85		240.85	hardwood faced frames, hardwood
	Reinstall	Ea.		21.03		21.03	door frames and drawer fronts,
	Clean	Ea.	1.09	11.20		12.29	hardwood veneer on raised door
	Paint	Ea.	1.03	28.50		29.53	panels, hinges, pulls.
	Minimum Charge	Job		163		163	
Premium Hardwood							
12″ Wide, 1 Door							
	Demolish	Ea.		6.55		6.55	Includes material and labor to install
	Install	Ea.	211	29.50		240.50	premium hardwood modular unit with
	Demolish and Install	Ea.	211	36.05		247.05	solid hardwood faced frames, drawer
	Reinstall	Ea.		23.71		23.71	fronts, and door panels, steel drawer
	Clean	Ea.	.44	4.53		4.97	guides, hinges, pulls, laminated
	Paint	Ea.	.41	13		13.41	interior.
	Minimum Charge	Job		163		163	
15″ Wide, 1 Door							
	Demolish	Ea.		8.20		8.20	Includes material and labor to install
	Install	Ea.	278	30.50		308.50	premium hardwood modular unit with
	Demolish and Install	Ea.	278	38.70		316.70	solid hardwood faced frames, drawer
	Reinstall	Ea.		24.37		24.37	fronts, and door panels, steel drawer
	Clean	Ea.	.55	5.60		6.15	guides, hinges, pulls, laminated
	Paint	Ea.	.52	15.90		16.42	interior.
	Minimum Charge	Job		163		163	
18″ Wide, 1 Door							
	Demolish	Ea.		9.90		9.90	Includes material and labor to install
	Install	Ea.	253	31		284	premium hardwood modular unit with
	Demolish and Install	Ea.	253	40.90		293.90	solid hardwood faced frames, drawer
	Reinstall	Ea.		24.96		24.96	fronts, and door panels, steel drawer
	Clean	Ea.	.66	6.75		7.41	guides, hinges, pulls, laminated
	Paint	Ea.	.62	19.05		19.67	interior.
	Minimum Charge	Job		163		163	

Cabinets and Countertops

Wall Cabinets		Unit	Material	Labor	Equip.	Total	Specification
21" Wide, 1 Door							
	Demolish	Ea.		11.40		11.40	Includes material and labor to install
	Install	Ea.	260	31.50		291.50	premium hardwood modular unit with
	Demolish and Install	Ea.	260	42.90		302.90	solid hardwood faced frames, drawer
	Reinstall	Ea.		25.32		25.32	fronts, and door panels, steel drawer
	Clean	Ea.	.77	7.85		8.62	guides, hinges, pulls, laminated
	Paint	Ea.	.73	22		22.73	interior.
	Minimum Charge	Job		163		163	
24" w / Blind Corner							
	Demolish	Ea.		13.10		13.10	Includes material and labor to install
	Install	Ea.	289	32		321	premium hardwood modular unit with
	Demolish and Install	Ea.	289	45.10		334.10	solid hardwood faced frames, drawer
	Reinstall	Ea.		25.69		25.69	fronts, and door panels, steel drawer
	Clean	Ea.	.88	9.05		9.93	guides, hinges, pulls, laminated
	Paint	Ea.	.83	26		26.83	interior.
	Minimum Charge	Job		163		163	
27" Wide, 2 Door							
	Demolish	Ea.		14.55		14.55	Includes material and labor to install
	Install	Ea.	490	33		523	premium hardwood modular unit with
	Demolish and Install	Ea.	490	47.55		537.55	solid hardwood faced frames, drawer
	Reinstall	Ea.		26.34		26.34	fronts, and door panels, steel drawer
	Clean	Ea.	.98	10.25		11.23	guides, hinges, pulls, laminated
	Paint	Ea.	.94	28.50		29.44	interior.
	Minimum Charge	Job		163		163	
30" Wide, 2 Door							
	Demolish	Ea.		16.35		16.35	Includes material and labor to install
	Install	Ea.	325	34		359	premium hardwood modular unit with
	Demolish and Install	Ea.	325	50.35		375.35	solid hardwood faced frames, drawer
	Reinstall	Ea.		27.03		27.03	fronts, and door panels, steel drawer
	Clean	Ea.	1.09	11.20		12.29	guides, hinges, pulls, laminated
	Paint	Ea.	1.03	32		33.03	interior.
	Minimum Charge	Job		163		163	
36" Wide, 2 Door							
	Demolish	Ea.		19.40		19.40	Includes material and labor to install
	Install	Ea.	370	34.50		404.50	premium hardwood modular unit with
	Demolish and Install	Ea.	370	53.90		423.90	solid hardwood faced frames, drawer
	Reinstall	Ea.		27.74		27.74	fronts, and door panels, steel drawer
	Clean	Ea.	1.31	13.85		15.16	guides, hinges, pulls, laminated
	Paint	Ea.	1.24	41		42.24	interior.
	Minimum Charge	Job		163		163	
48" Wide, 2 Door							
	Demolish	Ea.		26		26	Includes material and labor to install
	Install	Ea.	450	35.50		485.50	premium hardwood modular unit with
	Demolish and Install	Ea.	450	61.50		511.50	solid hardwood faced frames, drawer
	Reinstall	Ea.		28.35		28.35	fronts, and door panels, steel drawer
	Clean	Ea.	1.75	18.10		19.85	guides, hinges, pulls, laminated
	Paint	Ea.	1.65	47.50		49.15	interior.
	Minimum Charge	Job		163		163	
Above-appliance							
	Demolish	Ea.		16.35		16.35	Includes material and labor to install
	Install	Ea.	228	26.50		254.50	premium hardwood modular unit with
	Demolish and Install	Ea.	228	42.85		270.85	solid hardwood faced frames, drawer
	Reinstall	Ea.		21.03		21.03	fronts, and door panels, steel drawer
	Clean	Ea.	1.09	11.20		12.29	guides, hinges, pulls, laminated
	Paint	Ea.	1.03	32		33.03	interior.
	Minimum Charge	Job		163		163	

Cabinets and Countertops

Wall Cabinets	Unit	Material	Labor	Equip.	Total	Specification
Medicine Cabinet						
Good Grade						
Demolish	Ea.		10.90		10.90	Cost includes material and labor to
Install	Ea.	75	23.50		98.50	install 14" x 18" recessed medicine
Demolish and Install	Ea.	75	34.40		109.40	cabinet with stainless steel frame,
Reinstall	Ea.		18.63		18.63	mirror. Light not included.
Clean	Ea.	.02	10.60		10.62	
Paint	Ea.	1	17.90		18.90	
Minimum Charge	Job		163		163	
Good Grade Lighted						
Demolish	Ea.		10.90		10.90	Cost includes material and labor to
Install	Ea.	220	27		247	install 14" x 18" recessed medicine
Demolish and Install	Ea.	220	37.90		257.90	cabinet with stainless steel frame,
Reinstall	Ea.		21.73		21.73	mirror and light.
Clean	Ea.	.02	10.60		10.62	
Paint	Ea.	1	17.90		18.90	
Minimum Charge	Job		163		163	
Better Grade						
Demolish	Ea.		10.90		10.90	Includes material and labor to install
Install	Ea.	103	23.50		126.50	recessed medicine cabinet with
Demolish and Install	Ea.	103	34.40		137.40	polished chrome frame and mirror.
Reinstall	Ea.		18.63		18.63	Light not included.
Clean	Ea.	.02	10.60		10.62	
Paint	Ea.	1	17.90		18.90	
Minimum Charge	Job		163		163	
Better Grade Lighted						
Demolish	Ea.		10.90		10.90	Includes material and labor to install
Install	Ea.	249	27		276	recessed medicine cabinet with
Demolish and Install	Ea.	249	37.90		286.90	polished chrome frame, mirror and
Reinstall	Ea.		21.73		21.73	light.
Clean	Ea.	.02	10.60		10.62	
Paint	Ea.	1	17.90		18.90	
Minimum Charge	Job		163		163	
Premium Grade						
Demolish	Ea.		10.90		10.90	Cost includes material and labor to
Install	Ea.	132	23.50		155.50	install 14" x 24" recessed medicine
Demolish and Install	Ea.	132	34.40		166.40	cabinet with beveled mirror, frameless
Reinstall	Ea.		18.63		18.63	swing door. Light not included.
Clean	Ea.	.02	10.60		10.62	
Paint	Ea.	1	17.90		18.90	
Minimum Charge	Job		163		163	
Premium Grade Light						
Demolish	Ea.		10.90		10.90	Cost includes material and labor to
Install	Ea.	277	27		304	install 36" x 30" surface mounted
Demolish and Install	Ea.	277	37.90		314.90	medicine cabinet with beveled tri-view
Reinstall	Ea.		21.73		21.73	mirror, frameless swing door and light.
Clean	Ea.	.02	10.60		10.62	
Paint	Ea.	1	17.90		18.90	
Minimum Charge	Job		163		163	
Bath Mirror						
Demolish	Ea.		8.20		8.20	Cost includes material and labor to
Install	Ea.	72	16.30		88.30	install 18" x 24" surface mount mirror
Demolish and Install	Ea.	72	24.50		96.50	with stainless steel frame.
Reinstall	Ea.		13.04		13.04	
Clean	Ea.	.02	.71		.73	
Minimum Charge	Job		163		163	

Cabinets and Countertops

Wall Cabinets

Wall Cabinets		Unit	Material	Labor	Equip.	Total	Specification
Garage							
	Demolish	L.F.		6.55		6.55	Includes material and labor to install
	Install	L.F.	51	16.30		67.30	good modular unit with melamine
	Demolish and Install	L.F.	51	22.85		73.85	laminated to particle board, in
	Reinstall	L.F.		13.04		13.04	textured colors or wood grain print
	Clean	L.F.	.44	4.53		4.97	finish, hinges and pulls.
	Paint	L.F.	.41	10.20		10.61	
	Minimum Charge	Job		163		163	
Strip and Refinish							
	Install	L.F.	1.09	5.20		6.29	Includes labor and material to strip,
	Minimum Charge	Job		143		143	prep and refinish exterior of cabinets.

Tall Cabinets

Tall Cabinets		Unit	Material	Labor	Equip.	Total	Specification
Good Grade Laminated							
	Demolish	L.F.		11.65		11.65	Includes material and labor to install
	Install	L.F.	208	184		392	custom modular unit with solid
	Demolish and Install	L.F.	208	195.65		403.65	hardwood faced frames, hardwood
	Reinstall	L.F.		146.93		146.93	door frames and drawer fronts,
	Clean	L.F.	1.17	8.40		9.57	hardwood veneer on raised door
	Paint	L.F.	.96	20.50		21.46	panels, hinges and pulls.
	Minimum Charge	Job		163		163	
Better Grade Wood							
	Demolish	L.F.		11.65		11.65	Includes material and labor to install
	Install	L.F.	277	184		461	custom modular unit with solid
	Demolish and Install	L.F.	277	195.65		472.65	hardwood faced frames, hardwood
	Reinstall	L.F.		146.93		146.93	door frames and drawer fronts,
	Clean	L.F.	1.17	8.40		9.57	hardwood veneer on raised door
	Paint	L.F.	.96	22		22.96	panels, hinges and pulls.
	Minimum Charge	Job		163		163	
Premium Grade Wood							
	Demolish	L.F.		11.65		11.65	Includes material and labor to install
	Install	L.F.	370	184		554	premium modular unit with solid
	Demolish and Install	L.F.	370	195.65		565.65	hardwood faced frames, hardwood
	Reinstall	L.F.		146.93		146.93	door frames and drawer fronts,
	Clean	L.F.	1.17	8.40		9.57	hardwood veneer on raised door
	Paint	L.F.	.96	24		24.96	panels, hinges and pulls.
	Minimum Charge	Job		163		163	
Premium Hardwood							
	Demolish	L.F.		11.65		11.65	Includes material and labor to install
	Install	L.F.	485	184		669	premium hardwood modular unit with
	Demolish and Install	L.F.	485	195.65		680.65	solid hardwood faced frames, drawer
	Reinstall	L.F.		146.93		146.93	fronts, and door panels, steel drawer
	Clean	L.F.	1.17	8.40		9.57	guides, hinges and pulls, laminated
	Paint	L.F.	.96	26		26.96	interior.
	Minimum Charge	Job		163		163	

Cabinets and Countertops

Tall Cabinets	Unit	Material	Labor	Equip.	Total	Specification
Single Oven, 27″ Wide						
Good Grade Laminated						
Demolish	Ea.		26		26	Includes material and labor to install
Install	Ea.	470	81.50		551.50	good modular unit with melamine
Demolish and Install	Ea.	470	107.50		577.50	laminated to particle board, in
Reinstall	Ea.		65.20		65.20	textured colors or wood grain print
Clean	Ea.	3.64	14.75		18.39	finish, hinges and pulls.
Paint	Ea.	1.62	46		47.62	
Minimum Charge	Job		163		163	
Better Grade Veneer						
Demolish	Ea.		26		26	Includes material and labor to install
Install	Ea.	830	81.50		911.50	custom modular unit with solid
Demolish and Install	Ea.	830	107.50		937.50	hardwood faced frames, hardwood
Reinstall	Ea.		65.20		65.20	door frames and drawer fronts,
Clean	Ea.	3.64	14.75		18.39	hardwood veneer on raised door
Paint	Ea.	1.62	49.50		51.12	panels, hinges and pulls.
Minimum Charge	Job		163		163	
Premium Grade Veneer						
Demolish	Ea.		26		26	Includes material and labor to install
Install	Ea.	735	81.50		816.50	premium modular unit with solid
Demolish and Install	Ea.	735	107.50		842.50	hardwood faced frames, hardwood
Reinstall	Ea.		65.20		65.20	door frames and drawer fronts,
Clean	Ea.	3.64	14.75		18.39	hardwood veneer on raised door
Paint	Ea.	1.62	54		55.62	panels, hinges and pulls.
Minimum Charge	Job		163		163	
Premium Hardwood						
Demolish	Ea.		26		26	Includes material and labor to install
Install	Ea.	945	81.50		1026.50	premium hardwood modular unit with
Demolish and Install	Ea.	945	107.50		1052.50	solid hardwood faced frames, drawer
Reinstall	Ea.		65.20		65.20	fronts, and door panels, steel drawer
Clean	Ea.	3.64	14.75		18.39	guides, hinges and pulls, laminated
Paint	Ea.	1.62	58.50		60.12	interior.
Minimum Charge	Job		163		163	
Single Oven, 30″ Wide						
Good Grade Laminated						
Demolish	Ea.		26		26	Includes material and labor to install
Install	Ea.	425	81.50		506.50	good modular unit with melamine
Demolish and Install	Ea.	425	107.50		532.50	laminated to particle board, in
Reinstall	Ea.		65.20		65.20	textured colors or wood grain print
Clean	Ea.	3.64	14.75		18.39	finish, hinges and pulls.
Paint	Ea.	1.62	46		47.62	
Minimum Charge	Job		163		163	
Better Grade Veneer						
Demolish	Ea.		26		26	Includes material and labor to install
Install	Ea.	620	81.50		701.50	custom modular unit with solid
Demolish and Install	Ea.	620	107.50		727.50	hardwood faced frames, hardwood
Reinstall	Ea.		65.20		65.20	door frames and drawer fronts,
Clean	Ea.	3.64	14.75		18.39	hardwood veneer on raised door
Paint	Ea.	1.62	46		47.62	panels, hinges and pulls.
Minimum Charge	Job		163		163	

Cabinets and Countertops

Tall Cabinets	Unit	Material	Labor	Equip.	Total	Specification
Premium Grade Veneer						Includes material and labor to install
Demolish	Ea.		26		26	premium modular unit with solid
Install	Ea.	720	81.50		801.50	hardwood faced frames, hardwood
Demolish and Install	Ea.	720	107.50		827.50	door frames and drawer fronts,
Reinstall	Ea.		65.20		65.20	hardwood veneer on raised door
Clean	Ea.	3.64	14.75		18.39	panels, hinges and pulls.
Paint	Ea.	1.62	46		47.62	
Minimum Charge	Job		163		163	
Premium Hardwood						Includes material and labor to install
Demolish	Ea.		26		26	premium hardwood modular unit with
Install	Ea.	865	81.50		946.50	solid hardwood faced frames, drawer
Demolish and Install	Ea.	865	107.50		972.50	fronts, and door panels, steel drawer
Reinstall	Ea.		65.20		65.20	guides, hinges and pulls, laminated
Clean	Ea.	3.64	14.75		18.39	interior.
Paint	Ea.	1.62	46		47.62	
Minimum Charge	Job		163		163	
Utility 18″ W x 12″ D						
Good Grade Laminated						Includes material and labor to install
Demolish	Ea.		26		26	good modular unit with melamine
Install	Ea.	380	65		445	laminated to particle board, in
Demolish and Install	Ea.	380	91		471	textured colors or wood grain print
Reinstall	Ea.		52.16		52.16	finish, hinges and pulls.
Clean	Ea.	.86	7.15		8.01	
Paint	Ea.	.62	17.90		18.52	
Minimum Charge	Job		163		163	
Better Grade Veneer						Includes material and labor to install
Demolish	Ea.		26		26	custom modular unit with solid
Install	Ea.	415	65		480	hardwood faced frames, hardwood
Demolish and Install	Ea.	415	91		506	door frames and drawer fronts,
Reinstall	Ea.		52.16		52.16	hardwood veneer on raised door
Clean	Ea.	.86	7.15		8.01	panels, hinges and pulls.
Paint	Ea.	.62	19.05		19.67	
Minimum Charge	Job		163		163	
Premium Grade Veneer						Includes material and labor to install
Demolish	Ea.		26		26	premium modular unit with solid
Install	Ea.	530	65		595	hardwood faced frames, hardwood
Demolish and Install	Ea.	530	91		621	door frames and drawer fronts,
Reinstall	Ea.		52.16		52.16	hardwood veneer on raised door
Clean	Ea.	.86	7.15		8.01	panels, hinges and pulls.
Paint	Ea.	.62	22		22.62	
Minimum Charge	Job		163		163	
Premium Hardwood						Includes material and labor to install
Demolish	Ea.		26		26	premium hardwood modular unit with
Install	Ea.	670	65		735	solid hardwood faced frames, drawer
Demolish and Install	Ea.	670	91		761	fronts, and door panels, steel drawer
Reinstall	Ea.		52.16		52.16	guides, hinges and pulls, laminated
Clean	Ea.	.86	7.15		8.01	interior.
Paint	Ea.	.62	24		24.62	
Minimum Charge	Job		163		163	

Cabinets and Countertops

Tall Cabinets		Unit	Material	Labor	Equip.	Total	Specification
Utility 24″ W x 24″ D							
Good Grade Laminated							
	Demolish	Ea.		26		26	Includes material and labor to install
	Install	Ea.	460	81.50		541.50	good modular unit with melamine
	Demolish and Install	Ea.	460	107.50		567.50	laminated to particle board, in
	Reinstall	Ea.		65.20		65.20	textured colors or wood grain print
	Clean	Ea.	1.12	9.40		10.52	finish, hinges and pulls.
	Paint	Ea.	.83	24		24.83	
	Minimum Charge	Job		163		163	
Better Grade Veneer							
	Demolish	Ea.		26		26	Includes material and labor to install
	Install	Ea.	560	81.50		641.50	custom modular unit with solid
	Demolish and Install	Ea.	560	107.50		667.50	hardwood faced frames, hardwood
	Reinstall	Ea.		65.20		65.20	door frames and drawer fronts,
	Clean	Ea.	1.12	9.40		10.52	hardwood veneer on raised door
	Paint	Ea.	.83	26		26.83	panels, hinges and pulls.
	Minimum Charge	Job		163		163	
Premium Grade Veneer							
	Demolish	Ea.		26		26	Includes material and labor to install
	Install	Ea.	705	81.50		786.50	premium modular unit with solid
	Demolish and Install	Ea.	705	107.50		812.50	hardwood faced frames, hardwood
	Reinstall	Ea.		65.20		65.20	door frames and drawer fronts,
	Clean	Ea.	1.12	9.40		10.52	hardwood veneer on raised door
	Paint	Ea.	.83	28.50		29.33	panels, hinges and pulls.
	Minimum Charge	Job		163		163	
Premium Hardwood							
	Demolish	Ea.		26		26	Includes material and labor to install
	Install	Ea.	865	81.50		946.50	premium hardwood modular unit with
	Demolish and Install	Ea.	865	107.50		972.50	solid hardwood faced frames, drawer
	Reinstall	Ea.		65.20		65.20	fronts, and door panels, steel drawer
	Clean	Ea.	1.12	9.40		10.52	guides, hinges and pulls, laminated
	Paint	Ea.	.83	32		32.83	interior.
	Minimum Charge	Job		163		163	
Plain Shelves							
	Install	Ea.	30.50	20.50		51	Includes labor and material to install tall cabinet shelving, per set.
Rotating Shelves							
	Install	Ea.	76	16.30		92.30	Includes labor and material to install lazy susan shelving for wall cabinet.
Strip and Refinish							
	Install	L.F.	1.71	13		14.71	Includes labor and material to strip,
	Minimum Charge	Job		143		143	prep and refinish exterior of cabinets.
Stain							
	Install	L.F.	.15	2.60		2.75	Includes labor and material to replace
	Minimum Charge	Job		143		143	normal prep and stain on doors and exterior cabinets.

Cabinets and Countertops

Base Cabinets	Unit	Material	Labor	Equip.	Total	Specification
Good Grade Laminated						
12″ w, 1 Door, 1 Drawer						
Demolish	Ea.		6.55		6.55	Includes material and labor to install
Install	Ea.	109	26.50		135.50	good modular unit with melamine
Demolish and Install	Ea.	109	33.05		142.05	laminate, textured colors or wood
Reinstall	Ea.		21.03		21.03	grain print finish including hinges and
Clean	Ea.	.58	4.71		5.29	pulls.
Paint	Ea.	.41	11.90		12.31	
Minimum Charge	Job		163		163	
15″ w, 1 Door, 1 Drawer						
Demolish	Ea.		8.20		8.20	Includes material and labor to install
Install	Ea.	149	27		176	good modular unit with melamine
Demolish and Install	Ea.	149	35.20		184.20	laminate, textured colors or wood
Reinstall	Ea.		21.73		21.73	grain print finish including hinges and
Clean	Ea.	.73	5.90		6.63	pulls.
Paint	Ea.	.52	15.05		15.57	
Minimum Charge	Job		163		163	
18″ w, 1 Door, 1 Drawer						
Demolish	Ea.		9.70		9.70	Includes material and labor to install
Install	Ea.	163	28		191	good modular unit with melamine
Demolish and Install	Ea.	163	37.70		200.70	laminate, textured colors or wood
Reinstall	Ea.		22.39		22.39	grain print finish including hinges and
Clean	Ea.	.86	7.15		8.01	pulls.
Paint	Ea.	.62	17.90		18.52	
Minimum Charge	Job		163		163	
21″ w, 1 Door, 1 Drawer						
Demolish	Ea.		11.40		11.40	Includes material and labor to install
Install	Ea.	169	28.50		197.50	good modular unit with melamine
Demolish and Install	Ea.	169	39.90		208.90	laminate, textured colors or wood
Reinstall	Ea.		22.98		22.98	grain print finish including hinges and
Clean	Ea.	1.05	8.40		9.45	pulls.
Paint	Ea.	.73	20.50		21.23	
Minimum Charge	Job		163		163	
24″ w, 1 Door, 1 Drawer						
Demolish	Ea.		13.10		13.10	Includes material and labor to install
Install	Ea.	195	29		224	good modular unit with melamine
Demolish and Install	Ea.	195	42.10		237.10	laminate, textured colors or wood
Reinstall	Ea.		23.39		23.39	grain print finish including hinges and
Clean	Ea.	1.12	9.40		10.52	pulls.
Paint	Ea.	.83	24		24.83	
Minimum Charge	Job		163		163	
30″ Wide Sink Base						
Demolish	Ea.		16.35		16.35	Includes material and labor to install
Install	Ea.	190	30.50		220.50	good modular unit with melamine
Demolish and Install	Ea.	190	46.85		236.85	laminate, textured colors or wood
Reinstall	Ea.		24.37		24.37	grain print finish including hinges and
Clean	Ea.	1.46	11.80		13.26	pulls.
Paint	Ea.	1.03	28.50		29.53	
Minimum Charge	Job		163		163	
36″ Wide Sink Base						
Demolish	Ea.		20		20	Includes material and labor to install
Install	Ea.	213	32		245	good modular unit with melamine
Demolish and Install	Ea.	213	52		265	laminate, textured colors or wood
Reinstall	Ea.		25.69		25.69	grain print finish including hinges and
Clean	Ea.	1.62	13.85		15.47	pulls.
Paint	Ea.	1.24	36		37.24	
Minimum Charge	Job		163		163	

Cabinets and Countertops

Base Cabinets	Unit	Material	Labor	Equip.	Total	Specification
36" Blind Corner						
Demolish	Ea.		20		20	Includes material and labor to install
Install	Ea.	315	36		351	good modular unit with melamine
Demolish and Install	Ea.	315	56		371	laminate, textured colors or wood
Reinstall	Ea.		28.98		28.98	grain print finish including hinges and
Clean	Ea.	1.62	13.85		15.47	pulls.
Paint	Ea.	1.24	36		37.24	
Minimum Charge	Job		163		163	
42" w, 2 Door, 2 Drawer						
Demolish	Ea.		24		24	Includes material and labor to install
Install	Ea.	265	33		298	good modular unit with melamine
Demolish and Install	Ea.	265	57		322	laminate, textured colors or wood
Reinstall	Ea.		26.34		26.34	grain print finish including hinges and
Clean	Ea.	2.08	16.85		18.93	pulls.
Paint	Ea.	1.43	41		42.43	
Minimum Charge	Job		163		163	
Clean Interior						
Clean	L.F.	.58	3.57		4.15	Includes labor and materials to clean
Minimum Charge	Job		118		118	the interior of a 27" cabinet.
Better Grade Wood						
12" w, 1 Door, 1 Drawer						
Demolish	Ea.		6.55		6.55	Includes material and labor to install
Install	Ea.	145	26.50		171.50	custom modular unit including solid
Demolish and Install	Ea.	145	33.05		178.05	hardwood faced frames, hardwood
Reinstall	Ea.		21.03		21.03	door frames and drawer fronts,
Clean	Ea.	.58	4.71		5.29	hardwood veneer on raised door
Paint	Ea.	.41	13		13.41	panels, hinges and pulls.
Minimum Charge	Job		163		163	
15" w, 1 Door, 1 Drawer						
Demolish	Ea.		8.20		8.20	Includes material and labor to install
Install	Ea.	199	27		226	custom modular unit including solid
Demolish and Install	Ea.	199	35.20		234.20	hardwood faced frames, hardwood
Reinstall	Ea.		21.73		21.73	door frames and drawer fronts,
Clean	Ea.	.73	5.90		6.63	hardwood veneer on raised door
Paint	Ea.	.52	16.80		17.32	panels, hinges and pulls.
Minimum Charge	Job		163		163	
18" w, 1 Door, 1 Drawer						
Demolish	Ea.		9.70		9.70	Includes material and labor to install
Install	Ea.	217	28		245	custom modular unit including solid
Demolish and Install	Ea.	217	37.70		254.70	hardwood faced frames, hardwood
Reinstall	Ea.		22.39		22.39	door frames and drawer fronts,
Clean	Ea.	.86	7.15		8.01	hardwood veneer on raised door
Paint	Ea.	.62	19.05		19.67	panels, hinges and pulls.
Minimum Charge	Job		163		163	
21" w, 1 Door, 1 Drawer						
Demolish	Ea.		11.40		11.40	Includes material and labor to install
Install	Ea.	226	28.50		254.50	custom modular unit including solid
Demolish and Install	Ea.	226	39.90		265.90	hardwood faced frames, hardwood
Reinstall	Ea.		22.98		22.98	door frames and drawer fronts,
Clean	Ea.	1.05	8.40		9.45	hardwood veneer on raised door
Paint	Ea.	.73	22		22.73	panels, hinges and pulls.
Minimum Charge	Job		163		163	

Cabinets and Countertops

Base Cabinets	Unit	Material	Labor	Equip.	Total	Specification
24" w, 1 Door, 1 Drawer						
Demolish	Ea.		13.10		13.10	Includes material and labor to install
Install	Ea.	260	29		289	custom modular unit including solid
Demolish and Install	Ea.	260	42.10		302.10	hardwood faced frames, hardwood
Reinstall	Ea.		23.39		23.39	door frames and drawer fronts,
Clean	Ea.	1.12	9.40		10.52	hardwood veneer on raised door
Paint	Ea.	.83	26		26.83	panels, hinges and pulls.
Minimum Charge	Job		163		163	
30" Wide Sink Base						
Demolish	Ea.		16.35		16.35	Includes material and labor to install
Install	Ea.	254	30.50		284.50	custom modular unit including solid
Demolish and Install	Ea.	254	46.85		300.85	hardwood faced frames, hardwood
Reinstall	Ea.		24.37		24.37	door frames and drawer fronts,
Clean	Ea.	1.46	11.80		13.26	hardwood veneer on raised door
Paint	Ea.	1.03	32		33.03	panels, hinges and pulls.
Minimum Charge	Job		163		163	
36" Wide Sink Base						
Demolish	Ea.		20		20	Includes material and labor to install
Install	Ea.	285	32		317	custom modular unit including solid
Demolish and Install	Ea.	285	52		337	hardwood faced frames, hardwood
Reinstall	Ea.		25.69		25.69	door frames and drawer fronts,
Clean	Ea.	1.62	13.85		15.47	hardwood veneer on raised door
Paint	Ea.	1.24	41		42.24	panels, hinges and pulls.
Minimum Charge	Job		163		163	
36" Blind Corner						
Demolish	Ea.		20		20	Includes material and labor to install
Install	Ea.	420	36		456	custom modular unit including solid
Demolish and Install	Ea.	420	56		476	hardwood faced frames, hardwood
Reinstall	Ea.		28.98		28.98	door frames and drawer fronts,
Clean	Ea.	1.62	13.85		15.47	hardwood veneer on raised door
Paint	Ea.	1.24	41		42.24	panels, hinges and pulls.
Minimum Charge	Job		163		163	
42" w, 2 Door, 2 Drawer						
Demolish	Ea.		24		24	Includes material and labor to install
Install	Ea.	350	33		383	modular unit including solid hardwood
Demolish and Install	Ea.	350	57		407	faced frames, hardwood door frames
Reinstall	Ea.		26.34		26.34	and drawer fronts, hardwood veneer
Clean	Ea.	2.08	16.85		18.93	on raised door panels, hinges and
Paint	Ea.	1.43	47.50		48.93	pulls.
Minimum Charge	Job		163		163	
Clean Interior						
Clean	L.F.	.58	3.57		4.15	Includes labor and materials to clean
Minimum Charge	Job		118		118	the interior of a 27" cabinet.
Premium Grade Wood						
12" w, 1 Door, 1 Drawer						
Demolish	Ea.		6.55		6.55	Includes material and labor to install
Install	Ea.	251	13.15		264.15	premium hardwood modular unit with
Demolish and Install	Ea.	251	19.70		270.70	solid hardwood faced frames, drawer
Reinstall	Ea.		10.52		10.52	fronts, and door panels, steel drawer
Clean	Ea.	.58	4.71		5.29	guides, hinges, pulls and laminated
Paint	Ea.	.41	14.30		14.71	interior.
Minimum Charge	Job		163		163	

Cabinets and Countertops

Base Cabinets	Unit	Material	Labor	Equip.	Total	Specification
15″ w, 1 Door, 1 Drawer						
Demolish	Ea.		8.20		8.20	Includes material and labor to install
Install	Ea.	260	27		287	premium modular unit including solid
Demolish and Install	Ea.	260	35.20		295.20	hardwood faced frames, hardwood
Reinstall	Ea.		21.73		21.73	door frames and drawer fronts,
Clean	Ea.	.73	5.90		6.63	hardwood veneer on raised door
Paint	Ea.	.52	17.90		18.42	panels, hinges and pulls.
Minimum Charge	Job		163		163	
18″ w, 1 Door, 1 Drawer						
Demolish	Ea.		9.70		9.70	Includes material and labor to install
Install	Ea.	281	28		309	premium modular unit including solid
Demolish and Install	Ea.	281	37.70		318.70	hardwood faced frames, hardwood
Reinstall	Ea.		22.39		22.39	door frames and drawer fronts,
Clean	Ea.	.86	7.15		8.01	hardwood veneer on raised door
Paint	Ea.	.62	22		22.62	panels, hinges and pulls.
Minimum Charge	Job		163		163	
21″ w, 1 Door, 1 Drawer						
Demolish	Ea.		11.40		11.40	Includes material and labor to install
Install	Ea.	292	28.50		320.50	premium modular unit including solid
Demolish and Install	Ea.	292	39.90		331.90	hardwood faced frames, hardwood
Reinstall	Ea.		22.98		22.98	door frames and drawer fronts,
Clean	Ea.	1.05	8.40		9.45	hardwood veneer on raised door
Paint	Ea.	.73	24		24.73	panels, hinges and pulls.
Minimum Charge	Job		163		163	
24″ w, 1 Door, 1 Drawer						
Demolish	Ea.		13.10		13.10	Includes material and labor to install
Install	Ea.	330	29		359	premium modular unit including solid
Demolish and Install	Ea.	330	42.10		372.10	hardwood faced frames, hardwood
Reinstall	Ea.		23.39		23.39	door frames and drawer fronts,
Clean	Ea.	1.12	9.40		10.52	hardwood veneer on raised door
Paint	Ea.	.83	28.50		29.33	panels, hinges and pulls.
Minimum Charge	Job		163		163	
30″ Wide Sink Base						
Demolish	Ea.		16.35		16.35	Includes material and labor to install
Install	Ea.	296	30.50		326.50	premium modular unit including solid
Demolish and Install	Ea.	296	46.85		342.85	hardwood faced frames, hardwood
Reinstall	Ea.		24.37		24.37	door frames and drawer fronts,
Clean	Ea.	1.46	11.80		13.26	hardwood veneer on raised door
Paint	Ea.	1.03	36		37.03	panels, hinges and pulls.
Minimum Charge	Job		163		163	
36″ Wide Sink Base						
Demolish	Ea.		20		20	Includes material and labor to install
Install	Ea.	330	32		362	premium modular unit including solid
Demolish and Install	Ea.	330	52		382	hardwood faced frames, hardwood
Reinstall	Ea.		25.69		25.69	door frames and drawer fronts,
Clean	Ea.	1.62	13.85		15.47	hardwood veneer on raised door
Paint	Ea.	1.24	41		42.24	panels, hinges and pulls.
Minimum Charge	Job		163		163	
36″ Blind Corner						
Demolish	Ea.		20		20	Includes material and labor to install
Install	Ea.	695	36		731	premium modular unit including solid
Demolish and Install	Ea.	695	56		751	hardwood faced frames, hardwood
Reinstall	Ea.		28.98		28.98	door frames and drawer fronts,
Clean	Ea.	1.62	13.85		15.47	hardwood veneer on raised door
Paint	Ea.	1.24	41		42.24	panels, hinges and pulls.
Minimum Charge	Job		163		163	

Cabinets and Countertops

Base Cabinets	Unit	Material	Labor	Equip.	Total	Specification
42" w, 2 Door, 2 Drawer						Includes material and labor to install
Demolish	Ea.		24		24	premium modular unit including solid
Install	Ea.	475	33		508	hardwood faced frames, hardwood
Demolish and Install	Ea.	475	57		532	door frames and drawer fronts,
Reinstall	Ea.		26.34		26.34	hardwood veneer on raised door
Clean	Ea.	2.08	16.85		18.93	panels, hinges and pulls.
Paint	Ea.	1.43	47.50		48.93	
Minimum Charge	Job		163		163	
Clean Interior						Includes labor and materials to clean
Clean	L.F.	.58	3.57		4.15	the interior of a 27" cabinet.
Minimum Charge	Job		118		118	
Premium Hardwood						
12" w, 1 Door, 1 Drawer						Includes material and labor to install
Demolish	Ea.		6.55		6.55	premium hardwood modular unit with
Install	Ea.	305	26.50		331.50	solid hardwood faced frames, drawer
Demolish and Install	Ea.	305	33.05		338.05	fronts, and door panels, steel drawer
Reinstall	Ea.		21.03		21.03	guides, hinges, pulls and laminated
Clean	Ea.	.58	4.71		5.29	interior.
Paint	Ea.	.41	15.90		16.31	
Minimum Charge	Job		163		163	
15" w, 1 Door, 1 Drawer						Includes material and labor to install
Demolish	Ea.		8.20		8.20	premium hardwood modular unit with
Install	Ea.	310	27		337	solid hardwood faced frames, drawer
Demolish and Install	Ea.	310	35.20		345.20	fronts, and door panels, steel drawer
Reinstall	Ea.		21.73		21.73	guides, hinges, pulls and laminated
Clean	Ea.	.73	5.90		6.63	interior.
Paint	Ea.	.52	20.50		21.02	
Minimum Charge	Job		163		163	
18" w, 1 Door, 1 Drawer						Includes material and labor to install
Demolish	Ea.		9.70		9.70	premium hardwood modular unit with
Install	Ea.	335	28		363	solid hardwood faced frames, drawer
Demolish and Install	Ea.	335	37.70		372.70	fronts, and door panels, steel drawer
Reinstall	Ea.		22.39		22.39	guides, hinges, pulls and laminated
Clean	Ea.	.86	7.15		8.01	interior.
Paint	Ea.	.62	24		24.62	
Minimum Charge	Job		163		163	
21" w, 1 Door, 1 Drawer						Includes material and labor to install
Demolish	Ea.		11.40		11.40	premium hardwood modular unit with
Install	Ea.	350	28.50		378.50	solid hardwood faced frames, drawer
Demolish and Install	Ea.	350	39.90		389.90	fronts, and door panels, steel drawer
Reinstall	Ea.		22.98		22.98	guides, hinges, pulls and laminated
Clean	Ea.	1.05	8.40		9.45	interior.
Paint	Ea.	.73	26		26.73	
Minimum Charge	Job		163		163	
24" w, 1 Door, 1 Drawer						Includes material and labor to install
Demolish	Ea.		13.10		13.10	premium hardwood modular unit with
Install	Ea.	390	29		419	solid hardwood faced frames, drawer
Demolish and Install	Ea.	390	42.10		432.10	fronts, and door panels, steel drawer
Reinstall	Ea.		23.39		23.39	guides, hinges and pulls, laminated
Clean	Ea.	1.12	9.40		10.52	interior.
Paint	Ea.	.83	32		32.83	
Minimum Charge	Job		163		163	

Cabinets and Countertops

Base Cabinets	Unit	Material	Labor	Equip.	Total	Specification
30″ Wide Sink Base						
Demolish	Ea.		16.35		16.35	Includes material and labor to install
Install	Ea.	350	30.50		380.50	premium hardwood modular unit with
Demolish and Install	Ea.	350	46.85		396.85	solid hardwood faced frames, drawer
Reinstall	Ea.		24.37		24.37	fronts, and door panels, steel drawer
Clean	Ea.	1.46	11.80		13.26	guides, hinges, pulls and laminated
Paint	Ea.	1.03	36		37.03	interior.
Minimum Charge	Job		163		163	
36″ Wide Sink Base						
Demolish	Ea.		20		20	Includes material and labor to install
Install	Ea.	390	32		422	premium hardwood modular unit with
Demolish and Install	Ea.	390	52		442	solid hardwood faced frames, drawer
Reinstall	Ea.		25.69		25.69	fronts, and door panels, steel drawer
Clean	Ea.	1.62	13.85		15.47	guides, hinges, pulls and laminated
Paint	Ea.	1.24	47.50		48.74	interior.
Minimum Charge	Job		163		163	
36″ Blind Corner						
Demolish	Ea.		20		20	Includes material and labor to install
Install	Ea.	755	36		791	premium hardwood modular unit with
Demolish and Install	Ea.	755	56		811	solid hardwood faced frames, drawer
Reinstall	Ea.		28.98		28.98	fronts, and door panels, steel drawer
Clean	Ea.	1.62	13.85		15.47	guides, hinges, pulls and laminated
Paint	Ea.	1.24	47.50		48.74	interior.
Minimum Charge	Job		163		163	
42″ w, 2 Door, 2 Drawer						
Demolish	Ea.		24		24	Includes material and labor to install
Install	Ea.	620	33		653	premium hardwood modular unit with
Demolish and Install	Ea.	620	57		677	solid hardwood faced frames, drawer
Reinstall	Ea.		26.34		26.34	fronts, and door panels, steel drawer
Clean	Ea.	2.08	16.85		18.93	guides, hinges, pulls and laminated
Paint	Ea.	1.43	57		58.43	interior.
Minimum Charge	Job		163		163	
Clean Interior						
Clean	L.F.	.58	3.57		4.15	Includes labor and materials to clean
Minimum Charge	Job		118		118	the interior of a 27″ cabinet.
Island						
24″ Wide						
Demolish	Ea.		13.10		13.10	Includes material and labor to install
Install	Ea.	370	27		397	custom modular unit with solid
Demolish and Install	Ea.	370	40.10		410.10	hardwood faced frames, hardwood
Reinstall	Ea.		21.73		21.73	door frames and drawer fronts,
Clean	Ea.	1.62	13.85		15.47	hardwood veneer on raised door
Paint	Ea.	1.23	41		42.23	panels, hinges and pulls.
Minimum Charge	Job		163		163	
30″ Wide						
Demolish	Ea.		16.35		16.35	Includes material and labor to install
Install	Ea.	435	34		469	custom modular unit with solid
Demolish and Install	Ea.	435	50.35		485.35	hardwood faced frames, hardwood
Reinstall	Ea.		27.17		27.17	door frames and drawer fronts,
Clean	Ea.	2.08	16.85		18.93	hardwood veneer on raised door
Paint	Ea.	1.55	47.50		49.05	panels, hinges and pulls.
Minimum Charge	Job		163		163	

225

Cabinets and Countertops

Base Cabinets		Unit	Material	Labor	Equip.	Total	Specification
36″ Wide							
	Demolish	Ea.		20		20	Includes material and labor to install
	Install	Ea.	495	41		536	custom modular unit with solid
	Demolish and Install	Ea.	495	61		556	hardwood faced frames, hardwood
	Reinstall	Ea.		32.60		32.60	door frames and drawer fronts,
	Clean	Ea.	2.43	21.50		23.93	hardwood veneer on raised door
	Paint	Ea.	1.87	57		58.87	panels, hinges and pulls.
	Minimum Charge	Job		163		163	
48″ Wide							
	Demolish	Ea.		26		26	Includes material and labor to install
	Install	Ea.	735	54.50		789.50	custom modular unit with solid
	Demolish and Install	Ea.	735	80.50		815.50	hardwood faced frames, hardwood
	Reinstall	Ea.		43.47		43.47	door frames and drawer fronts,
	Clean	Ea.	2.92	26		28.92	hardwood veneer on raised door
	Paint	Ea.	2.15	71.50		73.65	panels, hinges and pulls.
	Minimum Charge	Job		163		163	
60″ Wide							
	Demolish	Ea.		29		29	Includes material and labor to install
	Install	Ea.	870	68		938	custom modular unit with solid
	Demolish and Install	Ea.	870	97		967	hardwood faced frames, hardwood
	Reinstall	Ea.		54.33		54.33	door frames and drawer fronts,
	Clean	Ea.	3.64	33.50		37.14	hardwood veneer on raised door
	Paint	Ea.	2.15	95.50		97.65	panels, hinges and pulls.
	Minimum Charge	Job		163		163	
Clean Interior							
	Clean	L.F.	.58	3.57		4.15	Includes labor and materials to clean
	Minimum Charge	Job		118		118	the interior of a 27″ cabinet.
Liquor Bar							
	Demolish	L.F.		6.55		6.55	Includes material and labor to install
	Install	L.F.	133	32.50		165.50	custom modular unit with solid
	Demolish and Install	L.F.	133	39.05		172.05	hardwood faced frames, hardwood
	Reinstall	L.F.		26.08		26.08	door frames and drawer fronts,
	Clean	L.F.	.58	4.91		5.49	hardwood veneer on raised door
	Paint	L.F.	.41	13		13.41	panels, hinges and pulls.
	Minimum Charge	Job		163		163	
Clean Interior							
	Clean	L.F.	.58	3.57		4.15	Includes labor and materials to clean
	Minimum Charge	Job		118		118	the interior of a 27″ cabinet.
Seal Interior							
	Install	L.F.	.29	4.69		4.98	Includes labor and materials to seal a
	Minimum Charge	Job		143		143	cabinet interior.
Back Bar							
	Demolish	L.F.		6.55		6.55	Includes material and labor to install
	Install	L.F.	133	32.50		165.50	custom modular unit with solid
	Demolish and Install	L.F.	133	39.05		172.05	hardwood faced frames, hardwood
	Reinstall	L.F.		26.08		26.08	door frames and drawer fronts,
	Clean	L.F.	.58	4.91		5.49	hardwood veneer on raised door
	Paint	L.F.	.41	13		13.41	panels, hinges and pulls.
	Minimum Charge	Job		163		163	
Clean Interior							
	Clean	L.F.	.58	3.57		4.15	Includes labor and materials to clean
	Minimum Charge	Job		118		118	the interior of a 27″ cabinet.

Cabinets and Countertops

Base Cabinets		Unit	Material	Labor	Equip.	Total	Specification
Seal Interior							
	Install	L.F.	.29	4.69		4.98	Includes labor and materials to seal a
	Minimum Charge	Job		143		143	cabinet interior.
Storage (garage)							
	Demolish	L.F.		6.55		6.55	Includes material and labor to install
	Install	L.F.	100	32.50		132.50	good modular unit with melamine
	Demolish and Install	L.F.	100	39.05		139.05	laminate, textured colors or wood
	Reinstall	L.F.		26.08		26.08	grain print finish including hinges and
	Clean	L.F.	.58	4.91		5.49	pulls.
	Paint	L.F.	.41	13		13.41	
	Minimum Charge	Job		163		163	
Clean Interior							
	Clean	L.F.	.58	3.57		4.15	Includes labor and materials to clean
	Minimum Charge	Job		118		118	the interior of a 27" cabinet.
Seal Interior							
	Install	L.F.	.29	4.69		4.98	Includes labor and materials to seal a
	Minimum Charge	Job		143		143	cabinet interior.
Built-in Desk							
	Demolish	L.F.		6.55		6.55	Includes material and labor to install
	Install	L.F.	133	32.50		165.50	custom modular unit with solid
	Demolish and Install	L.F.	133	39.05		172.05	hardwood faced frames, hardwood
	Reinstall	L.F.		26.08		26.08	door frames and drawer fronts,
	Clean	L.F.	.58	4.91		5.49	hardwood veneer on raised door
	Paint	L.F.	.41	13		13.41	panels, hinges and pulls.
	Minimum Charge	Job		163		163	
Built-in Bookcase							
	Demolish	L.F.		8.75		8.75	Includes material and labor to install
	Install	L.F.	22	32.50		54.50	custom modular bookcase unit with
	Demolish and Install	L.F.	22	41.25		63.25	clear pine faced frames, shelves 12"
	Reinstall	L.F.		26.08		26.08	O.C., 7' high, 8" deep.
	Clean	L.F.	1.17	8.40		9.57	
	Paint	L.F.	.96	22		22.96	
	Minimum Charge	Job		163		163	
Strip and Refinish							
	Install	L.F.	1.09	5.20		6.29	Includes labor and material to strip,
	Minimum Charge	Job		143		143	prep and refinish exterior of cabinets.
Stain							
	Install	L.F.	.35	1.33		1.68	Includes labor and material to stain
	Minimum Charge	Job		163		163	door and exterior of cabinets.

Countertop		Unit	Material	Labor	Equip.	Total	Specification
Laminated							
With Splash							
	Demolish	L.F.		4.36		4.36	Includes material and labor to install
	Install	L.F.	27.50	11.65		39.15	one piece laminated top with 4"
	Demolish and Install	L.F.	27.50	16.01		43.51	backsplash.
	Reinstall	L.F.		9.31		9.31	
	Clean	L.F.	.08	.59		.67	
	Minimum Charge	Job		163		163	

Cabinets and Countertops

Countertop		Unit	Material	Labor	Equip.	Total	Specification
Roll Top							
	Demolish	L.F.		4.36		4.36	Includes material and labor to install
	Install	L.F.	9.90	10.85		20.75	one piece laminated top with rolled
	Demolish and Install	L.F.	9.90	15.21		25.11	drip edge (post formed) and
	Reinstall	L.F.		8.69		8.69	backsplash.
	Clean	L.F.	.10	.89		.99	
	Minimum Charge	Job		163		163	
Ceramic Tile							
	Demolish	S.F.		2.18		2.18	Cost includes material and labor to
	Install	S.F.	14.30	6.50		20.80	install 4-1/4" x 4-1/4" to 6" x 6"
	Demolish and Install	S.F.	14.30	8.68		22.98	glazed tile set in mortar bed and grout
	Clean	S.F.	.06	.45		.51	on particle board substrate.
	Minimum Charge	Job		147		147	
Cultured Marble							
	Demolish	L.F.		4.36		4.36	Includes material and labor to install
	Install	L.F.	28	33		61	marble countertop, 24" wide, no
	Demolish and Install	L.F.	28	37.36		65.36	backsplash.
	Reinstall	L.F.		26.30		26.30	
	Clean	L.F.	.10	.89		.99	
	Minimum Charge	Job		147		147	
With Splash							
	Demolish	L.F.		4.36		4.36	Includes material and labor to install
	Install	L.F.	31.50	33		64.50	marble countertop, 24" wide, with
	Demolish and Install	L.F.	31.50	37.36		68.86	backsplash.
	Reinstall	L.F.		26.30		26.30	
	Clean	L.F.	.10	.89		.99	
	Minimum Charge	Job		147		147	
Quarry Tile							
	Demolish	S.F.		2.18		2.18	Includes material and labor to install
	Install	S.F.	7.80	6.50		14.30	quarry tile countertop with no
	Demolish and Install	S.F.	7.80	8.68		16.48	backsplash.
	Clean	S.F.	.06	.45		.51	
	Minimum Charge	Job		147		147	
With Splash							
	Demolish	S.F.		2.18		2.18	Includes material and labor to install
	Install	S.F.	9.90	7.25		17.15	quarry tile countertop with backsplash.
	Demolish and Install	S.F.	9.90	9.43		19.33	
	Clean	S.F.	.06	.45		.51	
	Minimum Charge	Job		147		147	
Butcher Block							
	Demolish	S.F.		2.18		2.18	Includes material and labor to install
	Install	S.F.	28.50	5.80		34.30	solid laminated maple countertop with
	Demolish and Install	S.F.	28.50	7.98		36.48	no backsplash.
	Reinstall	S.F.		4.66		4.66	
	Clean	S.F.	.02	.18		.20	
	Minimum Charge	Job		163		163	
Solid Surface							
	Demolish	L.F.		4.36		4.36	Includes material and labor to install
	Install	L.F.	48	23.50		71.50	solid surface counter top 22" deep.
	Demolish and Install	L.F.	48	27.86		75.86	
	Reinstall	L.F.		18.63		18.63	
	Clean	L.F.	.10	.89		.99	
	Minimum Charge	Job		163		163	

Cabinets and Countertops

Vanity Cabinets	Unit	Material	Labor	Equip.	Total	Specification
Good Grade Laminated						
24″ Wide, 2 Door						
Demolish	Ea.		13.10		13.10	Includes material and labor to install
Install	Ea.	146	32.50		178.50	modular unit with melamine laminated
Demolish and Install	Ea.	146	45.60		191.60	to particle board, textured colors or
Reinstall	Ea.		26.08		26.08	wood grain print finish, hinges and
Clean	Ea.	1.12	9.40		10.52	pulls.
Paint	Ea.	.83	24		24.83	
Minimum Charge	Job		163		163	
30″ Wide, 2 Door						
Demolish	Ea.		16.35		16.35	Includes material and labor to install
Install	Ea.	167	41		208	modular unit with melamine laminated
Demolish and Install	Ea.	167	57.35		224.35	to particle board, textured colors or
Reinstall	Ea.		32.60		32.60	wood grain print finish, hinges and
Clean	Ea.	1.46	11.80		13.26	pulls.
Paint	Ea.	1.03	28.50		29.53	
Minimum Charge	Job		163		163	
36″ Wide, 2 Door						
Demolish	Ea.		20		20	Includes material and labor to install
Install	Ea.	223	49		272	modular unit with melamine laminated
Demolish and Install	Ea.	223	69		292	to particle board, textured colors or
Reinstall	Ea.		39.13		39.13	wood grain print finish, hinges and
Clean	Ea.	1.62	13.85		15.47	pulls.
Paint	Ea.	1.24	36		37.24	
Minimum Charge	Job		163		163	
42″ w, 3 Doors / Drawer						
Demolish	Ea.		24		24	Includes material and labor to install
Install	Ea.	320	52.50		372.50	modular unit with melamine laminated
Demolish and Install	Ea.	320	76.50		396.50	to particle board, textured colors or
Reinstall	Ea.		42.13		42.13	wood grain print finish, hinges and
Clean	Ea.	2.08	16.85		18.93	pulls.
Paint	Ea.	1.43	41		42.43	
Minimum Charge	Job		163		163	
48″ w, 3 Doors / Drawer						
Demolish	Ea.		26		26	Includes material and labor to install
Install	Ea.	267	57		324	modular unit with melamine laminated
Demolish and Install	Ea.	267	83		350	to particle board, textured colors or
Reinstall	Ea.		45.63		45.63	wood grain print finish, hinges and
Clean	Ea.	2.92	26		28.92	pulls.
Paint	Ea.	2.15	71.50		73.65	
Minimum Charge	Job		163		163	
Clean Interior						
Install	L.F.	.58	4.91		5.49	Includes labor and materials to clean
Minimum Charge	Job		118		118	base cabinetry.
Better Grade Wood						
24″ Wide, 2 Door						
Demolish	Ea.		13.10		13.10	Includes material and labor to install
Install	Ea.	195	32.50		227.50	custom modular unit with solid
Demolish and Install	Ea.	195	45.60		240.60	hardwood faced frames, hardwood
Reinstall	Ea.		26.08		26.08	door frames and drawer fronts,
Clean	Ea.	1.12	9.40		10.52	hardwood veneer on raised door
Paint	Ea.	.83	26		26.83	panels, hinges and pulls.
Minimum Charge	Job		163		163	

Cabinets and Countertops

Vanity Cabinets	Unit	Material	Labor	Equip.	Total	Specification
30" Wide, 2 Door						
Demolish	Ea.		16.35		16.35	Includes material and labor to install
Install	Ea.	223	41		264	custom modular unit with solid
Demolish and Install	Ea.	223	57.35		280.35	hardwood faced frames, hardwood
Reinstall	Ea.		32.60		32.60	door frames and drawer fronts,
Clean	Ea.	1.46	11.80		13.26	hardwood veneer on raised door
Paint	Ea.	1.03	32		33.03	panels, hinges and pulls.
Minimum Charge	Job		163		163	
36" Wide, 2 Door						
Demolish	Ea.		20		20	Includes material and labor to install
Install	Ea.	298	49		347	custom modular unit with solid
Demolish and Install	Ea.	298	69		367	hardwood faced frames, hardwood
Reinstall	Ea.		39.13		39.13	door frames and drawer fronts,
Clean	Ea.	1.62	13.85		15.47	hardwood veneer on raised door
Paint	Ea.	1.24	41		42.24	panels, hinges and pulls.
Minimum Charge	Job		163		163	
42" w, 3 Doors / Drawer						
Demolish	Ea.		24		24	Includes material and labor to install
Install	Ea.	320	52.50		372.50	custom modular unit with solid
Demolish and Install	Ea.	320	76.50		396.50	hardwood faced frames, hardwood
Reinstall	Ea.		42.13		42.13	door frames and drawer fronts,
Clean	Ea.	2.08	16.85		18.93	hardwood veneer on raised door
Paint	Ea.	1.43	47.50		48.93	panels, hinges and pulls.
Minimum Charge	Job		163		163	
48" w, 3 Doors / Drawer						
Demolish	Ea.		26		26	Includes material and labor to install
Install	Ea.	360	57		417	custom modular unit with solid
Demolish and Install	Ea.	360	83		443	hardwood faced frames, hardwood
Reinstall	Ea.		45.63		45.63	door frames and drawer fronts,
Clean	Ea.	2.92	26		28.92	hardwood veneer on raised door
Paint	Ea.	2.15	71.50		73.65	panels, hinges and pulls.
Minimum Charge	Job		163		163	
Clean Interior						
Install	L.F.	.58	4.91		5.49	Includes labor and materials to clean
Minimum Charge	Job		118		118	base cabinetry.
Premium Grade Wood						
24" Wide, 2 Door						
Demolish	Ea.		13.10		13.10	Includes material and labor to install
Install	Ea.	231	32.50		263.50	premium modular unit with solid
Demolish and Install	Ea.	231	45.60		276.60	hardwood faced frames, hardwood
Reinstall	Ea.		26.08		26.08	door frames and drawer fronts,
Clean	Ea.	1.12	9.40		10.52	hardwood veneer on raised door
Paint	Ea.	.83	28.50		29.33	panels, hinges and pulls.
Minimum Charge	Job		163		163	
30" Wide, 2 Door						
Demolish	Ea.		16.35		16.35	Includes material and labor to install
Install	Ea.	265	41		306	premium modular unit with solid
Demolish and Install	Ea.	265	57.35		322.35	hardwood faced frames, hardwood
Reinstall	Ea.		32.60		32.60	door frames and drawer fronts,
Clean	Ea.	1.46	11.80		13.26	hardwood veneer on raised door
Paint	Ea.	1.03	36		37.03	panels, hinges and pulls.
Minimum Charge	Job		163		163	

Cabinets and Countertops

Vanity Cabinets	Unit	Material	Labor	Equip.	Total	Specification
36" Wide, 2 Door						
Demolish	Ea.		20		20	Includes material and labor to install
Install	Ea.	415	49		464	premium modular unit with solid
Demolish and Install	Ea.	415	69		484	hardwood faced frames, hardwood
Reinstall	Ea.		39.13		39.13	door frames and drawer fronts,
Clean	Ea.	1.62	13.85		15.47	hardwood veneer on raised door
Paint	Ea.	1.24	41		42.24	panels, hinges and pulls.
Minimum Charge	Job		163		163	
42" w, 3 Doors / Drawer						
Demolish	Ea.		24		24	Includes material and labor to install
Install	Ea.	395	52.50		447.50	premium modular unit with solid
Demolish and Install	Ea.	395	76.50		471.50	hardwood faced frames, hardwood
Reinstall	Ea.		42.13		42.13	door frames and drawer fronts,
Clean	Ea.	2.08	16.85		18.93	hardwood veneer on raised door
Paint	Ea.	1.43	47.50		48.93	panels, hinges and pulls.
Minimum Charge	Job		163		163	
48" w, 3 Doors / Drawer						
Demolish	Ea.		26		26	Includes material and labor to install
Install	Ea.	485	57		542	premium modular unit with solid
Demolish and Install	Ea.	485	83		568	hardwood faced frames, hardwood
Reinstall	Ea.		45.63		45.63	door frames and drawer fronts,
Clean	Ea.	2.92	26		28.92	hardwood veneer on raised door
Paint	Ea.	2.15	71.50		73.65	panels, hinges and pulls.
Minimum Charge	Job		163		163	
Clean Interior						
Install	L.F.	.58	4.91		5.49	Includes labor and materials to clean
Minimum Charge	Job		118		118	base cabinetry.
Premium Hardwood						
24" Wide, 2 Door						
Demolish	Ea.		13.10		13.10	Includes material and labor to install
Install	Ea.	289	32.50		321.50	premium modular unit with solid
Demolish and Install	Ea.	289	45.60		334.60	hardwood faced frames, hardwood
Reinstall	Ea.		26.08		26.08	door frames and drawer fronts,
Clean	Ea.	1.12	9.40		10.52	hardwood veneer on raised door
Paint	Ea.	.83	32		32.83	panels, hinges and pulls.
Minimum Charge	Job		163		163	
30" Wide, 2 Door						
Demolish	Ea.		16.35		16.35	Includes material and labor to install
Install	Ea.	325	41		366	premium modular unit with solid
Demolish and Install	Ea.	325	57.35		382.35	hardwood faced frames, hardwood
Reinstall	Ea.		32.60		32.60	door frames and drawer fronts,
Clean	Ea.	1.46	11.80		13.26	hardwood veneer on raised door
Paint	Ea.	1.03	36		37.03	panels, hinges and pulls.
Minimum Charge	Job		163		163	
36" Wide, 2 Door						
Demolish	Ea.		20		20	Includes material and labor to install
Install	Ea.	475	49		524	premium modular unit with solid
Demolish and Install	Ea.	475	69		544	hardwood faced frames, hardwood
Reinstall	Ea.		39.13		39.13	door frames and drawer fronts,
Clean	Ea.	1.62	13.85		15.47	hardwood veneer on raised door
Paint	Ea.	1.24	47.50		48.74	panels, hinges and pulls.
Minimum Charge	Job		163		163	

Cabinets and Countertops

Vanity Cabinets	Unit	Material	Labor	Equip.	Total	Specification
42" w, 3 Doors / Drawer						
Demolish	Ea.		24		24	Includes material and labor to install
Install	Ea.	440	52.50		492.50	premium modular unit with solid
Demolish and Install	Ea.	440	76.50		516.50	hardwood faced frames, hardwood
Reinstall	Ea.		42.13		42.13	door frames and drawer fronts,
Clean	Ea.	2.08	16.85		18.93	hardwood veneer on raised door
Paint	Ea.	1.43	57		58.43	panels, hinges and pulls.
Minimum Charge	Job		163		163	
48" w, 3 Doors / Drawer						
Demolish	Ea.		26		26	Includes material and labor to install
Install	Ea.	545	57		602	premium modular unit with solid
Demolish and Install	Ea.	545	83		628	hardwood faced frames, hardwood
Reinstall	Ea.		45.63		45.63	door frames and drawer fronts,
Clean	Ea.	2.92	26		28.92	hardwood veneer on raised door
Paint	Ea.	2.15	71.50		73.65	panels, hinges and pulls.
Minimum Charge	Job		163		163	
Clean Interior						
Install	L.F.	.58	4.91		5.49	Includes labor and materials to clean
Minimum Charge	Job		118		118	base cabinetry.
Strip and Refinish						
Install	L.F.	1.09	5.20		6.29	Includes labor and material to strip,
Minimum Charge	Job		143		143	prep and refinish exterior of cabinets.

Vanity Cabinet Tops	Unit	Material	Labor	Equip.	Total	Specification
Cultured Marble Top						
Demolish	L.F.		4.36		4.36	Includes material and labor to install
Install	L.F.	35	19.35		54.35	cultured marble vanity top with
Demolish and Install	L.F.	35	23.71		58.71	integral sink, 22" deep.
Reinstall	L.F.		15.47		15.47	
Clean	L.F.	.10	.89		.99	
Minimum Charge	Job		147		147	
Laminated Top						
Demolish	L.F.		2.62		2.62	Includes material and labor to install
Install	L.F.	27.50	11.65		39.15	one piece laminated top with 4"
Demolish and Install	L.F.	27.50	14.27		41.77	backsplash.
Reinstall	L.F.		9.31		9.31	
Clean	L.F.	.08	.59		.67	
Minimum Charge	Job		163		163	
Solid Surface						
Demolish	L.F.		4.36		4.36	Includes material and labor to install
Install	L.F.	67	24		91	solid surface vanity top with integral
Demolish and Install	L.F.	67	28.36		95.36	sink installed, 22" deep.
Reinstall	L.F.		19.32		19.32	
Clean	L.F.	.10	.89		.99	
Minimum Charge	Job		163		163	

Roller Handle

Paint Brush

Paint Preparation, Walls		Unit	Material	Labor	Equip.	Total	Specification
Cover / Protect Floors							
	Install	S.F.	.01	.11		.12	Includes material and labor to install
	Minimum Charge	Job		141		141	plastic masking sheet on large uninterrupted areas.
Cover / Protect Walls							
	Install	S.F.	.01	.11		.12	Includes material and labor to install
	Minimum Charge	Job		141		141	plastic masking sheet on walls.
Clean Walls							
Light							
	Clean	S.F.		.09		.09	Includes labor to wash gypsum
	Minimum Charge	Job		141		141	drywall or plaster wall surfaces.
Heavy							
	Clean	S.F.		.11		.11	Includes labor and material for heavy
	Minimum Charge	Job		141		141	cleaning (multiple applications) with detergent and solvent.
Prep Walls							
	Paint	S.F.	.03	.14		.17	Includes labor and material to prepare
	Minimum Charge	Job		141		141	for painting including scraping, patching and puttying.
Sand Walls							
	Paint	S.F.		.09		.09	Includes labor to sand gypsum drywall
	Minimum Charge	Job		141		141	wall surfaces.
Seal Walls							
Large Area							
	Paint	S.F.	.06	.14		.20	Includes labor and material to paint
	Minimum Charge	Job		141		141	with primer / sealer by roller.
Spot							
	Paint	S.F.	.06	.25		.31	Includes labor and material to seal
	Minimum Charge	Job		141		141	existing drywall with shellac-based material.

Paint Preparation, Ceilings		Unit	Material	Labor	Equip.	Total	Specification
Cover / Protect Floors							
	Install	S.F.	.01	.11		.12	Includes material and labor to install
	Minimum Charge	Job		141		141	plastic masking sheet on large uninterrupted areas.

Painting

Paint Preparation, Ceilings

Paint Preparation, Ceilings		Unit	Material	Labor	Equip.	Total	Specification
Cover / Protect Walls							
	Install	S.F.	.01	.11		.12	Includes material and labor to install
	Minimum Charge	Job		141		141	plastic masking sheet on walls.
Prep Ceiling							
	Paint	S.F.	.04	.53		.57	Includes labor and material to fill in
	Minimum Charge	Job		141		141	hairline cracks up to 1/8" with filler, sand and prep for paint.
Sand Ceiling							
	Paint	S.F.		.13		.13	Includes labor to sand gypsum drywall
	Minimum Charge	Job		141		141	ceiling surfaces.
Seal Ceiling							
Large Area							
	Paint	S.F.	.06	.22		.28	Includes labor and material to paint
	Minimum Charge	Job		143		143	ceiling, one coat flat latex.
Spot							
	Paint	S.F.	.06	.25		.31	Includes labor and material to seal
	Minimum Charge	Job		143		143	existing drywall with shellac-based material.

Paint / Texture, Walls

Paint / Texture, Walls		Unit	Material	Labor	Equip.	Total	Specification
Texture Walls							
Spray							
	Install	S.F.	.04	.36		.40	Includes labor and material to install
	Minimum Charge	Job		143		143	by spray, texture finish.
Trowel							
	Install	S.F.		.19		.19	Includes labor and material to apply
	Minimum Charge	Job		143		143	hand troweled texture.
Paint Walls 1 Coat							
	Paint	S.F.	.06	.22		.28	Includes labor and material to paint,
	Minimum Charge	Job		143		143	one coat flat latex.
Paint Walls 2 Coats							
	Paint	S.F.	.11	.36		.47	Includes labor and material to paint,
	Minimum Charge	Job		143		143	two coats flat latex.
Paint Walls 3 Coats							
	Paint	S.F.	.15	.44		.59	Includes labor and material to paint,
	Minimum Charge	Job		143		143	three coats flat latex.
Clean and Seal							
	Paint	S.F.	.22	.76		.98	Includes labor and material to do light
	Minimum Charge	Job		143		143	cleaning, sealing with 2 coats of latex paint.

Paint / Texture, Ceilings

Paint / Texture, Ceilings		Unit	Material	Labor	Equip.	Total	Specification
Texture Ceiling							
Spray							
	Install	S.F.	.04	.36		.40	Includes labor and material to install
	Minimum Charge	Job		143		143	by spray, texture finish.
Trowel							
	Install	S.F.		.19		.19	Includes labor and material to apply
	Minimum Charge	Job		143		143	hand troweled texture.

Painting

Paint / Texture, Ceilings		Unit	Material	Labor	Equip.	Total	Specification
Paint Ceiling, 1 Coat							
	Paint	S.F.	.06	.22		.28	Includes labor and material to paint
	Minimum Charge	Job		143		143	ceiling, one coat flat latex.
Paint Ceiling, 2 Coats							
	Paint	S.F.	.11	.36		.47	Includes labor and material to paint
	Minimum Charge	Job		143		143	two coats flat latex paint.
Paint Ceiling, 3 Coats							
	Paint	S.F.	.15	.42		.57	Includes labor and material to paint
	Minimum Charge	Job		143		143	ceiling, three coats flat latex.
Clean and Seal							
	Paint	S.F.	.15	.60		.75	Includes labor and material for light
							cleaning, sealing and paint 3 coats of
	Minimum Charge	Job		143		143	flat latex paint.

Flooring

Plank Flooring

Wood Strip Flooring

Tile Flooring

Wood Plank Flooring	Unit	Material	Labor	Equip.	Total	Specification
Maple						
2-1/4"						
Demolish	S.F.		1.05		1.05	Includes material and labor to install
Install	S.F.	4.19	1.63		5.82	unfinished T&G maple flooring,
Demolish and Install	S.F.	4.19	2.68		6.87	25/32" thick, 2-1/4" wide, random
Reinstall	S.F.		1.30		1.30	3' to 16' lengths including felt
Clean	S.F.		.29		.29	underlayment, nailed in place over
Paint	S.F.	1.14	1.81		2.95	prepared subfloor.
Minimum Charge	Job		163		163	
3-1/4"						
Demolish	S.F.		1.05		1.05	Includes material and labor to install
Install	S.F.	3.27	1.63		4.90	unfinished T&G maple flooring,
Demolish and Install	S.F.	3.27	2.68		5.95	33/32" thick, 3-1/4" wide, random
Reinstall	S.F.		1.30		1.30	3' to 16' lengths including felt
Clean	S.F.		.29		.29	underlayment, nailed in place over
Paint	S.F.	1.14	1.81		2.95	prepared subfloor.
Minimum Charge	Job		163		163	
Sand and Finish						
Install	S.F.	.76	.59		1.35	Includes labor and material to sand (3
Minimum Charge	Job		163		163	passes) and finish and two coats of
						urethane existing floor.
Oak						
2-1/4"						
Demolish	S.F.		1.05		1.05	Includes material and labor to install
Install	S.F.	4.02	1.63		5.65	unfinished T&G oak flooring, 25/32"
Demolish and Install	S.F.	4.02	2.68		6.70	thick, 2-1/4" wide, random 3' to 16'
Reinstall	S.F.		1.30		1.30	lengths including felt underlayment,
Clean	S.F.		.29		.29	nailed in place over prepared
Paint	S.F.	1.14	1.81		2.95	subfloor.
Minimum Charge	Job		163		163	
3-1/4"						
Demolish	S.F.		1.05		1.05	Includes material and labor to install
Install	S.F.	5.50	1.15		6.65	unfinished T&G oak flooring, 25/32"
Demolish and Install	S.F.	5.50	2.20		7.70	thick, 3-1/4" wide, random 3' to 16'
Reinstall	S.F.		.92		.92	lengths including felt underlayment,
Clean	S.F.		.29		.29	nailed in place over prepared
Paint	S.F.	1.14	1.81		2.95	subfloor.
Minimum Charge	Job		163		163	

Flooring

Wood Plank Flooring		Unit	Material	Labor	Equip.	Total	Specification
Sand and Finish							
	Install	S.F.	1.14	1.81		2.95	Includes labor and material to sand
	Minimum Charge	Job		163		163	and finish (3 passes) with two coats of
							urethane on a new floor.
Walnut							
3" to 7"							
	Demolish	S.F.		1.05		1.05	Includes material and labor to install
	Install	S.F.	9.05	1.15		10.20	unfinished T&G walnut flooring,
	Demolish and Install	S.F.	9.05	2.20		11.25	25/32" thick, random 3' to 16'
	Reinstall	S.F.		.92		.92	lengths including felt underlayment,
	Clean	S.F.		.29		.29	nailed in place over prepared
	Paint	S.F.	1.14	1.81		2.95	subfloor.
	Minimum Charge	Job		163		163	
Sand and Finish							
	Install	S.F.	1.14	1.81		2.95	Includes labor and material to sand
	Minimum Charge	Job		163		163	and finish (3 passes) with two coats of
							urethane on a new floor.
Pine							
2-1/4"							
	Demolish	S.F.		1.05		1.05	Includes material and labor to install
	Install	S.F.	7.35	1.23		8.58	unfinished T&G pine flooring, 25/32"
	Demolish and Install	S.F.	7.35	2.28		9.63	thick, 2-1/4" wide, random 3' to 16'
	Reinstall	S.F.		.98		.98	lengths including felt underlayment,
	Clean	S.F.		.29		.29	nailed in place over prepared
	Paint	S.F.	.76	.59		1.35	subfloor.
	Minimum Charge	Job		163		163	
Sand and Finish							
	Install	S.F.	1.14	1.81		2.95	Includes labor and material to sand
	Minimum Charge	Job		163		163	and finish (3 passes) with two coats of
							urethane on a new floor.
Prefinished Oak							
2-1/4"							
	Demolish	S.F.		1.05		1.05	Includes material and labor to install
	Install	S.F.	5.60	1.63		7.23	prefinished T&G oak flooring, 2-1/4"
	Demolish and Install	S.F.	5.60	2.68		8.28	wide, random 3' to 16' lengths
	Reinstall	S.F.		1.30		1.30	including felt underlayment, nailed in
	Clean	S.F.		.29		.29	place over prepared subfloor.
	Minimum Charge	Job		163		163	
3-1/4"							
	Demolish	S.F.		1.05		1.05	Includes material and labor to install
	Install	S.F.	7.15	1.49		8.64	prefinished T&G oak flooring, 25/32"
	Demolish and Install	S.F.	7.15	2.54		9.69	thick, 3" to 7" wide, random 3' to 16'
	Reinstall	S.F.		1.19		1.19	lengths including felt underlayment,
	Clean	S.F.		.29		.29	nailed in place over prepared
	Minimum Charge	Job		163		163	subfloor.
Sand and Finish New							
	Install	S.F.	.76	.80		1.56	Includes labor and material to sand (3
	Minimum Charge	Job		163		163	passes) and finish, two coats of
							urethane, new floor.
Refinish Existing							
	Install	S.F.	.76	.59		1.35	Includes labor and material to sand (3
	Minimum Charge	Job		163		163	passes) and finish and two coats of
							urethane existing floor.

Flooring

Wood Plank Flooring

	Unit	Material	Labor	Equip.	Total	Specification
Urethane Coat						
Install	S.F.	.14	.15		.29	Includes labor and material to install
Minimum Charge	Job		163		163	two coats of urethane, on existing floor.
Clean and Wax						
Clean	S.F.	.02	.18		.20	Includes labor and material to wash
Minimum Charge	Job		163		163	and wax a hardwood floor.
Underlayment Per S.F.						
Demolish	S.F.		.48		.48	Cost includes material and labor to
Install	S.F.	.86	.43		1.29	install 1/4" lauan subfloor, standard
Demolish and Install	S.F.	.86	.91		1.77	interior grade, nailed every 6".
Clean	S.F.		.23		.23	
Minimum Charge	Job		163		163	

Wood Parquet Tile Floor

	Unit	Material	Labor	Equip.	Total	Specification
Oak						
9" x 9"						
Demolish	S.F.		.80		.80	Includes material and labor to install
Install	S.F.	4.62	2.61		7.23	prefinished oak 9" x 9" parquet block
Demolish and Install	S.F.	4.62	3.41		8.03	flooring, 5/16" thick, installed in
Reinstall	S.F.		2.09		2.09	mastic.
Clean	S.F.		.29		.29	
Paint	S.F.	1.14	1.81		2.95	
Minimum Charge	Job		163		163	
13" x 13"						
Demolish	S.F.		.80		.80	Includes material and labor to install
Install	S.F.	8.75	2.04		10.79	prefinished oak 13" x 13" parquet
Demolish and Install	S.F.	8.75	2.84		11.59	block flooring, 5/16" thick, installed in
Reinstall	S.F.		1.63		1.63	mastic.
Clean	S.F.		.29		.29	
Paint	S.F.	1.14	1.81		2.95	
Minimum Charge	Job		163		163	
Cherry						
9" x 9"						
Demolish	S.F.		.80		.80	Includes material and labor to install
Install	S.F.	8.15	2.61		10.76	prefinished cherry 9" x 9" parquet
Demolish and Install	S.F.	8.15	3.41		11.56	block flooring, 5/16" thick, installed in
Reinstall	S.F.		2.61		2.61	mastic.
Clean	S.F.		.29		.29	
Paint	S.F.	1.14	1.81		2.95	
Minimum Charge	Job		163		163	
13" x 13"						
Demolish	S.F.		.80		.80	Includes material and labor to install
Install	S.F.	9.30	2.04		11.34	prefinished cherry 13" x 13" parquet
Demolish and Install	S.F.	9.30	2.84		12.14	block flooring, 5/16" thick, installed in
Reinstall	S.F.		2.04		2.04	mastic.
Clean	S.F.		.29		.29	
Paint	S.F.	1.14	1.81		2.95	
Minimum Charge	Job		163		163	

Flooring

Wood Parquet Tile Floor		Unit	Material	Labor	Equip.	Total	Specification
Walnut							
9″ x 9″							
	Demolish	S.F.		.80		.80	Includes material and labor to install
	Install	S.F.	5	2.61		7.61	prefinished walnut 9″ x 9″ parquet
	Demolish and Install	S.F.	5	3.41		8.41	flooring in mastic.
	Reinstall	S.F.		2.09		2.09	
	Clean	S.F.		.29		.29	
	Paint	S.F.	1.14	1.81		2.95	
	Minimum Charge	Job		163		163	
13″ x 13″							
	Demolish	S.F.		.80		.80	Includes material and labor to install
	Install	S.F.	9.35	2.04		11.39	prefinished walnut 13″ x 13″ parquet
	Demolish and Install	S.F.	9.35	2.84		12.19	block flooring, 5/16″ thick, installed in
	Reinstall	S.F.		2.04		2.04	mastic.
	Clean	S.F.		.29		.29	
	Paint	S.F.	1.14	1.81		2.95	
	Minimum Charge	Job		163		163	
Acrylic Impregnated							
Oak 12″ x 12″							
	Demolish	S.F.		.80		.80	Includes material and labor to install
	Install	S.F.	7.55	2.33		9.88	red oak 12″ x 12″ acrylic
	Demolish and Install	S.F.	7.55	3.13		10.68	impregnated parquet block flooring,
	Reinstall	S.F.		2.33		2.33	5/16″ thick, installed in mastic.
	Clean	S.F.		.29		.29	
	Minimum Charge	Job		163		163	
Cherry 12″ x 12″							
	Demolish	S.F.		.80		.80	Includes material and labor to install
	Install	S.F.	11.40	2.33		13.73	prefinished cherry 12″ x 12″ acrylic
	Demolish and Install	S.F.	11.40	3.13		14.53	impregnated parquet block flooring,
	Reinstall	S.F.		2.33		2.33	5/16″ thick, installed in mastic.
	Clean	S.F.		.29		.29	
	Minimum Charge	Job		163		163	
Ash 12″ x 12″							
	Demolish	S.F.		.80		.80	Includes material and labor to install
	Install	S.F.	10.25	2.33		12.58	ash 12″ x 12″ acrylic impregnated
	Demolish and Install	S.F.	10.25	3.13		13.38	parquet block flooring, 5/16″ thick in
	Reinstall	S.F.		2.33		2.33	mastic.
	Clean	S.F.		.29		.29	
	Minimum Charge	Job		163		163	
Teak							
	Demolish	S.F.		.80		.80	Includes material and labor to install
	Install	S.F.	7.20	2.04		9.24	prefinished teak parquet block
	Demolish and Install	S.F.	7.20	2.84		10.04	flooring, 5/16″ thick, and installed in
	Reinstall	S.F.		2.04		2.04	mastic.
	Clean	S.F.		.29		.29	
	Paint	S.F.	1.14	1.81		2.95	
	Minimum Charge	Job		163		163	
Clean and Wax							
	Clean	S.F.	.02	.18		.20	Includes labor and material to wash
	Minimum Charge	Job		163		163	and wax a hardwood floor.

Flooring

Ceramic Tile Flooring		Unit	Material	Labor	Equip.	Total	Specification
Economy Grade							
	Demolish	S.F.		.78		.78	Cost includes material and labor to
	Install	S.F.	4.29	5.80		10.09	install 4-1/4" x 4-1/4" ceramic tile in
	Demolish and Install	S.F.	4.29	6.58		10.87	mortar bed including grout.
	Clean	S.F.		.45		.45	
	Minimum Charge	Job		147		147	
Average Grade							
	Demolish	S.F.		.78		.78	Cost includes material and labor to
	Install	S.F.	5.70	5.90		11.60	install 4-1/4" x 4-1/4" ceramic tile in
	Demolish and Install	S.F.	5.70	6.68		12.38	mortar bed including grout.
	Clean	S.F.		.45		.45	
	Minimum Charge	Job		147		147	
Premium Grade							
	Demolish	S.F.		.78		.78	Cost includes material and labor to
	Install	S.F.	8.95	5.90		14.85	install 4-1/4" x 4-1/4" ceramic tile in
	Demolish and Install	S.F.	8.95	6.68		15.63	mortar bed including grout.
	Clean	S.F.		.45		.45	
	Minimum Charge	Job		147		147	
Re-grout							
	Install	S.F.	.14	2.36		2.50	Includes material and labor to regrout
	Minimum Charge	Job		147		147	tile floors.

Hard Tile Flooring		Unit	Material	Labor	Equip.	Total	Specification
Thick Set Paver							
Brick							
	Demolish	S.F.		1.64		1.64	Cost includes material and labor to
	Install	S.F.	.72	13.80		14.52	install 6" x 12" adobe brick paver with
	Demolish and Install	S.F.	.72	15.44		16.16	1/2" mortar joints.
	Clean	S.F.		.45		.45	
	Minimum Charge	Job		147		147	
Mexican Red							
	Demolish	S.F.		1.64		1.64	Cost includes material and labor to
	Install	S.F.	1.43	6.15		7.58	install 12" x 12" red Mexican paver
	Demolish and Install	S.F.	1.43	7.79		9.22	tile in mortar bed with grout.
	Clean	S.F.		.45		.45	
	Minimum Charge	Job		147		147	
Saltillo							
	Demolish	S.F.		.78		.78	Cost includes material and labor to
	Install	S.F.	2.24	6.15		8.39	install 12" x 12"saltillo tile in mortar
	Demolish and Install	S.F.	2.24	6.93		9.17	bed with grout.
	Clean	S.F.		.45		.45	
	Minimum Charge	Job		147		147	
Marble							
Premium Grade							
	Demolish	S.F.		.78		.78	Cost includes material and labor to
	Install	S.F.	18.90	12.80		31.70	install 3/8" x 12" x 12" marble, in
	Demolish and Install	S.F.	18.90	13.58		32.48	mortar bed with grout.
	Clean	S.F.		.45		.45	
	Minimum Charge	Job		147		147	
Terrazzo							
Gray Cement							
	Demolish	S.F.		1.04		1.04	Cost includes material and labor to
	Install	S.F.	2.74	4.07	1.77	8.58	install 1-3/4" terrazzo , #1 and #2
	Demolish and Install	S.F.	2.74	5.11	1.77	9.62	chips in gray Portland cement.
	Clean	S.F.		.45		.45	
	Minimum Charge	Job		265	115	380	

Flooring

Hard Tile Flooring

Hard Tile Flooring		Unit	Material	Labor	Equip.	Total	Specification
White Cement							
	Demolish	S.F.		1.04		1.04	Cost includes material and labor to
	Install	S.F.	3.11	4.07	1.77	8.95	install 1-3/4" terrazzo, #1 and #2
	Demolish and Install	S.F.	3.11	5.11	1.77	9.99	chips in white Portland cement.
	Clean	S.F.		.45		.45	
	Minimum Charge	Job		265	115	380	
Non-skid Gray							
	Demolish	S.F.		1.04		1.04	Cost includes material and labor to
	Install	S.F.	4.97	8.85	3.84	17.66	install 1-3/4" terrazzo, #1 and #2
	Demolish and Install	S.F.	4.97	9.89	3.84	18.70	chips in gray Portland cement with
	Clean	S.F.		.45		.45	light non-skid abrasive.
	Minimum Charge	Job		265	115	380	
Non-skid White							
	Demolish	S.F.		1.04		1.04	Cost includes material and labor to
	Install	S.F.	5.45	8.85	3.84	18.14	install 1-3/4" terrazzo, #1 and #2
	Demolish and Install	S.F.	5.45	9.89	3.84	19.18	chips in white Portland cement with
	Clean	S.F.		.45		.45	light non-skid abrasive.
	Minimum Charge	Job		265	115	380	
Gray w / Brass Divider							
	Demolish	S.F.		1.05		1.05	Cost includes material and labor to
	Install	S.F.	6.80	9.65	4.19	20.64	install 1-3/4" terrazzo, #1 and #2
	Demolish and Install	S.F.	6.80	10.70	4.19	21.69	chips in gray Portland cement with
	Clean	S.F.		.45		.45	brass strips 2' O.C. each way.
	Minimum Charge	Job		265	115	380	
Clean							
	Clean	S.F.	.12	.30		.42	Includes labor and material to clean
	Minimum Charge	Job		265	115	380	terrazzo flooring.
Slate							
	Demolish	S.F.		.94		.94	Includes material and labor to install
	Install	S.F.	4.55	2.91		7.46	slate tile flooring in thin set with grout.
	Demolish and Install	S.F.	4.55	3.85		8.40	
	Clean	S.F.		.45		.45	
	Minimum Charge	Job		147		147	
Granite							
	Demolish	S.F.		.77		.77	Includes material and labor to install
	Install	S.F.	18.55	24.50		43.05	granite tile flooring in thin set with
	Demolish and Install	S.F.	18.55	25.27		43.82	grout.
	Reinstall	S.F.		24.57		24.57	
	Clean	S.F.		.45		.45	
	Minimum Charge	Job		147		147	
Re-grout							
	Install	S.F.	.14	2.36		2.50	Includes material and labor to regrout
	Minimum Charge	Job		147		147	tile floors.

Sheet Vinyl Per S.F.

Sheet Vinyl Per S.F.		Unit	Material	Labor	Equip.	Total	Specification
Economy Grade							
	Demolish	S.F.		.37		.37	Includes material and labor to install
	Install	S.F.	1.52	.92		2.44	resilient sheet vinyl flooring.
	Demolish and Install	S.F.	1.52	1.29		2.81	
	Clean	S.F.		.29		.29	
	Minimum Charge	Job		147		147	

Flooring

Sheet Vinyl Per S.F.

		Unit	Material	Labor	Equip.	Total	Specification
Average Grade							
	Demolish	S.F.		.37		.37	Includes material and labor to install
	Install	S.F.	2.43	.92		3.35	resilient sheet vinyl flooring.
	Demolish and Install	S.F.	2.43	1.29		3.72	
	Clean	S.F.		.29		.29	
	Minimum Charge	Job		147		147	
Premium Grade							
	Demolish	S.F.		.37		.37	Includes material and labor to install
	Install	S.F.	3.84	.92		4.76	resilient sheet vinyl flooring.
	Demolish and Install	S.F.	3.84	1.29		5.13	
	Clean	S.F.		.29		.29	
	Minimum Charge	Job		147		147	
Clean and Wax							
	Clean	S.F.	.02	.15		.17	Includes labor and material to wash
	Minimum Charge	Job		147		147	and wax a vinyl tile floor.

Sheet Vinyl

		Unit	Material	Labor	Equip.	Total	Specification
Economy Grade							
	Demolish	S.Y.		3.38		3.38	Includes material and labor to install
	Install	S.Y.	13.60	8.30		21.90	no wax sheet vinyl flooring.
	Demolish and Install	S.Y.	13.60	11.68		25.28	
	Clean	S.Y.		2.05		2.05	
	Minimum Charge	Job		147		147	
Average Grade							
	Demolish	S.Y.		3.38		3.38	Includes material and labor to install
	Install	S.Y.	21.50	8.30		29.80	no wax sheet vinyl flooring.
	Demolish and Install	S.Y.	21.50	11.68		33.18	
	Clean	S.Y.		2.05		2.05	
	Minimum Charge	Job		147		147	
Premium Grade							
	Demolish	S.Y.		3.38		3.38	Includes material and labor to install
	Install	S.Y.	34.50	8.30		42.80	no wax sheet vinyl flooring.
	Demolish and Install	S.Y.	34.50	11.68		46.18	
	Clean	S.Y.		2.05		2.05	
	Minimum Charge	Job		147		147	
Clean and Wax							
	Clean	S.F.	.02	.15		.17	Includes labor and material to wash
	Minimum Charge	Job		147		147	and wax a vinyl tile floor.

Vinyl Tile

		Unit	Material	Labor	Equip.	Total	Specification
Economy Grade							
	Demolish	S.F.		.52		.52	Includes material and labor to install
	Install	S.F.	1.94	.59		2.53	no wax 12" x 12" vinyl tile.
	Demolish and Install	S.F.	1.94	1.11		3.05	
	Clean	S.F.		.29		.29	
	Minimum Charge	Job		147		147	
Average Grade							
	Demolish	S.F.		.52		.52	Includes material and labor to install
	Install	S.F.	5.30	.59		5.89	no wax 12" x 12" vinyl tile.
	Demolish and Install	S.F.	5.30	1.11		6.41	
	Clean	S.F.		.29		.29	
	Minimum Charge	Job		147		147	

Flooring

Vinyl Tile

Vinyl Tile	Unit	Material	Labor	Equip.	Total	Specification
Premium Grade						
Demolish	S.F.		.52		.52	Includes material and labor to install
Install	S.F.	9.25	.59		9.84	no wax 12" x 12" vinyl tile.
Demolish and Install	S.F.	9.25	1.11		10.36	
Clean	S.F.		.29		.29	
Minimum Charge	Job		147		147	
Clean and Wax						
Clean	S.F.	.02	.15		.17	Includes labor and material to wash
Minimum Charge	Job		147		147	and wax a vinyl tile floor.

Carpeting Per S.F.

Carpeting Per S.F.	Unit	Material	Labor	Equip.	Total	Specification
Economy Grade						
Demolish	S.F.		.06		.06	Includes material and labor to install
Install	S.F.	1.10	.57		1.67	carpet including tack strips and hot
Demolish and Install	S.F.	1.10	.63		1.73	melt tape on seams.
Clean	S.F.		.41		.41	
Minimum Charge	Job		147		147	
Average Grade						
Demolish	S.F.		.06		.06	Includes material and labor to install
Install	S.F.	1.30	.57		1.87	carpet including tack strips and hot
Demolish and Install	S.F.	1.30	.63		1.93	melt tape on seams.
Clean	S.F.		.41		.41	
Minimum Charge	Job		147		147	
Premium Grade						
Demolish	S.F.		.06		.06	Includes material and labor to install
Install	S.F.	2.31	.57		2.88	carpet including tack strips and hot
Demolish and Install	S.F.	2.31	.63		2.94	melt tape on seams.
Clean	S.F.		.41		.41	
Minimum Charge	Job		147		147	
Indoor/Outdoor						
Demolish	S.F.		.06		.06	Includes material and labor to install
Install	S.F.	1.08	.70		1.78	indoor-outdoor carpet.
Demolish and Install	S.F.	1.08	.76		1.84	
Clean	S.F.		.41		.41	
Minimum Charge	Job		147		147	
100% Wool						
Average Grade						
Demolish	S.F.		.06		.06	Includes material and labor to install
Install	S.F.	6.45	.58		7.03	carpet including tack strips and hot
Demolish and Install	S.F.	6.45	.64		7.09	melt tape on seams.
Clean	S.F.		.41		.41	
Minimum Charge	Job		147		147	
Premium Grade						
Demolish	S.F.		.06		.06	Includes material and labor to install
Install	S.F.	7.15	.58		7.73	carpet including tack strips and hot
Demolish and Install	S.F.	7.15	.64		7.79	melt tape on seams.
Clean	S.F.		.41		.41	
Minimum Charge	Job		147		147	
Luxury Grade						
Demolish	S.F.		.06		.06	Includes material and labor to install
Install	S.F.	10.10	.58		10.68	carpet including tack strips and hot
Demolish and Install	S.F.	10.10	.64		10.74	melt tape on seams.
Clean	S.F.		.41		.41	
Minimum Charge	Job		147		147	

Flooring

Carpeting Per S.F.

Carpeting Per S.F.		Unit	Material	Labor	Equip.	Total	Specification
Berber							
Average Grade							
	Demolish	S.F.		.06		.06	Includes material and labor to install
	Install	S.F.	2.88	.70		3.58	carpet including tack strips and hot
	Demolish and Install	S.F.	2.88	.76		3.64	melt tape on seams.
	Clean	S.F.		.41		.41	
	Minimum Charge	Job		147		147	
Premium Grade							
	Demolish	S.F.		.06		.06	Includes material and labor to install
	Install	S.F.	5.90	.70		6.60	carpet including tack strips and hot
	Demolish and Install	S.F.	5.90	.76		6.66	melt tape on seams.
	Clean	S.F.		.41		.41	
	Minimum Charge	Job		147		147	

Carpeting

Carpeting		Unit	Material	Labor	Equip.	Total	Specification
Indoor/Outdoor							
	Demolish	S.Y.		.87		.87	Includes material and labor to install
	Install	S.Y.	9.70	6.25		15.95	indoor-outdoor carpet.
	Demolish and Install	S.Y.	9.70	7.12		16.82	
	Reinstall	S.Y.		5.02		5.02	
	Clean	S.Y.		3.68		3.68	
	Minimum Charge	Job		147		147	
Economy Grade							
	Demolish	S.Y.		.87		.87	Includes material and labor to install
	Install	S.Y.	9.85	5.15		15	carpet including tack strips and hot
	Demolish and Install	S.Y.	9.85	6.02		15.87	melt tape on seams.
	Reinstall	S.Y.		4.14		4.14	
	Clean	S.Y.		3.68		3.68	
	Minimum Charge	Job		147		147	
Average Grade							
	Demolish	S.Y.		.87		.87	Includes material and labor to install
	Install	S.Y.	11.70	5.15		16.85	carpet including tack strips and hot
	Demolish and Install	S.Y.	11.70	6.02		17.72	melt tape on seams.
	Reinstall	S.Y.		4.14		4.14	
	Clean	S.Y.		3.68		3.68	
	Minimum Charge	Job		147		147	
Premium Grade							
	Demolish	S.Y.		.87		.87	Includes material and labor to install
	Install	S.Y.	21	5.15		26.15	carpet including tack strips and hot
	Demolish and Install	S.Y.	21	6.02		27.02	melt tape on seams.
	Reinstall	S.Y.		4.14		4.14	
	Clean	S.Y.		3.68		3.68	
	Minimum Charge	Job		147		147	
100% Wool							
Average Grade							
	Demolish	S.Y.		.87		.87	Includes material and labor to install
	Install	S.Y.	58	5.25		63.25	wool carpet including tack strips and
	Demolish and Install	S.Y.	58	6.12		64.12	hot melt tape on seams.
	Reinstall	S.Y.		4.21		4.21	
	Clean	S.Y.		3.68		3.68	
	Minimum Charge	Job		147		147	

Flooring

Carpeting

Carpeting		Unit	Material	Labor	Equip.	Total	Specification
Premium Grade							
	Demolish	S.Y.		.87		.87	Includes material and labor to install
	Install	S.Y.	64.50	5.25		69.75	carpet including tack strips and hot
	Demolish and Install	S.Y.	64.50	6.12		70.62	melt tape on seams.
	Reinstall	S.Y.		4.21		4.21	
	Clean	S.Y.		3.68		3.68	
	Minimum Charge	Job		147		147	
Luxury Grade							
	Demolish	S.Y.		.87		.87	Includes material and labor to install
	Install	S.Y.	91	5.25		96.25	carpet including tack strips and hot
	Demolish and Install	S.Y.	91	6.12		97.12	melt tape on seams.
	Reinstall	S.Y.		4.21		4.21	
	Clean	S.Y.		3.68		3.68	
	Minimum Charge	Job		147		147	
Berber							
Average Grade							
	Demolish	S.Y.		.87		.87	Includes material and labor to install
	Install	S.Y.	26	6.25		32.25	berber carpet including tack strips and
	Demolish and Install	S.Y.	26	7.12		33.12	hot melt tape on seams.
	Reinstall	S.Y.		5.02		5.02	
	Clean	S.Y.		3.68		3.68	
	Minimum Charge	Job		147		147	
Premium Grade							
	Demolish	S.Y.		.87		.87	Includes material and labor to install
	Install	S.Y.	53	6.25		59.25	carpet including tack strips and hot
	Demolish and Install	S.Y.	53	7.12		60.12	melt tape on seams.
	Reinstall	S.Y.		5.02		5.02	
	Clean	S.Y.		3.68		3.68	
	Minimum Charge	Job		147		147	

Carpet Pad Per S.F.

Carpet Pad Per S.F.		Unit	Material	Labor	Equip.	Total	Specification
Urethane							
	Demolish	S.F.		.02		.02	Includes material and labor to install
	Install	S.F.	.33	.22		.55	urethane carpet pad.
	Demolish and Install	S.F.	.33	.24		.57	
	Minimum Charge	Job		147		147	
Foam Rubber Slab							
	Demolish	S.F.		.02		.02	Includes material and labor to install
	Install	S.F.	.53	.22		.75	carpet pad, 3/8" thick foam rubber.
	Demolish and Install	S.F.	.53	.24		.77	
	Minimum Charge	Job		147		147	
Waffle							
	Demolish	S.F.		.02		.02	Includes material and labor to install
	Install	S.F.	.35	.02		.37	rubber waffle carpet pad.
	Demolish and Install	S.F.	.35	.04		.39	
	Minimum Charge	Job		147		147	
Jute							
	Demolish	S.F.		.02		.02	Includes material and labor to install
	Install	S.F.	.64	.19		.83	jute hair carpet pad.
	Demolish and Install	S.F.	.64	.21		.85	
	Minimum Charge	Job		147		147	
Rebound							
	Demolish	S.F.		.02		.02	Includes material and labor to install
	Install	S.F.	.30	.16		.46	rebound carpet pad.
	Demolish and Install	S.F.	.30	.18		.48	
	Minimum Charge	Job		147		147	

Flooring

Carpet Pad		Unit	Material	Labor	Equip.	Total	Specification
Urethane							
	Demolish	S.Y.		.15		.15	Includes material and labor to install
	Install	S.Y.	3	1.97		4.97	urethane carpet pad.
	Demolish and Install	S.Y.	3	2.12		5.12	
	Reinstall	S.Y.		1.57		1.57	
	Minimum Charge	Job		147		147	
Foam Rubber Slab							
	Demolish	S.Y.		.15		.15	Includes material and labor to install
	Install	S.Y.	4.75	1.97		6.72	rubber slab carpet pad.
	Demolish and Install	S.Y.	4.75	2.12		6.87	
	Reinstall	S.Y.		1.57		1.57	
	Minimum Charge	Job		147		147	
Waffle							
	Demolish	S.Y.		.15		.15	Includes material and labor to install
	Install	S.Y.	3.17	.22		3.39	rubber waffle carpet pad.
	Demolish and Install	S.Y.	3.17	.37		3.54	
	Reinstall	S.Y.		.17		.17	
	Minimum Charge	Job		147		147	
Jute							
	Demolish	S.Y.		.15		.15	Includes material and labor to install
	Install	S.Y.	5.70	1.73		7.43	jute hair carpet pad.
	Demolish and Install	S.Y.	5.70	1.88		7.58	
	Reinstall	S.Y.		1.39		1.39	
	Minimum Charge	Job		147		147	
Rebound							
	Demolish	S.Y.		.15		.15	Includes material and labor to install
	Install	S.Y.	2.69	1.47		4.16	rebound carpet pad.
	Demolish and Install	S.Y.	2.69	1.62		4.31	
	Reinstall	S.Y.		1.18		1.18	
	Minimum Charge	Job		147		147	

Stair Components		Unit	Material	Labor	Equip.	Total	Specification
Carpeting							
	Demolish	Ea.		1.59		1.59	Includes material and labor to install
	Install	Riser	32.50	7.75		40.25	carpet and pad on stairs.
	Demolish and Install	Riser	32.50	9.34		41.84	
	Reinstall	Riser		6.21		6.21	
	Clean	Ea.		3.68		3.68	
	Minimum Charge	Job		147		147	

Tub/Shower

Stainless Steel Double Sink

Water Closet

Commode	Unit	Material	Labor	Equip.	Total	Specification
Demolish	Ea.		35.50		35.50	Includes material and labor to install a
Install	Ea.	179	120		299	toilet with valve, seat and cover.
Demolish and Install	Ea.	179	155.50		334.50	
Reinstall	Ea.		96.24		96.24	
Clean	Ea.	.30	11.80		12.10	
Minimum Charge	Job		177		177	
Wall w / Flush Valve						
Demolish	Ea.		35.50		35.50	Includes material and labor to install a
Install	Ea.	380	110		490	wall hung toilet with flush valve, seat
Demolish and Install	Ea.	380	145.50		525.50	and cover.
Reinstall	Ea.		87.94		87.94	
Clean	Ea.	.30	11.80		12.10	
Minimum Charge	Job		177		177	
Designer Floor Mount						
Demolish	Ea.		35.50		35.50	Includes material and labor to install a
Install	Ea.	545	120		665	toilet with valve, seat and cover.
Demolish and Install	Ea.	545	155.50		700.50	
Reinstall	Ea.		96.24		96.24	
Clean	Ea.	.30	11.80		12.10	
Minimum Charge	Job		177		177	
Urinal						
Demolish	Ea.		35.50		35.50	Includes material and labor to install a
Install	Ea.	330	213		543	wall hung urinal with flush valve.
Demolish and Install	Ea.	330	248.50		578.50	
Reinstall	Ea.		170.03		170.03	
Clean	Ea.	.30	11.80		12.10	
Minimum Charge	Job		177		177	
Bidet						
Vitreous China						
Demolish	Ea.		35.50		35.50	Includes material and labor to install a
Install	Ea.	625	128		753	vitreous china bidet complete with
Demolish and Install	Ea.	625	163.50		788.50	trim.
Reinstall	Ea.		102.02		102.02	
Clean	Ea.	.30	11.80		12.10	
Minimum Charge	Job		177		177	

Finish Mechanical

Commode		Unit	Material	Labor	Equip.	Total	Specification
Chrome Fittings							
	Install	Ea.	152	44.50		196.50	Includes labor and materials to install chrome fittings for a bidet.
Brass Fittings							
	Install	Ea.	167	44.50		211.50	Includes labor and materials to install brass fittings for a bidet.
Seat							
	Demolish	Ea.		14.75		14.75	Includes material and labor to install a toilet seat and cover.
	Install	Ea.	22.50	14.75		37.25	
	Demolish and Install	Ea.	22.50	29.50		52	
	Reinstall	Ea.		11.81		11.81	
	Clean	Ea.	.14	4.91		5.05	
	Minimum Charge	Job		177		177	
Rough-in							
	Install	Ea.	185	225		410	Includes labor and materials to install water supply pipe with valves and drain, waste and vent pipe with all couplings, hangers and fasteners necessary.

Sink (assembly)		Unit	Material	Labor	Equip.	Total	Specification
Single							
Porcelain							
	Demolish	Ea.		22		22	Includes material and labor to install a single bowl enamel finished cast iron kitchen sink with faucet and drain.
	Install	Ea.	277	114		391	
	Demolish and Install	Ea.	277	136		413	
	Reinstall	Ea.		91.09		91.09	
	Clean	Ea.	.19	8.40		8.59	
	Minimum Charge	Job		177		177	
Stainless Steel							
	Demolish	Ea.		22		22	Includes material and labor to install a single bowl stainless steel kitchen sink with faucet and drain.
	Install	Ea.	405	114		519	
	Demolish and Install	Ea.	405	136		541	
	Reinstall	Ea.		91.09		91.09	
	Clean	Ea.	.19	8.40		8.59	
	Minimum Charge	Job		177		177	
Double							
Porcelain							
	Demolish	Ea.		25.50		25.50	Includes material and labor to install a double bowl enamel finished cast iron kitchen sink with faucet and drain.
	Install	Ea.	530	133		663	
	Demolish and Install	Ea.	530	158.50		688.50	
	Reinstall	Ea.		106.27		106.27	
	Clean	Ea.	.30	11.80		12.10	
	Minimum Charge	Job		177		177	
Stainless Steel							
	Demolish	Ea.		25.50		25.50	Includes material and labor to install a double bowl stainless steel kitchen sink with faucet and drain.
	Install	Ea.	257	133		390	
	Demolish and Install	Ea.	257	158.50		415.50	
	Reinstall	Ea.		106.27		106.27	
	Clean	Ea.	.30	11.80		12.10	
	Minimum Charge	Job		177		177	
Bar							
	Demolish	Ea.		22		22	Includes material and labor to install a small single bowl stainless steel bar / vegetable / kitchen sink with faucet and drain.
	Install	Ea.	78.50	88.50		167	
	Demolish and Install	Ea.	78.50	110.50		189	
	Reinstall	Ea.		70.88		70.88	
	Clean	Ea.	.19	8.40		8.59	
	Minimum Charge	Job		177		177	

Finish Mechanical

Sink (assembly)

Sink (assembly)		Unit	Material	Labor	Equip.	Total	Specification
Floor							Includes material and labor to install
	Demolish	Ea.		71		71	an enameled cast iron floor mounted
	Install	Ea.	585	145		730	corner service sink with faucet and
	Demolish and Install	Ea.	585	216		801	drain.
	Reinstall	Ea.		115.93		115.93	
	Clean	Ea.	.30	5.90		6.20	
	Minimum Charge	Job		177		177	
Pedestal							Includes material and labor to install a
	Demolish	Ea.		22		22	vitreous china pedestal lavatory with
	Install	Ea.	445	99.50		544.50	faucet set and pop-up drain.
	Demolish and Install	Ea.	445	121.50		566.50	
	Reinstall	Ea.		79.70		79.70	
	Clean	Ea.	.19	8.40		8.59	
	Minimum Charge	Job		177		177	
Vanity Lavatory							Includes material and labor to install a
	Demolish	Ea.		22		22	vitreous china lavatory with faucet set
	Install	Ea.	254	118		372	and pop-up drain.
	Demolish and Install	Ea.	254	140		394	
	Reinstall	Ea.		94.46		94.46	
	Clean	Ea.	.19	8.40		8.59	
	Minimum Charge	Job		177		177	
Rough-in							Includes labor and materials to install
	Install	Ea.	114	298		412	copper supply pipe with valves and
							drain, waste and vent pipe with all
							couplings, hangers and fasteners
							necessary.
Laundry							Includes material and labor to install a
	Demolish	Ea.		23.50		23.50	laundry sink (wall mounted or with
	Install	Ea.	101	98		199	legs) including faucet and pop up
	Demolish and Install	Ea.	101	121.50		222.50	drain.
	Clean	Ea.	.19	8.40		8.59	
	Minimum Charge	Job		177		177	

Sink Only

Sink Only		Unit	Material	Labor	Equip.	Total	Specification
Single							
Porcelain							Includes material and labor to install a
	Demolish	Ea.		22		22	single bowl enamel finished cast iron
	Install	Ea.	207	127		334	kitchen sink.
	Demolish and Install	Ea.	207	149		356	
	Reinstall	Ea.		101.26		101.26	
	Clean	Ea.	.19	8.40		8.59	
	Minimum Charge	Job		177		177	
Stainless Steel							Includes material and labor to install a
	Demolish	Ea.		22		22	single bowl stainless steel kitchen sink.
	Install	Ea.	335	88.50		423.50	
	Demolish and Install	Ea.	335	110.50		445.50	
	Reinstall	Ea.		70.88		70.88	
	Clean	Ea.	.19	8.40		8.59	
	Minimum Charge	Job		177		177	

Finish Mechanical

Sink Only		Unit	Material	Labor	Equip.	Total	Specification
Double							
Porcelain							
	Demolish	Ea.		25.50		25.50	Includes material and labor to install a
	Install	Ea.	234	148		382	double bowl enamel finished cast iron
	Demolish and Install	Ea.	234	173.50		407.50	kitchen sink.
	Reinstall	Ea.		118.13		118.13	
	Clean	Ea.	.30	11.80		12.10	
	Minimum Charge	Job		177		177	
Stainless Steel							
	Demolish	Ea.		25.50		25.50	Includes material and labor to install a
	Install	Ea.	500	114		614	double bowl stainless steel kitchen
	Demolish and Install	Ea.	500	139.50		639.50	sink.
	Reinstall	Ea.		91.46		91.46	
	Clean	Ea.	.30	11.80		12.10	
	Minimum Charge	Job		177		177	
Bar							
	Demolish	Ea.		22		22	Includes material and labor to install a
	Install	Ea.	42	88.50		130.50	small single bowl stainless steel bar /
	Demolish and Install	Ea.	42	110.50		152.50	vegetable / kitchen sink.
	Reinstall	Ea.		70.88		70.88	
	Clean	Ea.	.19	8.40		8.59	
	Minimum Charge	Job		177		177	
Floor							
	Demolish	Ea.		71		71	Includes material and labor to install
	Install	Ea.	490	161		651	an enameled cast iron floor mounted
	Demolish and Install	Ea.	490	232		722	corner service sink.
	Reinstall	Ea.		128.87		128.87	
	Clean	Ea.	.30	5.90		6.20	
	Minimum Charge	Job		177		177	
Pedestal							
	Demolish	Ea.		22		22	Includes material and labor to install a
	Install	Ea.	345	111		456	vitreous china pedestal lavatory.
	Demolish and Install	Ea.	345	133		478	
	Reinstall	Ea.		88.60		88.60	
	Clean	Ea.	.19	8.40		8.59	
	Minimum Charge	Job		177		177	
Vanity Lavatory							
	Demolish	Ea.		22		22	Includes material and labor to install a
	Install	Ea.	145	111		256	vitreous china lavatory.
	Demolish and Install	Ea.	145	133		278	
	Reinstall	Ea.		88.60		88.60	
	Clean	Ea.	.19	8.40		8.59	
	Minimum Charge	Job		177		177	
Rough-in							
	Install	Ea.	99.50	277		376.50	Includes labor and materials to install copper supply pipe with valves and drain, waste and vent pipe with all couplings, hangers and fasteners necessary.
Faucet Sink							
Good Grade							
	Demolish	Ea.		17.70		17.70	Includes material and labor to install a
	Install	Ea.	56.50	24.50		81	sink faucet and fittings.
	Demolish and Install	Ea.	56.50	42.20		98.70	
	Reinstall	Ea.		19.57		19.57	
	Clean	Ea.	.12	7.35		7.47	
	Minimum Charge	Job		177		177	

Finish Mechanical

Sink Only

Sink Only	Unit	Material	Labor	Equip.	Total	Specification
Better Grade						
Demolish	Ea.		17.70		17.70	Includes material and labor to install a
Install	Ea.	95	24.50		119.50	sink faucet and fittings.
Demolish and Install	Ea.	95	42.20		137.20	
Reinstall	Ea.		19.57		19.57	
Clean	Ea.	.12	7.35		7.47	
Minimum Charge	Job		177		177	
Premium Grade						
Demolish	Ea.		17.70		17.70	Includes material and labor to install a
Install	Ea.	186	30.50		216.50	sink faucet and fittings.
Demolish and Install	Ea.	186	48.20		234.20	
Reinstall	Ea.		24.40		24.40	
Clean	Ea.	.12	7.35		7.47	
Minimum Charge	Job		177		177	
Drain & Basket						
Demolish	Ea.		11.10		11.10	Includes material and labor to install a
Install	Ea.	13.35	22		35.35	sink drain and basket.
Demolish and Install	Ea.	13.35	33.10		46.45	
Reinstall	Ea.		17.72		17.72	
Clean	Ea.	.12	7.35		7.47	
Minimum Charge	Job		177		177	

Bathtub

Bathtub	Unit	Material	Labor	Equip.	Total	Specification
Enameled						
Steel						
Demolish	Ea.		64		64	Includes material and labor to install a
Install	Ea.	320	116		436	formed steel bath tub with spout,
Demolish and Install	Ea.	320	180		500	mixing valve, shower head and
Reinstall	Ea.		92.74		92.74	pop-up drain.
Clean	Ea.	.58	19.65		20.23	
Minimum Charge	Job		177		177	
Cast Iron						
Demolish	Ea.		64		64	Includes material and labor to install a
Install	Ea.	460	145		605	cast iron bath tub with spout, mixing
Demolish and Install	Ea.	460	209		669	valve, shower head and pop-up drain.
Reinstall	Ea.		115.93		115.93	
Clean	Ea.	.58	19.65		20.23	
Minimum Charge	Job		177		177	
Fiberglass						
Demolish	Ea.		64		64	Includes material and labor to install
Install	Ea.	885	116		1001	an acrylic soaking tub with spout,
Demolish and Install	Ea.	885	180		1065	mixing valve, shower head and
Reinstall	Ea.		92.74		92.74	pop-up drain.
Clean	Ea.	.58	19.65		20.23	
Minimum Charge	Job		177		177	
Institutional						
Demolish	Ea.		64		64	Includes material and labor to install a
Install	Ea.	980	213		1193	hospital / institutional type bathtub
Demolish and Install	Ea.	980	277		1257	with spout, mixing valve, shower head
Reinstall	Ea.		170.03		170.03	and pop-up drain.
Clean	Ea.	.58	19.65		20.23	
Minimum Charge	Job		177		177	

Finish Mechanical

Bathtub

	Unit	Material	Labor	Equip.	Total	Specification
Whirlpool (Acrylic)						
Demolish	Ea.		197		197	Includes material and labor to install a
Install	Ea.	2200	640		2840	molded fiberglass whirlpool tub.
Demolish and Install	Ea.	2200	837		3037	
Reinstall	Ea.		510.08		510.08	
Clean	Ea.	.58	19.65		20.23	
Minimum Charge	Job		177		177	
Pump / Motor						
Demolish	Ea.		49		49	Includes material and labor to install a
Install	Ea.	173	148		321	whirlpool tub pump / motor.
Demolish and Install	Ea.	173	197		370	
Reinstall	Ea.		118.13		118.13	
Clean	Ea.	.14	14.75		14.89	
Minimum Charge	Job		177		177	
Heater / Motor						
Demolish	Ea.		49		49	Includes material and labor to install a
Install	Ea.	230	158		388	whirlpool tub heater.
Demolish and Install	Ea.	230	207		437	
Reinstall	Ea.		126.01		126.01	
Clean	Ea.	.14	14.75		14.89	
Minimum Charge	Job		177		177	
Thermal Cover						
Demolish	Ea.		7.90		7.90	Includes material and labor to install a
Install	Ea.	155	22		177	whirlpool tub thermal cover.
Demolish and Install	Ea.	155	29.90		184.90	
Reinstall	Ea.		17.72		17.72	
Clean	Ea.	1.46	14.75		16.21	
Minimum Charge	Job		177		177	
Tub / Shower Combination						
Demolish	Ea.		39.50		39.50	Includes material and labor to install a
Install	Ea.	545	159		704	fiberglass module tub with shower
Demolish and Install	Ea.	545	198.50		743.50	surround with spout, mixing valve,
Reinstall	Ea.		127.52		127.52	shower head and pop-up drain.
Clean	Ea.	.73	29.50		30.23	
Minimum Charge	Job		177		177	
Accessories						
Sliding Door						
Demolish	Ea.		16.35		16.35	Includes material and labor to install a
Install	Ea.	380	75		455	48" aluminum framed tempered glass
Demolish and Install	Ea.	380	91.35		471.35	shower door.
Reinstall	Ea.		60.03		60.03	
Clean	Ea.	1.21	9.80		11.01	
Minimum Charge	Job		177		177	
Shower Head						
Demolish	Ea.		4.43		4.43	Includes material and labor to install a
Install	Ea.	71.50	14.75		86.25	water saving shower head.
Demolish and Install	Ea.	71.50	19.18		90.68	
Reinstall	Ea.		11.81		11.81	
Clean	Ea.	.01	2.95		2.96	
Minimum Charge	Job		177		177	
Faucet Set						
Demolish	Ea.		22		22	Includes material and labor to install a
Install	Ea.	76.50	44.50		121	combination spout / diverter for a
Demolish and Install	Ea.	76.50	66.50		143	bathtub.
Reinstall	Ea.		35.44		35.44	
Clean	Ea.	.12	7.35		7.47	
Minimum Charge	Job		177		177	

Finish Mechanical

Bathtub

Bathtub		Unit	Material	Labor	Equip.	Total	Specification
Shower Rod							
	Demolish	Ea.		2.73		2.73	Includes material and labor to install a
	Install	Ea.	13.20	25		38.20	chrome plated shower rod.
	Demolish and Install	Ea.	13.20	27.73		40.93	
	Reinstall	Ea.		20.06		20.06	
	Clean	Ea.		2.95		2.95	
	Minimum Charge	Job		177		177	
Rough-in							
	Install	Ea.	156	310		466	Includes labor and materials to install copper supply pipe with valves and drain, waste and vent pipe with all couplings, hangers and fasteners necessary.

Shower

Shower		Unit	Material	Labor	Equip.	Total	Specification
Fiberglass							
32″ x 32″							
	Demolish	Ea.		133		133	Includes material and labor to install a
	Install	Ea.	455	266		721	fiberglass shower stall with door,
	Demolish and Install	Ea.	455	399		854	mixing valve and shower head and
	Reinstall	Ea.		212.53		212.53	drain fitting.
	Clean	Ea.	.58	19.65		20.23	
	Minimum Charge	Job		177		177	
36″ x 36″							
	Demolish	Ea.		133		133	Includes material and labor to install a
	Install	Ea.	505	266		771	fiberglass shower stall with door,
	Demolish and Install	Ea.	505	399		904	mixing valve and shower head and
	Reinstall	Ea.		212.53		212.53	drain fitting.
	Clean	Ea.	.58	19.65		20.23	
	Minimum Charge	Job		177		177	
35″ x 60″							
	Demolish	Ea.		133		133	Includes material and labor to install a
	Install	Ea.	725	320		1045	handicap fiberglass shower stall with
	Demolish and Install	Ea.	725	453		1178	door & seat, mixing valve and shower
	Reinstall	Ea.		255.04		255.04	head and drain fitting.
	Clean	Ea.	.58	19.65		20.23	
	Minimum Charge	Job		177		177	
Accessories							
Single Door							
	Demolish	Ea.		16.35		16.35	Includes material and labor to install a
	Install	Ea.	114	39.50		153.50	shower door.
	Demolish and Install	Ea.	114	55.85		169.85	
	Reinstall	Ea.		31.68		31.68	
	Clean	Ea.	1.21	9.80		11.01	
	Minimum Charge	Job		177		177	
Shower Head							
	Demolish	Ea.		4.43		4.43	Includes material and labor to install a
	Install	Ea.	71.50	14.75		86.25	water saving shower head.
	Demolish and Install	Ea.	71.50	19.18		90.68	
	Reinstall	Ea.		11.81		11.81	
	Clean	Ea.	.01	2.95		2.96	
	Minimum Charge	Job		177		177	

Finish Mechanical

Shower		Unit	Material	Labor	Equip.	Total	Specification
Faucet Set							
	Demolish	Ea.		22		22	Includes material and labor to install a
	Install	Ea.	76.50	44.50		121	combination spout / diverter for a
	Demolish and Install	Ea.	76.50	66.50		143	bathtub.
	Reinstall	Ea.		35.44		35.44	
	Clean	Ea.	.12	7.35		7.47	
	Minimum Charge	Job		177		177	
Shower Pan							
	Demolish	Ea.		16.40		16.40	Includes material and labor to install
	Install	Ea.	177	44.50		221.50	the base portion of a fiberglass
	Demolish and Install	Ea.	177	60.90		237.90	shower unit.
	Reinstall	Ea.		35.44		35.44	
	Clean	Ea.	.58	19.65		20.23	
	Minimum Charge	Job		177		177	
Rough-in							
	Install	Ea.	107	310		417	Includes labor and materials to install copper supply pipe with valves and drain, waste and vent pipe with all couplings, hangers and fasteners necessary.

Ductwork		Unit	Material	Labor	Equip.	Total	Specification
Return Grill							
Under 10" Wide							
	Demolish	Ea.		7.90		7.90	Includes material and labor to install a
	Install	Ea.	34	19.80		53.80	10" x 10" air return register.
	Demolish and Install	Ea.	34	27.70		61.70	
	Reinstall	Ea.		15.84		15.84	
	Clean	Ea.	.14	6.55		6.69	
	Minimum Charge	Job		178		178	
12" to 20" Wide							
	Demolish	Ea.		7.90		7.90	Includes material and labor to install a
	Install	Ea.	62.50	21		83.50	16" x 16" air return register.
	Demolish and Install	Ea.	62.50	28.90		91.40	
	Reinstall	Ea.		16.77		16.77	
	Clean	Ea.	.14	6.55		6.69	
	Minimum Charge	Job		178		178	
Over 20" Wide							
	Demolish	Ea.		12		12	Includes material and labor to install a
	Install	Ea.	123	32.50		155.50	24" x 24" air return register.
	Demolish and Install	Ea.	123	44.50		167.50	
	Reinstall	Ea.		25.92		25.92	
	Clean	Ea.	.14	7.35		7.49	
	Minimum Charge	Job		178		178	
Supply Grill							
Baseboard Type							
	Demolish	Ea.		5.65		5.65	Includes material and labor to install a
	Install	Ea.	17.60	17.80		35.40	14" x 6" baseboard air supply
	Demolish and Install	Ea.	17.60	23.45		41.05	register.
	Reinstall	Ea.		14.26		14.26	
	Clean	Ea.	.14	6.55		6.69	
	Minimum Charge	Job		178		178	

Finish Mechanical

Ductwork		Unit	Material	Labor	Equip.	Total	Specification
10" Wide							
	Demolish	Ea.		6.60		6.60	Includes material and labor to install a
	Install	Ea.	26	17.80		43.80	10" x 6" air supply register.
	Demolish and Install	Ea.	26	24.40		50.40	
	Reinstall	Ea.		14.26		14.26	
	Clean	Ea.	.14	6.55		6.69	
	Minimum Charge	Job		178		178	
12" to 15" Wide							
	Demolish	Ea.		6.60		6.60	Includes material and labor to install a
	Install	Ea.	36.50	21		57.50	14" x 8" air supply register.
	Demolish and Install	Ea.	36.50	27.60		64.10	
	Reinstall	Ea.		16.77		16.77	
	Clean	Ea.	.14	6.55		6.69	
	Minimum Charge	Job		178		178	
18" to 24" Wide							
	Demolish	Ea.		7.90		7.90	Includes material and labor to install a
	Install	Ea.	50	27.50		77.50	24" x 8" air supply register.
	Demolish and Install	Ea.	50	35.40		85.40	
	Reinstall	Ea.		21.93		21.93	
	Clean	Ea.	.14	7.35		7.49	
	Minimum Charge	Job		178		178	
30" to 36" Wide							
	Demolish	Ea.		9.90		9.90	Includes material and labor to install a
	Install	Ea.	66	25.50		91.50	30" x 8" air supply register.
	Demolish and Install	Ea.	66	35.40		101.40	
	Reinstall	Ea.		20.37		20.37	
	Clean	Ea.	.14	14.75		14.89	
	Minimum Charge	Job		178		178	
Diffuser							
12" Louver							
	Demolish	Ea.		7.90		7.90	Includes material and labor to install a
	Install	Ea.	36	25.50		61.50	12" aluminum louvered diffuser.
	Demolish and Install	Ea.	36	33.40		69.40	
	Reinstall	Ea.		20.37		20.37	
	Clean	Ea.	.14	6.55		6.69	
	Minimum Charge	Job		178		178	
14" to 20" Louver							
	Demolish	Ea.		7.90		7.90	Includes material and labor to install a
	Install	Ea.	41.50	35.50		77	14" to 20" aluminum louvered diffuser.
	Demolish and Install	Ea.	41.50	43.40		84.90	
	Reinstall	Ea.		28.51		28.51	
	Clean	Ea.	.14	7.35		7.49	
	Minimum Charge	Job		178		178	
25" to 32" Louver							
	Demolish	Ea.		12		12	Includes material and labor to install a
	Install	Ea.	50.50	51		101.50	25" to 32" aluminum louvered diffuser.
	Demolish and Install	Ea.	50.50	63		113.50	
	Reinstall	Ea.		40.73		40.73	
	Clean	Ea.	.14	9.80		9.94	
	Minimum Charge	Job		178		178	
12" Round							
	Demolish	Ea.		11		11	Includes material and labor to install a
	Install	Ea.	32.50	29.50		62	12" diameter aluminum diffuser with a
	Demolish and Install	Ea.	32.50	40.50		73	butterfly damper.
	Reinstall	Ea.		23.76		23.76	
	Clean	Ea.	.14	6.55		6.69	
	Minimum Charge	Job		178		178	

Finish Mechanical

Ductwork		Unit	Material	Labor	Equip.	Total	Specification
14″ to 20″ Round							
	Demolish	Ea.		14.65		14.65	Includes material and labor to install a
	Install	Ea.	119	39.50		158.50	20″ diameter aluminum diffuser with a
	Demolish and Install	Ea.	119	54.15		173.15	butterfly damper.
	Reinstall	Ea.		31.68		31.68	
	Clean	Ea.	.14	7.35		7.49	
	Minimum Charge	Job		178		178	
Flush-out / Sanitize							
	Install	L.F.	.58	2.95		3.53	Includes labor and material costs to
	Minimum Charge	Job		178		178	clean ductwork.

Finish Electrical

Lighting Fixture

Duplex Receptacle

Electrical Per S.F.	Unit	Material	Labor	Equip.	Total	Specification
Residential						
Light Fixtures						
Install	S.F.	.23	.18		.41	Includes labor and material to replace house light fixtures. Apply S.F. cost to floor area including garages but not basements.

Circuits	Unit	Material	Labor	Equip.	Total	Specification
Outlet with Wiring						
Demolish	Ea.		19.75		19.75	Includes material and labor to install
Install	Ea.	7.10	24.50		31.60	general purpose outlet with cover, up
Demolish and Install	Ea.	7.10	44.25		51.35	to 20 feet of #12/2 wire.
Reinstall	Ea.		19.55		19.55	
Clean	Ea.		1.47		1.47	
Minimum Charge	Job		178		178	
GFCI Outlet w / Wiring						
Demolish	Ea.		19.75		19.75	Cost includes material and labor to
Install	Ea.	38.50	29		67.50	install 120V GFCI receptacle and wall
Demolish and Install	Ea.	38.50	48.75		87.25	plate, with up to 20 feet of #12/2
Reinstall	Ea.		23.11		23.11	wire.
Clean	Ea.		1.47		1.47	
Minimum Charge	Job		178		178	
Exterior Outlet w / Wiring						
Demolish	Ea.		19.75		19.75	Cost includes material and labor to
Install	Ea.	27.50	29.50		57	install 110V waterproof external
Demolish and Install	Ea.	27.50	49.25		76.75	receptacle and cover plate with up to
Reinstall	Ea.		23.71		23.71	20' of #12/2 wire.
Clean	Ea.		1.47		1.47	
Minimum Charge	Job		178		178	
240V Outlet w / Wiring						
Demolish	Ea.		19.75		19.75	Cost includes material and labor to
Install	Ea.	59.50	55.50		115	install 240V receptacle and cover
Demolish and Install	Ea.	59.50	75.25		134.75	plate with up to 20' of #12/2 wire.
Reinstall	Ea.		44.38		44.38	
Clean	Ea.		1.47		1.47	
Minimum Charge	Job		178		178	

Finish Electrical

Circuits

	Unit	Material	Labor	Equip.	Total	Specification
TV / Cable Jack w / Wiring						Includes material and labor to install
Demolish	Ea.		19.75		19.75	TV/cable outlet box and cover plate
Install	Ea.	13.90	22		35.90	including 20' of wiring.
Demolish and Install	Ea.	13.90	41.75		55.65	
Reinstall	Ea.		17.78		17.78	
Clean	Ea.		1.47		1.47	
Minimum Charge	Job		178		178	
Phone Jack with Wiring						Includes material and labor to install
Demolish	Ea.		19.75		19.75	phone jack and cover plate including
Install	Ea.	8.15	13.70		21.85	up to 20' of telephone cable.
Demolish and Install	Ea.	8.15	33.45		41.60	
Reinstall	Ea.		10.94		10.94	
Clean	Ea.		.92		.92	
Minimum Charge	Job		178		178	

Light Switch

	Unit	Material	Labor	Equip.	Total	Specification
Single						
Switch						
Demolish	Ea.		3.95		3.95	Includes material and labor to install
Install	Ea.	5.15	16.95		22.10	single switch and wall plate. Wiring is
Demolish and Install	Ea.	5.15	20.90		26.05	not included.
Reinstall	Ea.		13.55		13.55	
Clean	Ea.		1.47		1.47	
Minimum Charge	Job		178		178	
Plate						
Demolish	Ea.		1.98		1.98	Includes material and labor to install a
Install	Ea.	.34	4.45		4.79	one gang brown wall switch plate.
Demolish and Install	Ea.	.34	6.43		6.77	
Reinstall	Ea.		3.56		3.56	
Clean	Ea.	.06	.29		.35	
Minimum Charge	Job		178		178	
Double						
Switch						
Demolish	Ea.		7.90		7.90	Includes material and labor to install
Install	Ea.	7.45	25.50		32.95	double switch and wall plate. Wiring
Demolish and Install	Ea.	7.45	33.40		40.85	is not included.
Reinstall	Ea.		20.32		20.32	
Clean	Ea.	.06	4.71		4.77	
Minimum Charge	Job		178		178	
Plate						
Demolish	Ea.		1.98		1.98	Includes material and labor to install a
Install	Ea.	.85	6.70		7.55	two gang brown wall switch plate.
Demolish and Install	Ea.	.85	8.68		9.53	
Reinstall	Ea.		5.37		5.37	
Clean	Ea.	.06	.29		.35	
Minimum Charge	Job		178		178	
Triple						
Switch						
Demolish	Ea.		11.95		11.95	Includes material and labor to install
Install	Ea.	22.50	29.50		52	triple switch and wall plate. Wiring is
Demolish and Install	Ea.	22.50	41.45		63.95	not included.
Reinstall	Ea.		23.71		23.71	
Clean	Ea.	.06	7.05		7.11	
Minimum Charge	Job		178		178	

Finish Electrical

Light Switch		Unit	Material	Labor	Equip.	Total	Specification
Plate							
	Demolish	Ea.		1.98		1.98	Includes material and labor to install a
	Install	Ea.	1.25	11.10		12.35	three gang brown wall switch plate.
	Demolish and Install	Ea.	1.25	13.08		14.33	
	Reinstall	Ea.		8.89		8.89	
	Clean	Ea.	.06	.29		.35	
	Minimum Charge	Job		178		178	
Dimmer							
Switch							
	Demolish	Ea.		3.95		3.95	Includes material and labor to install
	Install	Ea.	17.80	21		38.80	dimmer switch. Wiring is not included.
	Demolish and Install	Ea.	17.80	24.95		42.75	
	Reinstall	Ea.		16.73		16.73	
	Clean	Ea.		1.47		1.47	
	Minimum Charge	Job		178		178	
Plate							
	Demolish	Ea.		1.98		1.98	Includes material and labor to install a
	Install	Ea.	.34	4.45		4.79	one gang brown wall switch plate.
	Demolish and Install	Ea.	.34	6.43		6.77	
	Reinstall	Ea.		3.56		3.56	
	Clean	Ea.	.06	.29		.35	
	Minimum Charge	Job		178		178	
With Wiring							
Single							
	Demolish	Ea.		19.75		19.75	Includes material and labor to install
	Install	Ea.	8.90	21		29.90	single switch and plate including up to
	Demolish and Install	Ea.	8.90	40.75		49.65	20' of romex cable.
	Reinstall	Ea.		16.64		16.64	
	Clean	Ea.		1.47		1.47	
	Minimum Charge	Job		178		178	
Double							
	Demolish	Ea.		19.75		19.75	Includes material and labor to install
	Install	Ea.	21	35.50		56.50	double switch and plate including up
	Demolish and Install	Ea.	21	55.25		76.25	to 20' of romex cable.
	Reinstall	Ea.		28.45		28.45	
	Clean	Ea.	.06	4.71		4.77	
	Minimum Charge	Job		178		178	
Triple							
	Demolish	Ea.		29.50		29.50	Includes material and labor to install
	Install	Ea.	29	40		69	triple switch and plate including up to
	Demolish and Install	Ea.	29	69.50		98.50	20' of romex cable.
	Reinstall	Ea.		32		32	
	Clean	Ea.	.06	7.05		7.11	
	Minimum Charge	Job		178		178	
Dimmer							
	Demolish	Ea.		19.75		19.75	Includes material and labor to install
	Install	Ea.	21.50	24.50		46	dimmer switch and cover plate
	Demolish and Install	Ea.	21.50	44.25		65.75	including up to 20' of romex cable.
	Reinstall	Ea.		19.55		19.55	
	Clean	Ea.		1.47		1.47	
	Minimum Charge	Job		178		178	

261

Finish Electrical

Cover for Outlet / Switch

	Unit	Material	Labor	Equip.	Total	Specification
High Grade						
Demolish	Ea.		1.98		1.98	Includes material and labor to install
Install	Ea.	1.98	4.45		6.43	wall plate, stainless steel, 1 gang.
Demolish and Install	Ea.	1.98	6.43		8.41	
Reinstall	Ea.		3.56		3.56	
Clean	Ea.	.06	.29		.35	
Minimum Charge	Job		178		178	
Deluxe Grade						
Demolish	Ea.		1.98		1.98	Includes material and labor to install
Install	Ea.	4.24	4.45		8.69	wall plate, brushed brass, 1 gang.
Demolish and Install	Ea.	4.24	6.43		10.67	
Reinstall	Ea.		3.56		3.56	
Clean	Ea.	.06	.29		.35	
Minimum Charge	Job		178		178	

Residential Light Fixture

	Unit	Material	Labor	Equip.	Total	Specification
Ceiling						
Good Quality						
Demolish	Ea.		8.25		8.25	Includes material and labor to install
Install	Ea.	28.50	8.90		37.40	standard quality incandescent ceiling
Demolish and Install	Ea.	28.50	17.15		45.65	light fixture. Wiring is not included.
Reinstall	Ea.		7.11		7.11	
Clean	Ea.	.14	9.80		9.94	
Minimum Charge	Job		178		178	
Better Quality						
Demolish	Ea.		8.25		8.25	Includes material and labor to install
Install	Ea.	86	18.70		104.70	custom quality incandescent ceiling
Demolish and Install	Ea.	86	26.95		112.95	light fixture. Wiring is not included.
Reinstall	Ea.		14.97		14.97	
Clean	Ea.	.14	9.80		9.94	
Minimum Charge	Job		178		178	
Premium Quality						
Demolish	Ea.		8.25		8.25	Includes material and labor to install
Install	Ea.	253	18.70		271.70	deluxe quality incandescent ceiling
Demolish and Install	Ea.	253	26.95		279.95	light fixture. Wiring is not included.
Reinstall	Ea.		14.97		14.97	
Clean	Ea.	.14	9.80		9.94	
Minimum Charge	Job		178		178	
Recessed Ceiling						
Demolish	Ea.		16.45		16.45	Includes material and labor to install
Install	Ea.	39.50	11.85		51.35	standard quality recessed eyeball
Demolish and Install	Ea.	39.50	28.30		67.80	spotlight with housing. Wiring is not
Reinstall	Ea.		9.48		9.48	included.
Clean	Ea.	.03	2.45		2.48	
Minimum Charge	Job		178		178	
Recessed Eyeball						
Demolish	Ea.		16.45		16.45	Includes material and labor to install
Install	Ea.	59.50	12.70		72.20	custom quality recessed eyeball
Demolish and Install	Ea.	59.50	29.15		88.65	spotlight with housing. Wiring is not
Reinstall	Ea.		10.16		10.16	included.
Clean	Ea.	.03	2.45		2.48	
Minimum Charge	Job		178		178	

Finish Electrical

Residential Light Fixture

	Unit	Material	Labor	Equip.	Total	Specification
Recessed Shower Type						
Demolish	Ea.		16.45		16.45	Includes material and labor to install
Install	Ea.	49.50	11.85		61.35	deluxe quality recessed eyeball
Demolish and Install	Ea.	49.50	28.30		77.80	spotlight with housing. Wiring is not
Reinstall	Ea.		9.48		9.48	included.
Clean	Ea.	.03	2.45		2.48	
Minimum Charge	Job		178		178	
Run Wiring						
Install	Ea.	10.35	14.20		24.55	Includes labor and material to install a
Minimum Charge	Job		178		178	electric fixture utility box with
						non-metallic cable.
Fluorescent						
Good Quality						
Demolish	Ea.		17.95		17.95	Includes material and labor to install
Install	Ea.	48.50	51		99.50	standard quality 12" wide by 48" long
Demolish and Install	Ea.	48.50	68.95		117.45	decorative surface mounted fluorescent
Reinstall	Ea.		40.64		40.64	fixture w/acrylic diffuser.
Clean	Ea.	.03	2.45		2.48	
Minimum Charge	Job		178		178	
Better Quality						
Demolish	Ea.		17.95		17.95	Includes material and labor to install
Install	Ea.	99.50	57.50		157	an interior fluorescent lighting fixture.
Demolish and Install	Ea.	99.50	75.45		174.95	
Reinstall	Ea.		45.88		45.88	
Clean	Ea.	.03	2.45		2.48	
Minimum Charge	Job		178		178	
Premium Quality						
Demolish	Ea.		17.95		17.95	Includes material and labor to install
Install	Ea.	103	67		170	deluxe quality 24" wide by 48" long
Demolish and Install	Ea.	103	84.95		187.95	decorative surface mounted fluorescent
Reinstall	Ea.		53.68		53.68	fixture w/acrylic diffuser.
Clean	Ea.	.03	2.45		2.48	
Minimum Charge	Job		178		178	
Run Wiring						
Install	Ea.	10.35	14.20		24.55	Includes labor and material to install a
Minimum Charge	Job		178		178	electric fixture utility box with
						non-metallic cable.
Exterior						
Good Quality						
Demolish	Ea.		15.80		15.80	Includes material and labor to install
Install	Ea.	34	22		56	standard quality incandescent outdoor
Demolish and Install	Ea.	34	37.80		71.80	light fixture. Wiring is not included.
Reinstall	Ea.		17.78		17.78	
Clean	Ea.	.14	9.80		9.94	
Minimum Charge	Job		178		178	
Better Quality						
Demolish	Ea.		15.80		15.80	Includes material and labor to install
Install	Ea.	86	22		108	custom quality incandescent outdoor
Demolish and Install	Ea.	86	37.80		123.80	light fixture. Wiring is not included.
Reinstall	Ea.		17.78		17.78	
Clean	Ea.	.14	9.80		9.94	
Minimum Charge	Job		178		178	
Premium Quality						
Demolish	Ea.		15.80		15.80	Includes material and labor to install
Install	Ea.	164	89		253	deluxe quality incandescent outdoor
Demolish and Install	Ea.	164	104.80		268.80	light fixture. Wiring is not included.
Reinstall	Ea.		71.12		71.12	
Clean	Ea.	.14	9.80		9.94	
Minimum Charge	Job		178		178	

Finish Electrical

Residential Light Fixture

	Unit	Material	Labor	Equip.	Total	Specification
Outdoor w / Pole						
Demolish	Ea.		15.80		15.80	Includes material and labor to install
Install	Ea.	121	89		210	post lantern type incandescent outdoor
Demolish and Install	Ea.	121	104.80		225.80	light fixture. Wiring is not included.
Reinstall	Ea.		71.12		71.12	
Clean	Ea.	.14	9.80		9.94	
Minimum Charge	Job		178		178	
Run Wiring						
Install	Ea.	10.35	14.20		24.55	Includes labor and material to install a
Minimum Charge	Job		178		178	electric fixture utility box with
						non-metallic cable.
Track Lighting						
Demolish	L.F.		5.50		5.50	Includes material and labor to install
Install	L.F.	12.30	14.80		27.10	track lighting.
Demolish and Install	L.F.	12.30	20.30		32.60	
Reinstall	L.F.		11.85		11.85	
Clean	L.F.	.14	4.71		4.85	
Minimum Charge	Job		178		178	

Commercial Light Fixture

	Unit	Material	Labor	Equip.	Total	Specification
Fluorescent						
2 Bulb, 2'						
Demolish	Ea.		15.80		15.80	Includes material and labor to install a
Install	Ea.	33	44.50		77.50	2' long surface mounted fluorescent
Demolish and Install	Ea.	33	60.30		93.30	light fixture with two 40 watt bulbs.
Reinstall	Ea.		35.56		35.56	
Clean	Ea.	.73	16.85		17.58	
Minimum Charge	Job		178		178	
2 Bulb, 4'						
Demolish	Ea.		15.80		15.80	Includes material and labor to install a
Install	Ea.	61.50	51		112.50	4' long surface mounted fluorescent
Demolish and Install	Ea.	61.50	66.80		128.30	light fixture with two 40 watt bulbs.
Reinstall	Ea.		40.64		40.64	
Clean	Ea.	.73	16.85		17.58	
Minimum Charge	Job		178		178	
3 Bulb, 4'						
Demolish	Ea.		15.80		15.80	Includes material and labor to install a
Install	Ea.	103	51		154	4' long pendant mounted fluorescent
Demolish and Install	Ea.	103	66.80		169.80	light fixture with three 34 watt bulbs.
Reinstall	Ea.		40.64		40.64	
Clean	Ea.	.73	16.85		17.58	
Minimum Charge	Job		178		178	
4 Bulb, 4'						
Demolish	Ea.		15.80		15.80	Includes material and labor to install a
Install	Ea.	93.50	54.50		148	4' long pendant mounted fluorescent
Demolish and Install	Ea.	93.50	70.30		163.80	light fixture with four 34 watt bulbs.
Reinstall	Ea.		43.77		43.77	
Clean	Ea.	.73	16.85		17.58	
Minimum Charge	Job		178		178	
2 Bulb, 8'						
Demolish	Ea.		19.75		19.75	Includes material and labor to install a
Install	Ea.	95	67		162	8' long surface mounted fluorescent
Demolish and Install	Ea.	95	86.75		181.75	light fixture with two 110 watt bulbs.
Reinstall	Ea.		53.68		53.68	
Clean	Ea.	.73	16.85		17.58	
Minimum Charge	Job		178		178	

Finish Electrical

Commercial Light Fixture	Unit	Material	Labor	Equip.	Total	Specification
Run Wiring						
Install	Ea.	10.35	14.20		24.55	Includes labor and material to install a
Minimum Charge	Job		178		178	electric fixture utility box with
						non-metallic cable.
Ceiling, Recessed						
Eyeball Type						
Demolish	Ea.		16.45		16.45	Includes material and labor to install
Install	Ea.	59.50	12.70		72.20	custom quality recessed eyeball
Demolish and Install	Ea.	59.50	29.15		88.65	spotlight with housing. Wiring is not
Reinstall	Ea.		10.16		10.16	included.
Clean	Ea.	.03	2.45		2.48	
Minimum Charge	Job		178		178	
Shower Type						
Demolish	Ea.		16.45		16.45	Includes material and labor to install
Install	Ea.	49.50	11.85		61.35	deluxe quality recessed eyeball
Demolish and Install	Ea.	49.50	28.30		77.80	spotlight with housing. Wiring is not
Reinstall	Ea.		9.48		9.48	included.
Clean	Ea.	.03	2.45		2.48	
Minimum Charge	Job		178		178	
Run Wiring						
Install	Ea.	10.35	14.20		24.55	Includes labor and material to install a
Minimum Charge	Job		178		178	electric fixture utility box with
						non-metallic cable.
Ceiling Grid Type						
2 Tube						
Demolish	Ea.		23		23	Cost includes material and labor to
Install	Ea.	56.50	67		123.50	install 2 lamp fluorescent light fixture
Demolish and Install	Ea.	56.50	90		146.50	in 2' x 4' ceiling grid system including
Reinstall	Ea.		53.68		53.68	lens and tubes.
Clean	Ea.	.73	16.85		17.58	
Minimum Charge	Job		178		178	
4 Tube						
Demolish	Ea.		26.50		26.50	Cost includes material and labor to
Install	Ea.	64.50	75.50		140	install 4 lamp fluorescent light fixture
Demolish and Install	Ea.	64.50	102		166.50	in 2' x 4' ceiling grid system including
Reinstall	Ea.		60.53		60.53	lens and tubes.
Clean	Ea.	.73	16.85		17.58	
Minimum Charge	Job		178		178	
Acrylic Diffuser						
Demolish	Ea.		4.12		4.12	Includes material and labor to install
Install	Ea.	38.50	10.80		49.30	fluorescent light fixture w/acrylic
Demolish and Install	Ea.	38.50	14.92		53.42	diffuser.
Reinstall	Ea.		8.62		8.62	
Clean	Ea.	.46	2.95		3.41	
Minimum Charge	Job		178		178	
Louver Diffuser						
Demolish	Ea.		4.94		4.94	Includes material and labor to install
Install	Ea.	42	10.80		52.80	fluorescent light fixture w/louver
Demolish and Install	Ea.	42	15.74		57.74	diffuser.
Reinstall	Ea.		8.62		8.62	
Clean	Ea.	.46	2.95		3.41	
Minimum Charge	Job		178		178	
Run Wiring						
Install	Ea.	10.35	14.20		24.55	Includes labor and material to install a
Minimum Charge	Job		178		178	electric fixture utility box with
						non-metallic cable.

265

Finish Electrical

Commercial Light Fixture	Unit	Material	Labor	Equip.	Total	Specification
Track Lighting						
Demolish	L.F.		5.50		5.50	Includes material and labor to install
Install	L.F.	12.30	14.80		27.10	track lighting.
Demolish and Install	L.F.	12.30	20.30		32.60	
Reinstall	L.F.		11.85		11.85	
Clean	L.F.	.14	4.71		4.85	
Minimum Charge	Job		178		178	
Emergency						
Demolish	Ea.		28		28	Includes material and labor to install
Install	Ea.	125	89		214	battery-powered emergency lighting
Demolish and Install	Ea.	125	117		242	unit with two (2) lights, including
Reinstall	Ea.		71.12		71.12	wiring.
Clean	Ea.	.36	16.85		17.21	
Minimum Charge	Job		178		178	
Run Wiring						
Install	Ea.	10.35	14.20		24.55	Includes labor and material to install a
Minimum Charge	Job		178		178	electric fixture utility box with
						non-metallic cable.

Fan	Unit	Material	Labor	Equip.	Total	Specification
Ceiling Paddle						
Good Quality						
Demolish	Ea.		22		22	Includes material and labor to install
Install	Ea.	85	35.50		120.50	standard ceiling fan.
Demolish and Install	Ea.	85	57.50		142.50	
Reinstall	Ea.		28.45		28.45	
Clean	Ea.	.30	14.75		15.05	
Minimum Charge	Job		178		178	
Better Quality						
Demolish	Ea.		22		22	Includes material and labor to install
Install	Ea.	141	35.50		176.50	better ceiling fan.
Demolish and Install	Ea.	141	57.50		198.50	
Reinstall	Ea.		28.45		28.45	
Clean	Ea.	.30	14.75		15.05	
Minimum Charge	Job		178		178	
Premium Quality						
Demolish	Ea.		22		22	Includes material and labor to install
Install	Ea.	330	44.50		374.50	deluxe ceiling fan.
Demolish and Install	Ea.	330	66.50		396.50	
Reinstall	Ea.		35.56		35.56	
Clean	Ea.	.30	14.75		15.05	
Minimum Charge	Job		178		178	
Light Kit						
Install	Ea.	30.50	44.50		75	Includes labor and material to install a
Minimum Charge	Job		178		178	ceiling fan light kit.
Bathroom Exhaust						
Ceiling Mounted						
Demolish	Ea.		18.10		18.10	Cost includes material and labor to
Install	Ea.	42	35.50		77.50	install 60 CFM ceiling mounted
Demolish and Install	Ea.	42	53.60		95.60	bathroom exhaust fan, up to 10' of
Reinstall	Ea.		28.36		28.36	wiring with one wall switch.
Clean	Ea.	.14	5.90		6.04	
Minimum Charge	Job		178		178	

Finish Electrical

Fan

Fan	Unit	Material	Labor	Equip.	Total	Specification
Wall Mounted						
Demolish	Ea.		18.10		18.10	Cost includes material and labor to
Install	Ea.	56	35.50		91.50	install 110 CFM wall mounted
Demolish and Install	Ea.	56	53.60		109.60	bathroom exhaust fan, up to 10' of
Reinstall	Ea.		28.36		28.36	wiring with one wall switch.
Clean	Ea.	.14	5.90		6.04	
Minimum Charge	Job		178		178	
Lighted						
Demolish	Ea.		18.10		18.10	Cost includes material and labor to
Install	Ea.	40	35.50		75.50	install 50 CFM ceiling mounted lighted
Demolish and Install	Ea.	40	53.60		93.60	exhaust fan, up to 10' of wiring with
Reinstall	Ea.		28.36		28.36	one wall switch..
Clean	Ea.	.14	5.90		6.04	
Minimum Charge	Job		178		178	
Blower Heater w / Light						
Demolish	Ea.		18.10		18.10	Includes material and labor to install
Install	Ea.	89	59		148	ceiling blower type ventilator with
Demolish and Install	Ea.	89	77.10		166.10	switch, light and snap-on grill. Includes
Reinstall	Ea.		47.27		47.27	1300 watt/120 volt heater.
Clean	Ea.	.14	5.90		6.04	
Minimum Charge	Job		178		178	
Fan Heater w / Light						
Demolish	Ea.		18.10		18.10	Includes material and labor to install
Install	Ea.	112	59		171	circular ceiling fan heater unit, 1500
Demolish and Install	Ea.	112	77.10		189.10	watt, surface mounted.
Reinstall	Ea.		47.27		47.27	
Clean	Ea.	.14	5.90		6.04	
Minimum Charge	Job		178		178	

Whole House Exhaust

36" Ceiling Mounted	Unit	Material	Labor	Equip.	Total	Specification
Install	Ea.	505	88.50		593.50	Includes material and labor to install a
Clean	Ea.	.04	4.24		4.28	whole house exhaust fan.
Minimum Charge	Job		178		178	

Kitchen Exhaust

Standard Model	Unit	Material	Labor	Equip.	Total	Specification
Install	Ea.	64	23.50		87.50	Includes labor and materials to install
Clean	Ea.	.04	4.24		4.28	a bathroom or kitchen vent fan.
Minimum Charge	Job		178		178	
Low Noise Model						
Install	Ea.	84.50	23.50		108	Includes labor and materials to install
Clean	Ea.	.04	4.24		4.28	a bathroom or kitchen vent fan, low
Minimum Charge	Job		178		178	noise model.

Low Voltage Systems

Smoke Detector	Unit	Material	Labor	Equip.	Total	Specification
Pre-wired						
Demolish	Ea.		8.25		8.25	Includes material and labor to install
Install	Ea.	19.65	66.50		86.15	AC type smoke detector including
Demolish and Install	Ea.	19.65	74.75		94.40	wiring.
Reinstall	Ea.		53.37		53.37	
Clean	Ea.	.06	7.35		7.41	
Minimum Charge	Job		178		178	

Finish Electrical

Low Voltage Systems

	Unit	Material	Labor	Equip.	Total	Specification
Battery-powered						
Demolish	Ea.		8.25		8.25	Includes material and labor to install
Install	Ea.	24	22		46	battery-powered type smoke detector.
Demolish and Install	Ea.	24	30.25		54.25	
Reinstall	Ea.		17.78		17.78	
Clean	Ea.	.06	7.35		7.41	
Minimum Charge	Job		178		178	
Run Wiring						
Install	Ea.	10.35	14.20		24.55	Includes labor and material to install a
Minimum Charge	Job		178		178	electric fixture utility box with
						non-metallic cable.
Door Bell / Chimes						
Good Quality						
Demolish	Ea.		24.50		24.50	Includes material and labor to install
Install	Ea.	45.50	89		134.50	surface mounted 2 note door bell.
Demolish and Install	Ea.	45.50	113.50		159	Wiring is not included.
Reinstall	Ea.		71.12		71.12	
Clean	Ea.	.14	7.35		7.49	
Minimum Charge	Job		178		178	
Better Quality						
Demolish	Ea.		24.50		24.50	Includes material and labor to install
Install	Ea.	60	89		149	standard quality surface mounted 2
Demolish and Install	Ea.	60	113.50		173.50	note door bell. Wiring is not included.
Reinstall	Ea.		71.12		71.12	
Clean	Ea.	.14	7.35		7.49	
Minimum Charge	Job		178		178	
Premium Quality						
Demolish	Ea.		24.50		24.50	Includes material and labor to install
Install	Ea.	68	89		157	deluxe quality surface mounted 2 note
Demolish and Install	Ea.	68	113.50		181.50	door bell. Wiring is not included.
Reinstall	Ea.		71.12		71.12	
Clean	Ea.	.14	7.35		7.49	
Minimum Charge	Job		178		178	
Run Wiring						
Install	Ea.	10.35	14.20		24.55	Includes labor and material to install a
Minimum Charge	Job		178		178	electric fixture utility box with
						non-metallic cable.
Intercom System						
Intercom (master)						
Demolish	Ea.		66		66	Includes material and labor to install
Install	Ea.	510	178		688	master intercom station with AM/FM
Demolish and Install	Ea.	510	244		754	music and room monitoring. Wiring is
Reinstall	Ea.		142.24		142.24	not included.
Clean	Ea.	.14	19.65		19.79	
Minimum Charge	Job		178		178	
Intercom (remote)						
Demolish	Ea.		39.50		39.50	Includes material and labor to install
Install	Ea.	46	44.50		90.50	remote station for master intercom
Demolish and Install	Ea.	46	84		130	system. Wiring is not included.
Reinstall	Ea.		35.56		35.56	
Clean	Ea.	.14	7.35		7.49	
Minimum Charge	Job		178		178	

Finish Electrical

Low Voltage Systems	Unit	Material	Labor	Equip.	Total	Specification
Security System						
Alarm Panel						
Demolish	Ea.		79		79	Includes material and labor to install a
Install	Ea.	168	178		346	6 zone burglar alarm panel.
Demolish and Install	Ea.	168	257		425	
Reinstall	Ea.		142.24		142.24	
Clean	Ea.	.14	7.35		7.49	
Minimum Charge	Job		178		178	
Motion Detector						
Demolish	Ea.		56.50		56.50	Includes material and labor to install a
Install	Ea.	231	155		386	ultrasonic motion detector for a
Demolish and Install	Ea.	231	211.50		442.50	security system.
Reinstall	Ea.		123.69		123.69	
Clean	Ea.	.14	7.35		7.49	
Minimum Charge	Job		178		178	
Door Switch						
Demolish	Ea.		24.50		24.50	Includes material and labor to install a
Install	Ea.	35	67		102	door / window contact for a security
Demolish and Install	Ea.	35	91.50		126.50	system.
Reinstall	Ea.		53.68		53.68	
Clean	Ea.	.14	2.36		2.50	
Minimum Charge	Job		178		178	
Window Switch						
Demolish	Ea.		24.50		24.50	Includes material and labor to install a
Install	Ea.	35	67		102	door / window contact for a security
Demolish and Install	Ea.	35	91.50		126.50	system.
Reinstall	Ea.		53.68		53.68	
Clean	Ea.	.14	2.36		2.50	
Minimum Charge	Job		178		178	
Satellite System						
Dish						
Demolish	Ea.		110		110	Includes material, labor and
Install	Ea.	660	296		956	equipment to install 10' mesh dish.
Demolish and Install	Ea.	660	406		1066	
Reinstall	Ea.		237.07		237.07	
Clean	Ea.	2.92	29.50		32.42	
Minimum Charge	Job		178		178	
Motor						
Demolish	Ea.		55		55	Includes material and labor to install a
Install	Ea.	330	148		478	television satellite dish motor unit.
Demolish and Install	Ea.	330	203		533	
Reinstall	Ea.		118.53		118.53	
Clean	Ea.	.14	7.35		7.49	
Minimum Charge	Job		178		178	
Television Antenna						
40' Pole						
Demolish	Ea.		61.50		61.50	Includes material and labor to install
Install	Ea.	24	111		135	stand-alone 1"-2" metal pole, 10' long.
Demolish and Install	Ea.	24	172.50		196.50	Cost does not include antenna.
Reinstall	Ea.		88.90		88.90	
Clean	Ea.	.58	29.50		30.08	
Paint	Ea.	6.80	36		42.80	
Minimum Charge	Job		178		178	
Rotor Unit						
Demolish	Ea.		49.50		49.50	Includes material and labor to install a
Install	Ea.	118	44.50		162.50	television antenna rotor.
Demolish and Install	Ea.	118	94		212	
Reinstall	Ea.		35.56		35.56	
Clean	Ea.	.58	29.50		30.08	
Minimum Charge	Job		178		178	

Finish Electrical

Low Voltage Systems	Unit	Material	Labor	Equip.	Total	Specification
Single Booster						Includes material and labor to install
Demolish	Ea.		49.50		49.50	standard grade VHF television signal
Install	Ea.	15.80	44.50		60.30	booster unit and all hardware required
Demolish and Install	Ea.	15.80	94		109.80	for connection to TV cable.
Reinstall	Ea.		35.56		35.56	
Clean	Ea.	.03	3.68		3.71	
Minimum Charge	Job		178		178	

Gazebo	Gutter	Shutter

Window Treatment	Unit	Material	Labor	Equip.	Total	Specification
Drapery						
Good Grade						
Install	L.F.	34	4.71		38.71	Includes material and labor to install
Reinstall	L.F.		3.77		3.77	good quality drapery.
Clean	L.F.	2.31	.12		2.43	
Minimum Charge	Job		118		118	
Premium Grade						
Install	L.F.	59.50	4.71		64.21	Includes material and labor to install
Reinstall	L.F.		3.77		3.77	premium quality drapery.
Clean	L.F.	2.31	.13		2.44	
Minimum Charge	Job		118		118	
Lining						
Install	L.F.	14.75	4.71		19.46	Includes material and labor to install
Reinstall	L.F.		3.77		3.77	drapery lining.
Clean	L.F.	1.16	.07		1.23	
Minimum Charge	Job		118		118	
Valance Board						
Demolish	L.F.		1.31		1.31	Includes material and labor to install a
Install	L.F.	4.43	9.40		13.83	drapery valance board.
Demolish and Install	L.F.	4.43	10.71		15.14	
Reinstall	L.F.		7.54		7.54	
Clean	L.F.	1.16	.24		1.40	
Minimum Charge	Job		118		118	
Rod						
Demolish	L.F.		3.27		3.27	Includes material and labor to install
Install	L.F.	2.12	5.55		7.67	standard traverse drapery hardware.
Demolish and Install	L.F.	2.12	8.82		10.94	
Reinstall	L.F.		4.42		4.42	
Clean	L.F.	.06	.21		.27	
Paint	L.F.	.06	.55		.61	
Minimum Charge	Job		118		118	
Blinds						
Mini-type						
Demolish	S.F.		.71		.71	Includes material and labor to install
Install	S.F.	2.51	.55		3.06	horizontal 1″ slat blinds.
Demolish and Install	S.F.	2.51	1.26		3.77	
Reinstall	S.F.		.44		.44	
Clean	S.F.	.44	.02		.46	
Minimum Charge	Job		118		118	

Window Treatment		Unit	Material	Labor	Equip.	Total	Specification
Vertical Type							
	Demolish	S.F.		.71		.71	Includes material and labor to install
	Install	S.F.	8.65	.71		9.36	vertical 3" - 5" cloth or PVC blinds.
	Demolish and Install	S.F.	8.65	1.42		10.07	
	Reinstall	S.F.		.57		.57	
	Clean	S.F.	.23	.02		.25	
	Minimum Charge	Job		118		118	
Vertical Type per L.F.							
	Demolish	L.F.		1.09		1.09	Includes material and labor to install
	Install	L.F.	51.50	3.14		54.64	vertical 3" - 5" cloth or PVC blinds x
	Demolish and Install	L.F.	51.50	4.23		55.73	72" long.
	Reinstall	L.F.		2.51		2.51	
	Clean	L.F.	1.64	.16		1.80	
	Minimum Charge	Job		118		118	
Venetian Type							
	Demolish	S.F.		.71		.71	Includes material and labor to install
	Install	S.F.	4.18	.55		4.73	horizontal 2" slat blinds.
	Demolish and Install	S.F.	4.18	1.26		5.44	
	Reinstall	S.F.		.44		.44	
	Clean	S.F.	.23	.02		.25	
	Minimum Charge	Job		118		118	
Wood Shutter							
27" x 36", 4 Panel							
	Demolish	Ea.		22.50		22.50	Includes material and labor to install
	Install	Pr.	80	19.20		99.20	moveable louver, up to 33" x 36", 4
	Demolish and Install	Pr.	80	41.70		121.70	panel shutter.
	Reinstall	Pr.		15.34		15.34	
	Clean	Ea.	.06	7.05		7.11	
	Paint	Ea.	1.09	20.50		21.59	
	Minimum Charge	Job		163		163	
33" x 36", 4 Panel							
	Demolish	Ea.		22.50		22.50	Includes material and labor to install
	Install	Pr.	80	19.20		99.20	moveable louver, up to 33" x 36", 4
	Demolish and Install	Pr.	80	41.70		121.70	panel shutter.
	Reinstall	Pr.		15.34		15.34	
	Clean	Ea.	.06	7.05		7.11	
	Paint	Ea.	1.13	22		23.13	
	Minimum Charge	Job		163		163	
39" x 36", 4 Panel							
	Demolish	Ea.		22.50		22.50	Includes material and labor to install
	Install	Pr.	90	19.20		109.20	moveable louver, up to 47" x 36", 4
	Demolish and Install	Pr.	90	41.70		131.70	panel shutter.
	Reinstall	Pr.		15.34		15.34	
	Clean	Ea.	.06	7.05		7.11	
	Paint	Ea.	1.13	22		23.13	
	Minimum Charge	Job		163		163	
47" x 36", 4 Panel							
	Demolish	Ea.		22.50		22.50	Includes material and labor to install
	Install	Pr.	90	19.20		109.20	moveable louver, up to 47" x 36", 4
	Demolish and Install	Pr.	90	41.70		131.70	panel shutter.
	Reinstall	Pr.		15.34		15.34	
	Clean	Ea.	.06	7.05		7.11	
	Paint	Ea.	1.18	24		25.18	
	Minimum Charge	Job		163		163	

Improvements / Appliances / Treatments

Stairs		Unit	Material	Labor	Equip.	Total	Specification
Concrete Steps							
Precast							
	Demolish	Ea.		43.50	29.50	73	Includes material and labor to install
	Install	Flight	360	111	39	510	precast concrete front entrance stairs,
	Demolish and Install	Flight	360	154.50	68.50	583	4' wide, with 48" platform and 2
	Clean	Ea.		3.27		3.27	risers.
	Paint	Ea.	2.33	36		38.33	
	Minimum Charge	Job		290		290	
Metal							
Concrete Filled							
	Demolish	Ea.		13.10		13.10	Includes material and labor to install
	Install	Riser	240	53.50	2.98	296.48	steel, cement filled metal pan stairs
	Demolish and Install	Riser	240	66.60	2.98	309.58	with picket rail.
	Clean	Ea.	.09	2.03		2.12	
	Paint	Ea.	.34	2.12		2.46	
	Minimum Charge	Job		197		197	
Landing							
	Demolish	S.F.		3.27		3.27	Includes material and labor to install
	Install	S.F.	49.50	6.30	.35	56.15	pre-erected conventional steel pan
	Demolish and Install	S.F.	49.50	9.57	.35	59.42	landing.
	Clean	S.F.	.09	.42		.51	
	Paint	S.F.	.12	.61		.73	
	Minimum Charge	Job		197		197	
Railing							
	Demolish	L.F.		1.81		1.81	Includes material and labor to install
	Install	L.F.	8.25	9.15	.51	17.91	primed steel pipe.
	Demolish and Install	L.F.	8.25	10.96	.51	19.72	
	Clean	L.F.	.58	1.47		2.05	
	Paint	L.F.	.65	4.77		5.42	
	Minimum Charge	Job		197		197	

Elevator		Unit	Material	Labor	Equip.	Total	Specification
2 Stop Residential							
	Install	Ea.	9350	3950		13300	Includes labor, material and
	Clean	Ea.	.67	14.75		15.42	equipment to install 2 stop elevator for
	Minimum Charge	Job		395		395	residential use.
2 Stop Commercial							
	Install	Ea.	33800	7925		41725	Includes labor, material and
	Clean	Ea.	.67	14.75		15.42	equipment to install 2 stop elevator for
	Minimum Charge	Job		395		395	apartment or commercial building with
							2,000 lb/13 passenger capacity.
3 Stop Commercial							
	Install	Ea.	38900	12200		51100	Includes labor, material and
	Clean	Ea.	.67	14.75		15.42	equipment to install 3 stop elevator for
	Minimum Charge	Job		395		395	apartment or commercial building with
							2,000 lb/13 passenger capacity.
4 Stop Commercial							
	Install	Ea.	44000	16500		60500	Includes labor, material and
	Clean	Ea.	.67	14.75		15.42	equipment to install 4 stop elevator for
	Minimum Charge	Job		395		395	apartment or commercial building with
							2,000 lb/13 passenger capacity.

273

Improvements / Appliances / Treatments

Elevator	Unit	Material	Labor	Equip.	Total	Specification
5 Stop Commercial						Includes labor, material and
Install	Ea.	49100	20800		69900	equipment to install 5 stop elevator for
Clean	Ea.	.67	14.75		15.42	apartment or commercial building with
Minimum Charge	Job		395		395	2,000 lb/13 passenger capacity.

Rangetop	Unit	Material	Labor	Equip.	Total	Specification
4-burner						
Demolish	Ea.		24		24	Cost includes material and labor to
Install	Ea.	205	59.50		264.50	install 4-burner electric surface unit.
Demolish and Install	Ea.	205	83.50		288.50	
Reinstall	Ea.		47.41		47.41	
Clean	Ea.	.04	21		21.04	
Minimum Charge	Job		59		59	
6-burner						
Demolish	Ea.		24		24	Includes material and labor to install a
Install	Ea.	450	71		521	6-burner cooktop unit.
Demolish and Install	Ea.	450	95		545	
Reinstall	Ea.		56.90		56.90	
Clean	Ea.	.04	21		21.04	
Minimum Charge	Job		59		59	
Premium Brand w / Grill						
Demolish	Ea.		24		24	Includes material and labor to install
Install	Ea.	885	71		956	electric downdraft cooktop with grill.
Demolish and Install	Ea.	885	95		980	
Reinstall	Ea.		56.90		56.90	
Clean	Ea.	.04	21		21.04	
Minimum Charge	Job		59		59	

Rangehood	Unit	Material	Labor	Equip.	Total	Specification
Vented						
30"						
Demolish	Ea.		22		22	Includes material and labor to install a
Install	Ea.	41.50	71		112.50	vented range hood.
Demolish and Install	Ea.	41.50	93		134.50	
Reinstall	Ea.		56.90		56.90	
Clean	Ea.	.04	6.05		6.09	
Minimum Charge	Job		59		59	
36"						
Demolish	Ea.		22		22	Includes material and labor to install a
Install	Ea.	88.50	71		159.50	vented range hood.
Demolish and Install	Ea.	88.50	93		181.50	
Reinstall	Ea.		56.90		56.90	
Clean	Ea.	.04	6.05		6.09	
Minimum Charge	Job		59		59	
42"						
Demolish	Ea.		22		22	Includes material and labor to install a
Install	Ea.	248	83		331	vented range hood.
Demolish and Install	Ea.	248	105		353	
Reinstall	Ea.		66.38		66.38	
Clean	Ea.	.04	6.05		6.09	
Minimum Charge	Job		59		59	

Improvements / Appliances / Treatments

Rangehood

	Unit	Material	Labor	Equip.	Total	Specification
Ventless						
30"						
Demolish	Ea.		22		22	Includes material and labor to install a
Install	Ea.	41.50	44.50		86	ventless range hood.
Demolish and Install	Ea.	41.50	66.50		108	
Reinstall	Ea.		35.56		35.56	
Clean	Ea.	.04	6.05		6.09	
Minimum Charge	Job		59		59	
36"						
Demolish	Ea.		22		22	Includes material and labor to install a
Install	Ea.	43.50	71		114.50	ventless range hood.
Demolish and Install	Ea.	43.50	93		136.50	
Reinstall	Ea.		56.90		56.90	
Clean	Ea.	.04	6.05		6.09	
Minimum Charge	Job		59		59	
42"						
Demolish	Ea.		22		22	Includes material and labor to install a
Install	Ea.	221	59.50		280.50	ventless range hood.
Demolish and Install	Ea.	221	81.50		302.50	
Reinstall	Ea.		47.41		47.41	
Clean	Ea.	.04	6.05		6.09	
Minimum Charge	Job		59		59	

Oven

	Unit	Material	Labor	Equip.	Total	Specification
Free Standing						
Demolish	Ea.		24		24	Cost includes material and labor to
Install	Ea.	430	47		477	install 30" free standing electric range
Demolish and Install	Ea.	430	71		501	with self-cleaning oven.
Reinstall	Ea.		37.70		37.70	
Clean	Ea.	.04	21		21.04	
Minimum Charge	Job		59		59	
Single Wall Oven						
Demolish	Ea.		24		24	Includes material and labor to install a
Install	Ea.	425	89		514	built-in oven.
Demolish and Install	Ea.	425	113		538	
Reinstall	Ea.		71.12		71.12	
Clean	Ea.	.04	21		21.04	
Minimum Charge	Job		59		59	
Double Wall Oven						
Demolish	Ea.		24		24	Includes material and labor to install a
Install	Ea.	1025	163		1188	conventional double oven.
Demolish and Install	Ea.	1025	187		1212	
Reinstall	Ea.		130.40		130.40	
Clean	Ea.	.04	21		21.04	
Minimum Charge	Job		59		59	
Drop-in Type w / Range						
Demolish	Ea.		24		24	Includes material and labor to install a
Install	Ea.	490	59.50		549.50	built-in cooking range with oven.
Demolish and Install	Ea.	490	83.50		573.50	
Reinstall	Ea.		47.41		47.41	
Clean	Ea.	.04	21		21.04	
Minimum Charge	Job		59		59	

Improvements / Appliances / Treatments

Oven	Unit	Material	Labor	Equip.	Total	Specification
Hi/Lo w / Range, Microwave						
Demolish	Ea.		24		24	Includes material and labor to install a
Install	Ea.	1625	325		1950	double oven, one conventional, one
Demolish and Install	Ea.	1625	349		1974	microwave.
Reinstall	Ea.		260.80		260.80	
Clean	Ea.	.04	21		21.04	
Minimum Charge	Job		59		59	
Hi/Lo w/ Range / Double Oven						
Demolish	Ea.		24		24	Includes material and labor to install a
Install	Ea.	1025	163		1188	conventional double oven.
Demolish and Install	Ea.	1025	187		1212	
Reinstall	Ea.		130.40		130.40	
Clean	Ea.	.04	21		21.04	
Minimum Charge	Job		59		59	
Countertop Microwave						
Demolish	Ea.		8.20		8.20	Includes material and labor to install a
Install	Ea.	116	29.50		145.50	1.0 cubic foot microwave oven.
Demolish and Install	Ea.	116	37.70		153.70	
Reinstall	Ea.		23.56		23.56	
Clean	Ea.	.04	10.60		10.64	
Minimum Charge	Job		59		59	
Cabinet / Wall Mounted Microwave						
Demolish	Ea.		16.35		16.35	Includes material and labor to install a
Install	Ea.	550	89		639	space saver microwave oven.
Demolish and Install	Ea.	550	105.35		655.35	
Reinstall	Ea.		71.12		71.12	
Clean	Ea.	.04	10.60		10.64	
Minimum Charge	Job		59		59	

Dryer	Unit	Material	Labor	Equip.	Total	Specification
Basic Grade						
Demolish	Ea.		10.90		10.90	Includes material and labor to install
Install	Ea.	310	188		498	electric or gas dryer. Vent kit not
Demolish and Install	Ea.	310	198.90		508.90	included.
Reinstall	Ea.		150.40		150.40	
Clean	Ea.	.04	21		21.04	
Minimum Charge	Job		59		59	
Good Grade						
Demolish	Ea.		10.90		10.90	Includes material and labor to install
Install	Ea.	385	188		573	electric or gas dryer. Vent kit not
Demolish and Install	Ea.	385	198.90		583.90	included.
Reinstall	Ea.		150.40		150.40	
Clean	Ea.	.04	21		21.04	
Minimum Charge	Job		59		59	
Premium Grade						
Demolish	Ea.		10.90		10.90	Includes material and labor to install
Install	Ea.	870	282		1152	electric or gas dryer. Vent kit not
Demolish and Install	Ea.	870	292.90		1162.90	included.
Reinstall	Ea.		225.60		225.60	
Clean	Ea.	.04	21		21.04	
Minimum Charge	Job		59		59	

Improvements / Appliances / Treatments

Dryer

Dryer	Unit	Material	Labor	Equip.	Total	Specification
Washer / Dryer Combination Unit						Includes material and labor to install
Demolish	Ea.		10.90		10.90	electric washer/dryer combination
Install	Ea.	860	47		907	unit, 27" wide, stackable or unitized.
Demolish and Install	Ea.	860	57.90		917.90	
Reinstall	Ea.		37.70		37.70	
Clean	Ea.	.04	21		21.04	
Minimum Charge	Job		59		59	

Washer

Washer	Unit	Material	Labor	Equip.	Total	Specification
Basic Grade						Includes material and labor to install
Demolish	Ea.		10.90		10.90	washing machine. New hoses not
Install	Ea.	305	118		423	included.
Demolish and Install	Ea.	305	128.90		433.90	
Reinstall	Ea.		94.51		94.51	
Clean	Ea.	.04	21		21.04	
Minimum Charge	Job		59		59	
Good Grade						Includes material and labor to install
Demolish	Ea.		10.90		10.90	washing machine. New hoses not
Install	Ea.	500	177		677	included.
Demolish and Install	Ea.	500	187.90		687.90	
Reinstall	Ea.		141.76		141.76	
Clean	Ea.	.04	21		21.04	
Minimum Charge	Job		59		59	
Premium Grade						Includes material and labor to install
Demolish	Ea.		10.90		10.90	washing machine with digital
Install	Ea.	1125	355		1480	computer readouts and large capacity.
Demolish and Install	Ea.	1125	365.90		1490.90	New hoses not included.
Reinstall	Ea.		283.52		283.52	
Clean	Ea.	.04	21		21.04	
Minimum Charge	Job		59		59	
Washer / Dryer Combination Unit						Includes material and labor to install
Demolish	Ea.		10.90		10.90	electric washer/dryer combination
Install	Ea.	860	47		907	unit, 27" wide, stackable or unitized.
Demolish and Install	Ea.	860	57.90		917.90	
Reinstall	Ea.		37.70		37.70	
Clean	Ea.	.04	21		21.04	
Minimum Charge	Job		59		59	

Refrigerator

Refrigerator	Unit	Material	Labor	Equip.	Total	Specification
12 Cubic Foot						Includes material and labor to install a
Demolish	Ea.		32.50		32.50	refrigerator.
Install	Ea.	485	47		532	
Demolish and Install	Ea.	485	79.50		564.50	
Reinstall	Ea.		37.70		37.70	
Clean	Ea.	.04	21		21.04	
Minimum Charge	Job		59		59	
16 Cubic Foot						Includes material and labor to install a
Demolish	Ea.		32.50		32.50	refrigerator.
Install	Ea.	490	52.50		542.50	
Demolish and Install	Ea.	490	85		575	
Reinstall	Ea.		41.88		41.88	
Clean	Ea.	.04	21		21.04	
Minimum Charge	Job		59		59	

Improvements / Appliances / Treatments

Refrigerator		Unit	Material	Labor	Equip.	Total	Specification
18 Cubic Foot							
	Demolish	Ea.		32.50		32.50	Includes material and labor to install a
	Install	Ea.	585	59		644	refrigerator.
	Demolish and Install	Ea.	585	91.50		676.50	
	Reinstall	Ea.		47.12		47.12	
	Clean	Ea.	.04	21		21.04	
	Minimum Charge	Job		59		59	
21 Cubic Foot							
	Demolish	Ea.		32.50		32.50	Includes material and labor to install a
	Install	Ea.	710	67.50		777.50	refrigerator.
	Demolish and Install	Ea.	710	100		810	
	Reinstall	Ea.		53.85		53.85	
	Clean	Ea.	.04	21		21.04	
	Minimum Charge	Job		59		59	
24 Cubic Foot							
	Demolish	Ea.		32.50		32.50	Includes material and labor to install a
	Install	Ea.	945	67.50		1012.50	refrigerator.
	Demolish and Install	Ea.	945	100		1045	
	Reinstall	Ea.		53.85		53.85	
	Clean	Ea.	.04	21		21.04	
	Minimum Charge	Job		59		59	
27 Cubic Foot							
	Demolish	Ea.		32.50		32.50	Includes material and labor to install a
	Install	Ea.	1675	47		1722	refrigerator.
	Demolish and Install	Ea.	1675	79.50		1754.50	
	Reinstall	Ea.		37.70		37.70	
	Clean	Ea.	.04	21		21.04	
	Minimum Charge	Job		59		59	
Ice Maker							
	Install	Ea.	66	44.50		110.50	Includes labor and material to install
	Minimum Charge	Job		59		59	an automatic ice maker in a
							refrigerator.

Freezer		Unit	Material	Labor	Equip.	Total	Specification
15 Cubic Foot							
	Demolish	Ea.		32.50		32.50	Cost includes material and labor to
	Install	Ea.	430	47		477	install 15 cubic foot freezer.
	Demolish and Install	Ea.	430	79.50		509.50	
	Reinstall	Ea.		37.70		37.70	
	Clean	Ea.	.04	21		21.04	
	Minimum Charge	Job		59		59	
18 Cubic Foot							
	Demolish	Ea.		32.50		32.50	Cost includes material and labor to
	Install	Ea.	570	47		617	install 18 cubic foot freezer.
	Demolish and Install	Ea.	570	79.50		649.50	
	Reinstall	Ea.		37.70		37.70	
	Clean	Ea.	.04	21		21.04	
	Minimum Charge	Job		59		59	
21 Cubic Foot							
	Demolish	Ea.		32.50		32.50	Includes material and labor to install a
	Install	Ea.	510	47		557	15 to 23 cubic foot deep freezer.
	Demolish and Install	Ea.	510	79.50		589.50	
	Reinstall	Ea.		37.70		37.70	
	Clean	Ea.	.04	21		21.04	
	Minimum Charge	Job		59		59	

Improvements / Appliances / Treatments

Freezer

Freezer	Unit	Material	Labor	Equip.	Total	Specification
24 Cubic Foot						
Demolish	Ea.		32.50		32.50	Includes material and labor to install a
Install	Ea.	565	94		659	24 cubic foot deep freezer.
Demolish and Install	Ea.	565	126.50		691.50	
Reinstall	Ea.		75.39		75.39	
Clean	Ea.	.04	21		21.04	
Minimum Charge	Job		59		59	
Chest-type						
Demolish	Ea.		32.50		32.50	Cost includes material and labor to
Install	Ea.	430	47		477	install 15 cubic foot freezer.
Demolish and Install	Ea.	430	79.50		509.50	
Clean	Ea.	.04	21		21.04	
Minimum Charge	Job		59		59	

Dishwasher

Dishwasher	Unit	Material	Labor	Equip.	Total	Specification
4 Cycle						
Demolish	Ea.		26		26	Includes material and labor to install
Install	Ea.	288	111		399	good grade automatic dishwasher.
Demolish and Install	Ea.	288	137		425	
Reinstall	Ea.		88.66		88.66	
Clean	Ea.	.04	21		21.04	
Minimum Charge	Job		59		59	

Garbage Disposal

Garbage Disposal	Unit	Material	Labor	Equip.	Total	Specification
In Sink Unit						
Demolish	Ea.		24.50		24.50	Cost includes material and labor to
Install	Ea.	88.50	44.50		133	install 1/2 HP custom disposal unit.
Demolish and Install	Ea.	88.50	69		157.50	
Reinstall	Ea.		35.46		35.46	
Clean	Ea.	.02	6.05		6.07	
Minimum Charge	Job		59		59	

Trash Compactor

Trash Compactor	Unit	Material	Labor	Equip.	Total	Specification
4 to 1 Compaction						
Demolish	Ea.		24.50		24.50	Includes material and labor to install a
Install	Ea.	605	65		670	trash compactor.
Demolish and Install	Ea.	605	89.50		694.50	
Reinstall	Ea.		52.16		52.16	
Clean	Ea.	.04	14.75		14.79	
Minimum Charge	Job		59		59	

Kitchenette Unit

Kitchenette Unit	Unit	Material	Labor	Equip.	Total	Specification
Range, Refrigerator and Sink						
Demolish	Ea.		99		99	Includes material and labor to install
Install	Ea.	3000	370		3370	range/refrigerator/sink unit, 72"
Demolish and Install	Ea.	3000	469		3469	wide, base unit only.
Reinstall	Ea.		295.53		295.53	
Clean	Ea.	.73	10.60		11.33	
Minimum Charge	Job		59		59	

Improvements / Appliances / Treatments

Bath Accessories	Unit	Material	Labor	Equip.	Total	Specification
Towel Bar						
Demolish	Ea.		5.45		5.45	Includes material and labor to install
Install	Ea.	31.50	13.60		45.10	towel bar up to 24″ long.
Demolish and Install	Ea.	31.50	19.05		50.55	
Reinstall	Ea.		10.87		10.87	
Clean	Ea.		2.95		2.95	
Minimum Charge	Job		163		163	
Toothbrush Holder						
Demolish	Ea.		5.45		5.45	Includes material and labor to install a
Install	Ea.	23.50	16.30		39.80	surface mounted tumbler and
Demolish and Install	Ea.	23.50	21.75		45.25	toothbrush holder.
Reinstall	Ea.		13.04		13.04	
Clean	Ea.		2.95		2.95	
Minimum Charge	Job		163		163	
Grab Rail						
Demolish	Ea.		8.20		8.20	Includes material and labor to install a
Install	Ea.	55.50	14.15		69.65	24″ long grab bar. Blocking is not
Demolish and Install	Ea.	55.50	22.35		77.85	included.
Reinstall	Ea.		11.34		11.34	
Clean	Ea.	.30	4.91		5.21	
Minimum Charge	Job		163		163	
Soap Dish						
Demolish	Ea.		5.45		5.45	Includes material and labor to install a
Install	Ea.	21	16.30		37.30	surface mounted soap dish.
Demolish and Install	Ea.	21	21.75		42.75	
Reinstall	Ea.		13.04		13.04	
Clean	Ea.		2.95		2.95	
Minimum Charge	Job		163		163	
Toilet Paper Roller						
Demolish	Ea.		5.45		5.45	Includes material and labor to install a
Install	Ea.	11.90	10.85		22.75	surface mounted toilet tissue dispenser.
Demolish and Install	Ea.	11.90	16.30		28.20	
Reinstall	Ea.		8.69		8.69	
Clean	Ea.	.30	7.35		7.65	
Minimum Charge	Job		163		163	
Dispenser						
Soap						
Demolish	Ea.		7.25		7.25	Includes material and labor to install a
Install	Ea.	48.50	16.30		64.80	surface mounted soap dispenser.
Demolish and Install	Ea.	48.50	23.55		72.05	
Reinstall	Ea.		13.04		13.04	
Clean	Ea.	.30	7.35		7.65	
Minimum Charge	Job		163		163	
Towel						
Demolish	Ea.		8.20		8.20	Includes material and labor to install a
Install	Ea.	45	20.50		65.50	surface mounted towel dispenser.
Demolish and Install	Ea.	45	28.70		73.70	
Reinstall	Ea.		16.30		16.30	
Clean	Ea.	.30	9.80		10.10	
Minimum Charge	Job		163		163	
Seat Cover						
Demolish	Ea.		5.45		5.45	Includes material and labor to install a
Install	Ea.	28.50	21.50		50	surface mounted toilet seat cover
Demolish and Install	Ea.	28.50	26.95		55.45	dispenser.
Reinstall	Ea.		17.39		17.39	
Clean	Ea.	.30	7.35		7.65	
Minimum Charge	Job		163		163	

Improvements / Appliances / Treatments

Bath Accessories		Unit	Material	Labor	Equip.	Total	Specification
Sanitary Napkin							
	Demolish	Ea.		13.80		13.80	Includes material and labor to install a
	Install	Ea.	104	34.50		138.50	surface mounted sanitary napkin
	Demolish and Install	Ea.	104	48.30		152.30	dispenser.
	Reinstall	Ea.		27.45		27.45	
	Clean	Ea.	.14	9.80		9.94	
	Minimum Charge	Job		163		163	
Partition							
Urinal							
	Demolish	Ea.		16.35		16.35	Includes material and labor to install a
	Install	Ea.	189	81.50		270.50	urinal screen.
	Demolish and Install	Ea.	189	97.85		286.85	
	Reinstall	Ea.		65.20		65.20	
	Clean	Ea.	.44	6.55		6.99	
	Minimum Charge	Job		163		163	
Toilet							
	Demolish	Ea.		48.50		48.50	Includes material and labor to install
	Install	Ea.	455	130		585	baked enamel standard size partition,
	Demolish and Install	Ea.	455	178.50		633.50	floor or ceiling mounted with one wall
	Reinstall	Ea.		104.32		104.32	and one door.
	Clean	Ea.	2.92	14.75		17.67	
	Minimum Charge	Job		163		163	
Handicap							
	Demolish	Ea.		48.50		48.50	Includes material and labor to install
	Install	Ea.	775	130		905	baked enamel handicap partition,
	Demolish and Install	Ea.	775	178.50		953.50	floor or ceiling mounted with one wall
	Reinstall	Ea.		104.32		104.32	and one door.
	Clean	Ea.	2.92	14.75		17.67	
	Minimum Charge	Job		163		163	

Fire Prevention / Protection		Unit	Material	Labor	Equip.	Total	Specification
Fire Extinguisher							
5# Carbon Dioxide							
	Install	Ea.	143			143	Includes labor and material to install one wall mounted 5 lb. carbon dioxide factory charged unit, complete with hose, horn and wall mounting bracket.
10# Carbon Dioxide							
	Install	Ea.	176			176	Includes labor and material to install one wall mounted 10 lb. carbon dioxide factory charged unit, complete with hose, horn and wall mounting bracket.
15# Carbon Dioxide							
	Install	Ea.	209			209	Includes labor and material to install one wall mounted 15 lb. carbon dioxide factory charged unit, complete with hose, horn and wall mounting bracket.
5# Dry Chemical							
	Install	Ea.	66			66	Includes labor and material to install one wall mounted 5 lb. dry chemical factory charged unit, complete with hose, horn and wall mounting bracket.

281

Improvements / Appliances / Treatments

Fire Prevention / Protection	Unit	Material	Labor	Equip.	Total	Specification
20# Dry Chemical						
Install	Ea.	121			121	Includes labor and material to install one wall mounted 20 lb. dry chemical factory charged unit, complete with hose, horn and wall mounting bracket.

Gutters	Unit	Material	Labor	Equip.	Total	Specification
Galvanized						
Demolish	L.F.		1.09		1.09	Cost includes material and labor to install 5″ galvanized steel gutter with enamel finish, fittings, hangers, and corners.
Install	L.F.	1.23	2.97		4.20	
Demolish and Install	L.F.	1.23	4.06		5.29	
Reinstall	L.F.		2.38		2.38	
Clean	L.F.		.31		.31	
Paint	L.F.	.18	.88		1.06	
Minimum Charge	Job		178		178	
Aluminum						
Demolish	L.F.		1.09		1.09	Cost includes material and labor to install 5″ aluminum gutter with enamel finish, fittings, hangers, and corners.
Install	L.F.	1.36	2.97		4.33	
Demolish and Install	L.F.	1.36	4.06		5.42	
Reinstall	L.F.		2.38		2.38	
Clean	L.F.		.31		.31	
Paint	L.F.	.18	.88		1.06	
Minimum Charge	Job		178		178	
Copper						
Demolish	L.F.		1.09		1.09	Cost includes material and labor to install 5″ half-round copper gutter with fittings, hangers, and corners.
Install	L.F.	5.90	2.97		8.87	
Demolish and Install	L.F.	5.90	4.06		9.96	
Reinstall	L.F.		2.38		2.38	
Clean	L.F.		.31		.31	
Minimum Charge	Job		178		178	
Plastic						
Demolish	L.F.		1.09		1.09	Cost includes material and labor to install 4″ white vinyl gutter fittings, hangers, and corners.
Install	L.F.	.94	2.96		3.90	
Demolish and Install	L.F.	.94	4.05		4.99	
Reinstall	L.F.		2.37		2.37	
Clean	L.F.		.31		.31	
Paint	L.F.	.18	.88		1.06	
Minimum Charge	Job		178		178	

Downspouts	Unit	Material	Labor	Equip.	Total	Specification
Galvanized						
Demolish	L.F.		.75		.75	Cost includes material and labor to install 4″ diameter downspout with enamel finish.
Install	L.F.	1.12	2.46		3.58	
Demolish and Install	L.F.	1.12	3.21		4.33	
Reinstall	L.F.		1.97		1.97	
Clean	L.F.		.31		.31	
Paint	L.F.	.18	.88		1.06	
Minimum Charge	Job		178		178	
Aluminum						
Demolish	L.F.		.75		.75	Includes material and labor to install aluminum enameled downspout.
Install	L.F.	1.90	2.55		4.45	
Demolish and Install	L.F.	1.90	3.30		5.20	
Reinstall	L.F.		2.04		2.04	
Clean	L.F.		.31		.31	
Paint	L.F.	.18	.88		1.06	
Minimum Charge	Job		178		178	

Improvements / Appliances / Treatments

Downspouts	Unit	Material	Labor	Equip.	Total	Specification
Copper						
Demolish	L.F.		.75		.75	Includes material and labor to install
Install	L.F.	4.95	1.88		6.83	round copper downspout.
Demolish and Install	L.F.	4.95	2.63		7.58	
Reinstall	L.F.		1.50		1.50	
Clean	L.F.		.31		.31	
Paint	L.F.	.18	.88		1.06	
Minimum Charge	Job		178		178	
Plastic						
Demolish	L.F.		.75		.75	Cost includes material and labor to
Install	L.F.	.79	1.70		2.49	install 2" x 3" vinyl rectangular
Demolish and Install	L.F.	.79	2.45		3.24	downspouts.
Reinstall	L.F.		1.36		1.36	
Clean	L.F.		.31		.31	
Paint	L.F.	.18	.88		1.06	
Minimum Charge	Job		178		178	

Shutters	Unit	Material	Labor	Equip.	Total	Specification
Exterior						
25" High						
Demolish	Ea.		8.20		8.20	Includes material and labor to install
Install	Pr.	89	32.50		121.50	pair of exterior shutters 25" x 16",
Demolish and Install	Pr.	89	40.70		129.70	unpainted, installed as fixed shutters
Reinstall	Pr.		32.60		32.60	on concrete or wood frame.
Clean	Ea.	.06	7.05		7.11	
Paint	Ea.	1.44	15.40		16.84	
Minimum Charge	Job		163		163	
39" High						
Demolish	Ea.		8.20		8.20	Includes material and labor to install
Install	Pr.	96.50	32.50		129	pair of shutters 39" x 16", unpainted,
Demolish and Install	Pr.	96.50	40.70		137.20	installed as fixed shutters on concrete
Reinstall	Pr.		32.60		32.60	or wood frame.
Clean	Ea.	.06	7.05		7.11	
Paint	Ea.	2.23	20.50		22.73	
Minimum Charge	Job		163		163	
51" High						
Demolish	Ea.		8.20		8.20	Includes material and labor to install
Install	Pr.	107	32.50		139.50	pair of shutters 51" x 16", unpainted,
Demolish and Install	Pr.	107	40.70		147.70	installed as fixed shutters on concrete
Reinstall	Pr.		32.60		32.60	or wood frame.
Clean	Ea.	.06	7.05		7.11	
Paint	Ea.	2.92	21.50		24.42	
Minimum Charge	Job		163		163	
59" High						
Demolish	Ea.		8.20		8.20	Includes material and labor to install
Install	Pr.	151	32.50		183.50	pair of shutters 59" x 16", unpainted,
Demolish and Install	Pr.	151	40.70		191.70	installed as fixed shutters on concrete
Reinstall	Pr.		32.60		32.60	or wood frame.
Clean	Ea.	.06	7.05		7.11	
Paint	Ea.	3.36	23.50		26.86	
Minimum Charge	Job		163		163	

Improvements / Appliances / Treatments

Shutters	Unit	Material	Labor	Equip.	Total	Specification
67" High						
Demolish	Ea.		8.20		8.20	Includes material and labor to install
Install	Pr.	152	36		188	pair of shutters 67" x 16", unpainted,
Demolish and Install	Pr.	152	44.20		196.20	installed as fixed shutters on concrete
Reinstall	Pr.		36.22		36.22	or wood frame.
Clean	Ea.	.06	7.05		7.11	
Paint	Ea.	3.84	26.50		30.34	
Minimum Charge	Job		163		163	
Interior Wood						
23" x 24", 2 Panel						
Demolish	Ea.		8.20		8.20	Cost includes material and labor to
Install	Ea.	81	19.20		100.20	install 23" x 24", 4 panel interior
Demolish and Install	Ea.	81	27.40		108.40	moveable pine shutter, 1-1/4" louver,
Reinstall	Ea.		15.34		15.34	knobs and hooks and hinges.
Clean	Ea.	.06	7.05		7.11	
Paint	Ea.	1.09	20.50		21.59	
Minimum Charge	Job		163		163	
27" x 24", 4 Panel						
Demolish	Ea.		8.20		8.20	Cost includes material and labor to
Install	Ea.	94	19.20		113.20	install 27" x 24", 4 panel interior
Demolish and Install	Ea.	94	27.40		121.40	moveable pine shutter, 1-1/4" louver,
Reinstall	Ea.		15.34		15.34	knobs and hooks and hinges.
Clean	Ea.	.06	7.05		7.11	
Paint	Ea.	1.09	20.50		21.59	
Minimum Charge	Job		163		163	
31" x 24", 4 Panel						
Demolish	Ea.		8.20		8.20	Cost includes material and labor to
Install	Ea.	108	19.20		127.20	install 31" x 24", 4 panel interior
Demolish and Install	Ea.	108	27.40		135.40	moveable pine shutter, 1-1/4" louver,
Reinstall	Ea.		15.34		15.34	knobs and hooks and hinges.
Clean	Ea.	.06	7.05		7.11	
Paint	Ea.	1.09	20.50		21.59	
Minimum Charge	Job		163		163	
35" x 24", 4 Panel						
Demolish	Ea.		8.20		8.20	Cost includes material and labor to
Install	Ea.	108	19.20		127.20	install 35" x 24", 4 panel interior
Demolish and Install	Ea.	108	27.40		135.40	moveable pine shutter, 1-1/4" louver,
Reinstall	Ea.		15.34		15.34	knobs and hooks and hinges.
Clean	Ea.	.06	7.05		7.11	
Paint	Ea.	1.09	20.50		21.59	
Minimum Charge	Job		163		163	
39" x 24", 4 Panel						
Demolish	Ea.		8.20		8.20	Cost includes material and labor to
Install	Ea.	124	19.20		143.20	install 39" x 24", 4 panel interior
Demolish and Install	Ea.	124	27.40		151.40	moveable pine shutter, 1-1/4" louver,
Reinstall	Ea.		15.34		15.34	knobs and hooks and hinges.
Clean	Ea.	.06	7.05		7.11	
Paint	Ea.	1.09	20.50		21.59	
Minimum Charge	Job		163		163	
47" x 24", 4 Panel						
Demolish	Ea.		8.20		8.20	Cost includes material and labor to
Install	Ea.	138	19.20		157.20	install 47" x 24", 4 panel interior
Demolish and Install	Ea.	138	27.40		165.40	moveable pine shutter, 1-1/4" louver,
Reinstall	Ea.		15.34		15.34	knobs and hooks and hinges.
Clean	Ea.	.06	7.05		7.11	
Paint	Ea.	1.13	22		23.13	
Minimum Charge	Job		163		163	

Awning	Unit	Material	Labor	Equip.	Total	Specification
Standard Aluminum						
36" Wide x 30" Deep						
Demolish	Ea.		8.20		8.20	Includes material and labor to install
Install	Ea.	169	27		196	one aluminum awning with
Demolish and Install	Ea.	169	35.20		204.20	weather-resistant finish.
Reinstall	Ea.		21.73		21.73	
Clean	Ea.		9.80		9.80	
Minimum Charge	Job		163		163	
48" Wide x 30" Deep						
Demolish	Ea.		8.20		8.20	Includes material and labor to install
Install	Ea.	195	32.50		227.50	one aluminum awning with
Demolish and Install	Ea.	195	40.70		235.70	weather-resistant finish.
Reinstall	Ea.		26.08		26.08	
Clean	Ea.		9.80		9.80	
Minimum Charge	Job		163		163	
60" Wide x 30" Deep						
Demolish	Ea.		16.35		16.35	Includes material and labor to install
Install	Ea.	219	36		255	one aluminum awning with
Demolish and Install	Ea.	219	52.35		271.35	weather-resistant finish.
Reinstall	Ea.		28.98		28.98	
Clean	Ea.		9.80		9.80	
Minimum Charge	Job		163		163	
72" Wide x 30" Deep						
Demolish	Ea.		16.35		16.35	Includes material and labor to install
Install	Ea.	243	41		284	one aluminum awning with
Demolish and Install	Ea.	243	57.35		300.35	weather-resistant finish.
Reinstall	Ea.		32.60		32.60	
Clean	Ea.		9.80		9.80	
Minimum Charge	Job		163		163	
Roll-up Type						
Demolish	S.F.		.59		.59	Includes material and labor to install
Install	S.F.	6.70	3.26		9.96	an aluminum roll-up awning.
Demolish and Install	S.F.	6.70	3.85		10.55	
Reinstall	S.F.		2.61		2.61	
Clean	S.F.	.03	.49		.52	
Minimum Charge	Job		163		163	
Canvas						
30" Wide						
Demolish	Ea.		4.91		4.91	Includes material and labor to install a
Install	Ea.	183	27		210	canvas window awning.
Demolish and Install	Ea.	183	31.91		214.91	
Reinstall	Ea.		21.73		21.73	
Clean	Ea.		2.95		2.95	
Minimum Charge	Job		163		163	
36" Wide						
Demolish	Ea.		4.91		4.91	Includes material and labor to install a
Install	Ea.	195	32.50		227.50	canvas window awning.
Demolish and Install	Ea.	195	37.41		232.41	
Reinstall	Ea.		26.08		26.08	
Clean	Ea.		3.23		3.23	
Minimum Charge	Job		163		163	
42" Wide						
Demolish	Ea.		4.91		4.91	Includes material and labor to install a
Install	Ea.	207	41		248	canvas window awning.
Demolish and Install	Ea.	207	45.91		252.91	
Reinstall	Ea.		32.60		32.60	
Clean	Ea.		3.83		3.83	
Minimum Charge	Job		163		163	

Improvements / Appliances / Treatments

Awning

Awning	Unit	Material	Labor	Equip.	Total	Specification
48″ Wide						
Demolish	Ea.		4.91		4.91	Includes material and labor to install a
Install	Ea.	219	46.50		265.50	canvas window awning.
Demolish and Install	Ea.	219	51.41		270.41	
Reinstall	Ea.		37.26		37.26	
Clean	Ea.		4.45		4.45	
Minimum Charge	Job		163		163	
Cloth Canopy Cover						
8′ x 10′						
Demolish	Ea.		270		270	Includes material and labor to install
Install	Ea.	340	163		503	one 8′ x 10′ cloth canopy patio cover
Demolish and Install	Ea.	340	433		773	with front bar and tension support
Reinstall	Ea.		130.40		130.40	rafters and 9″ valance.
Clean	Ea.	2.33	39.50		41.83	
Minimum Charge	Job		325		325	
8′ x 15′						
Demolish	Ea.		270		270	Includes material and labor to install
Install	Ea.	460	163		623	one 8′ x 15′ cloth canopy patio cover
Demolish and Install	Ea.	460	433		893	with front bar and tension support
Reinstall	Ea.		130.40		130.40	rafters and 9″ valance.
Clean	Ea.	2.33	59		61.33	
Minimum Charge	Job		325		325	

Vent

Vent	Unit	Material	Labor	Equip.	Total	Specification
Small Roof Turbine						
Demolish	Ea.		8.20		8.20	Includes material and labor to install
Install	Ea.	57	32		89	spinner ventilator, wind driven,
Demolish and Install	Ea.	57	40.20		97.20	galvanized, 4″ neck diam, 180 CFM.
Reinstall	Ea.		25.66		25.66	
Clean	Ea.		5.25		5.25	
Minimum Charge	Job		177		177	
Large Roof Turbine						
Demolish	Ea.		6.55		6.55	Includes material and labor to install
Install	Ea.	66.50	46		112.50	spinner ventilator, wind driven,
Demolish and Install	Ea.	66.50	52.55		119.05	galvanized, 8″ neck diam, 360 CFM.
Reinstall	Ea.		36.66		36.66	
Clean	Ea.		5.90		5.90	
Minimum Charge	Job		177		177	
Roof Ventilator						
Demolish	Ea.		8.20		8.20	Includes material and labor to install a
Install	Ea.	163	40		203	roof ventilator.
Demolish and Install	Ea.	163	48.20		211.20	
Reinstall	Ea.		32.08		32.08	
Clean	Ea.	.58	5.90		6.48	
Paint	Ea.	1.36	7.15		8.51	
Minimum Charge	Job		177		177	
Gable Louvered						
Demolish	Ea.		16.35		16.35	Includes material and labor to install a
Install	Ea.	32.50	10.85		43.35	vinyl gable end vent.
Demolish and Install	Ea.	32.50	27.20		59.70	
Reinstall	Ea.		8.69		8.69	
Clean	Ea.		3.14		3.14	
Paint	Ea.	.90	4.40		5.30	
Minimum Charge	Job		163		163	

Improvements / Appliances / Treatments

Vent		Unit	Material	Labor	Equip.	Total	Specification
Ridge Vent							
	Demolish	L.F.		.84		.84	Includes material and labor to install a
	Install	L.F.	2.78	2.30		5.08	mill finish aluminum ridge vent strip.
	Demolish and Install	L.F.	2.78	3.14		5.92	
	Minimum Charge	Job		163		163	
Foundation							
	Demolish	Ea.		5.80		5.80	Includes material and labor to install
	Install	Ea.	19.80	10.95		30.75	galvanized foundation block vent.
	Demolish and Install	Ea.	19.80	16.75		36.55	
	Reinstall	Ea.		8.77		8.77	
	Clean	Ea.	.14	5.90		6.04	
	Minimum Charge	Job		163		163	
Frieze							
	Demolish	L.F.		.58		.58	Includes material and labor to install
	Install	L.F.	.85	.74		1.59	pine frieze board.
	Demolish and Install	L.F.	.85	1.32		2.17	
	Reinstall	L.F.		.59		.59	
	Clean	L.F.	.03	.20		.23	
	Minimum Charge	Job		163		163	
Water Heater Cap							
	Demolish	Ea.		12.30		12.30	Includes material and labor to install a
	Install	Ea.	24	44.50		68.50	galvanized steel water heater cap.
	Demolish and Install	Ea.	24	56.80		80.80	
	Reinstall	Ea.		35.64		35.64	
	Clean	Ea.	1.02	14.75		15.77	

Cupolas		Unit	Material	Labor	Equip.	Total	Specification
22" x 22"							
	Demolish	Ea.		157		157	Includes material and labor to install a
	Install	Ea.	287	81.50		368.50	cedar cupola with aluminum roof
	Demolish and Install	Ea.	287	238.50		525.50	covering.
	Reinstall	Ea.		65.20		65.20	
	Minimum Charge	Job		163		163	
29" x 29"							
	Demolish	Ea.		157		157	Includes material and labor to install a
	Install	Ea.	380	81.50		461.50	cedar cupola with aluminum roof
	Demolish and Install	Ea.	380	238.50		618.50	covering.
	Reinstall	Ea.		65.20		65.20	
	Minimum Charge	Job		163		163	
35" x 35"							
	Demolish	Ea.		157		157	Includes material and labor to install a
	Install	Ea.	650	109		759	cedar cupola with aluminum roof
	Demolish and Install	Ea.	650	266		916	covering.
	Reinstall	Ea.		86.93		86.93	
	Minimum Charge	Job		163		163	
47" x 47"							
	Demolish	Ea.		157		157	Includes material and labor to install a
	Install	Ea.	1050	163		1213	cedar cupola with aluminum roof
	Demolish and Install	Ea.	1050	320		1370	covering.
	Reinstall	Ea.		130.40		130.40	
	Minimum Charge	Job		163		163	
Weather Vane							
	Install	Ea.	48.50	41		89.50	Includes minimum labor and
	Minimum Charge	Job		163		163	equipment to install residential type weathervane.

Aluminum Carport		Unit	Material	Labor	Equip.	Total	Specification
Natural Finish							
	Demolish	S.F.		1.64		1.64	Includes material, labor and
	Install	S.F.	21.50	5.35	.97	27.82	equipment to install an aluminum
	Demolish and Install	S.F.	21.50	6.99	.97	29.46	carport. Foundations are not included.
	Reinstall	S.F.		4.29	.78	5.07	
	Clean	S.F.	.03	.18		.21	
	Minimum Charge	Job		325		325	
Enamel Finish							
	Demolish	S.F.		1.64		1.64	Includes material, labor and
	Install	S.F.	21.50	5.35	.97	27.82	equipment to install an aluminum
	Demolish and Install	S.F.	21.50	6.99	.97	29.46	carport. Foundations are not included.
	Reinstall	S.F.		4.29	.78	5.07	
	Clean	S.F.	.03	.18		.21	
	Minimum Charge	Job		325		325	
20' x 20' Complete							
	Demolish	Ea.		164		164	Includes material, labor and
	Install	Ea.	6800	535	97	7432	equipment to install an aluminum
	Demolish and Install	Ea.	6800	699	97	7596	carport. Foundations are not included.
	Reinstall	Ea.		429.31	77.57	506.88	
	Clean	Ea.	.15	59		59.15	
	Minimum Charge	Job		325		325	
24' x 24' Complete							
	Demolish	Ea.		164		164	Includes material, labor and
	Install	Ea.	9250	715	129	10094	equipment to install an aluminum
	Demolish and Install	Ea.	9250	879	129	10258	carport. Foundations are not included.
	Reinstall	Ea.		572.42	103.42	675.84	
	Clean	Ea.	.15	118		118.15	
	Minimum Charge	Job		325		325	
Metal Support Posts							
	Demolish	Ea.		2.18		2.18	Includes material and labor to install
	Install	Ea.	15.15	41		56.15	metal support posts for a carport.
	Demolish and Install	Ea.	15.15	43.18		58.33	
	Reinstall	Ea.		32.60		32.60	
	Clean	Ea.	.15	2.45		2.60	
	Paint	Ea.	.12	5.70		5.82	
	Minimum Charge	Job		325		325	

Storage Shed		Unit	Material	Labor	Equip.	Total	Specification
Aluminum							
Pre-fab (8' x 10')							
	Demolish	Ea.		47.50		47.50	Includes material and labor to install
	Install	Ea.	273	470		743	an aluminum storage shed.
	Demolish and Install	Ea.	273	517.50		790.50	
	Clean	Ea.	.97	29.50		30.47	
	Minimum Charge	Ea.		395		395	
Pre-fab (10' x 12')							
	Demolish	Ea.		90		90	Includes material and labor to install
	Install	Ea.	460	470		930	an aluminum storage shed.
	Demolish and Install	Ea.	460	560		1020	
	Clean	Ea.	.97	32.50		33.47	
	Minimum Charge	Ea.		395		395	

Improvements / Appliances / Treatments

Storage Shed

	Unit	Material	Labor	Equip.	Total	Specification
Wood						
Pre-fab (8' x 10')						
Demolish	Ea.		47.50		47.50	Includes material and labor to install a
Install	Ea.	760	470		1230	pre-fabricated wood storage shed.
Demolish and Install	Ea.	760	517.50		1277.50	
Clean	Ea.	.97	29.50		30.47	
Minimum Charge	Ea.		395		395	
Pre-fab (10' x 12')						
Demolish	Ea.		90		90	Includes material and labor to install a
Install	Ea.	990	470		1460	pre-fabricated wood storage shed.
Demolish and Install	Ea.	990	560		1550	
Clean	Ea.	.97	32.50		33.47	
Minimum Charge	Ea.		395		395	

Gazebo

	Unit	Material	Labor	Equip.	Total	Specification
Good Grade						
Demolish	S.F.		2.62		2.62	Includes material and labor to install a
Install	S.F.	18.90	24.50		43.40	gazebo.
Demolish and Install	S.F.	18.90	27.12		46.02	
Reinstall	S.F.		19.56		19.56	
Clean	S.F.	.12	1.57		1.69	
Minimum Charge	Job		163		163	
Custom Grade						
Demolish	S.F.		2.62		2.62	Includes material and labor to install a
Install	S.F.	30.50	24.50		55	gazebo.
Demolish and Install	S.F.	30.50	27.12		57.62	
Reinstall	S.F.		19.56		19.56	
Clean	S.F.	.12	1.57		1.69	
Minimum Charge	Job		163		163	
Bench Seating						
Demolish	L.F.		4.36		4.36	Includes material and labor to install
Install	L.F.	2.23	16.30		18.53	pressure treated bench seating.
Demolish and Install	L.F.	2.23	20.66		22.89	
Reinstall	L.F.		13.04		13.04	
Clean	L.F.	.12	1.18		1.30	
Paint	L.F.	.51	2.29		2.80	
Minimum Charge	Job		163		163	
Re-screen						
Install	S.F.	.53	1.30		1.83	Includes labor and material to
Minimum Charge	Job		163		163	re-screen wood frame.

Pool

	Unit	Material	Labor	Equip.	Total	Specification
Screen Enclosure						
Re-screen						
Install	SF Wall		3.94		3.94	Includes material and labor to install a
Minimum Charge	Job		163		163	screen enclosure for a pool.
Screen Door						
Demolish	Ea.		6.45		6.45	Includes material and labor to install
Install	Ea.	133	54.50		187.50	wood screen door with aluminum cloth
Demolish and Install	Ea.	133	60.95		193.95	screen. Frame and hardware not
Reinstall	Ea.		43.47		43.47	included.
Clean	Ea.	.11	8.50		8.61	
Paint	Ea.	3.03	11.45		14.48	
Minimum Charge	Job		163		163	

Improvements / Appliances / Treatments

Pool		Unit	Material	Labor	Equip.	Total	Specification
Heater / Motor							
	Demolish	Ea.		31		31	Includes material and labor to install
	Install	Ea.	2025	400		2425	swimming pool heater, not incl. base
	Demolish and Install	Ea.	2025	431		2456	or pad, elec., 12 kW, 4,800 gal
	Reinstall	Ea.		318.57		318.57	pool, incl. pump.
	Clean	Ea.	2.92	29.50		32.42	
	Minimum Charge	Job		178		178	
Filter System							
	Demolish	Ea.		52.50		52.50	Includes material, labor and
	Install	Total	595	284		879	equipment to install sand filter system
	Demolish and Install	Total	595	336.50		931.50	tank including hook-up to existing
	Reinstall	Total		226.82		226.82	equipment. Pump and motor not
	Minimum Charge	Job		177		177	included.
Pump / Motor							
	Demolish	Ea.		31		31	Includes material and labor to install a
	Install	Ea.	515	71		586	22 GPM pump.
	Demolish and Install	Ea.	515	102		617	
	Reinstall	Ea.		56.70		56.70	
	Clean	Ea.	2.92	29.50		32.42	
	Minimum Charge	Job		178		178	
Pump Water							
	Install	Day		315	69	384	Includes labor and equipment to pump
	Minimum Charge	Job		148		148	water from a swimming pool.
Replaster							
	Install	SF Surf	.97	7.80		8.77	Includes labor and material to
	Minimum Charge	Job		148		148	replaster gunite pool with a surface area of 500 to 600 S.F.
Solar Panels							
	Demolish	S.F.		1.31		1.31	Includes material and labor to install
	Install	S.F.	18.75	2.33		21.08	solar panel.
	Demolish and Install	S.F.	18.75	3.64		22.39	
	Reinstall	S.F.		1.87		1.87	
	Minimum Charge	Ea.		395		395	

Construction Clean-up	Unit	Material	Labor	Equip.	Total	Specification
Final						
Clean	S.F.		.21		.21	Includes labor for final site/job
Minimum Charge	Job		118		118	clean-up including detail cleaning.

Costs shown in *Means cost data publications* are based on National Averages for materials and installation. To adjust these costs to a specific location, simply multiply the base cost by the factor for that city. The data is arranged alphabetically by state and postal zip code numbers. For a city not listed, use the factor for a nearby city with similar economic characteristics.

STATE	CITY	Residential
ALABAMA		
350-352	Birmingham	.87
354	Tuscaloosa	.73
355	Jasper	.71
356	Decatur	.77
357-358	Huntsville	.84
359	Gadsden	.73
360-361	Montgomery	.76
362	Anniston	.68
363	Dothan	.75
364	Evergreen	.71
365-366	Mobile	.80
367	Selma	.72
368	Phenix City	.73
369	Butler	.71
ALASKA		
995-996	Anchorage	1.26
997	Fairbanks	1.28
998	Juneau	1.26
999	Ketchikan	1.27
ARIZONA		
850,853	Phoenix	.87
852	Mesa/Tempe	.84
855	Globe	.80
856-857	Tucson	.84
859	Show Low	.83
860	Flagstaff	.85
863	Prescott	.82
864	Kingman	.81
865	Chambers	.81
ARKANSAS		
716	Pine Bluff	.75
717	Camden	.65
718	Texarkana	.70
719	Hot Springs	.64
720-722	Little Rock	.81
723	West Memphis	.74
724	Jonesboro	.73
725	Batesville	.71
726	Harrison	.72
727	Fayetteville	.68
728	Russellville	.70
729	Fort Smith	.76
CALIFORNIA		
900-902	Los Angeles	1.06
903-905	Inglewood	1.04
906-908	Long Beach	1.03
910-912	Pasadena	1.04
913-916	Van Nuys	1.07
917-918	Alhambra	1.08
919-921	San Diego	1.04
922	Palm Springs	1.02
923-924	San Bernardino	1.04
925	Riverside	1.06
926-927	Santa Ana	1.04
928	Anaheim	1.07
930	Oxnard	1.07
931	Santa Barbara	1.07
932-933	Bakersfield	1.04
934	San Luis Obispo	1.07
935	Mojave	1.05
936-938	Fresno	1.10
939	Salinas	1.12
940-941	San Francisco	1.21
942,956-958	Sacramento	1.11
943	Palo Alto	1.16
944	San Mateo	1.19
945	Vallejo	1.13
946	Oakland	1.18
947	Berkeley	1.21
948	Richmond	1.22
949	San Rafael	1.20
950	Santa Cruz	1.14
951	San Jose	1.18

STATE	CITY	Residential
CALIFORNIA (CONT'D)		
952	Stockton	1.09
953	Modesto	1.08
954	Santa Rosa	1.14
955	Eureka	1.09
959	Marysville	1.10
960	Redding	1.10
961	Susanville	1.11
COLORADO		
800-802	Denver	.96
803	Boulder	.94
804	Golden	.92
805	Fort Collins	.91
806	Greeley	.80
807	Fort Morgan	.94
808-809	Colorado Springs	.92
810	Pueblo	.93
811	Alamosa	.90
812	Salida	.91
813	Durango	.93
814	Montrose	.88
815	Grand Junction	.93
816	Glenwood Springs	.92
CONNECTICUT		
060	New Britain	1.06
061	Hartford	1.05
062	Willimantic	1.06
063	New London	1.05
064	Meriden	1.05
065	New Haven	1.06
066	Bridgeport	1.06
067	Waterbury	1.06
068	Norwalk	1.06
069	Stamford	1.07
D.C.		
200-205	Washington	.93
DELAWARE		
197	Newark	1.00
198	Wilmington	1.01
199	Dover	1.00
FLORIDA		
320,322	Jacksonville	.79
321	Daytona Beach	.86
323	Tallahassee	.72
324	Panama City	.67
325	Pensacola	.75
326,344	Gainesville	.77
327-328,347	Orlando	.85
329	Melbourne	.87
330-332,340	Miami	.83
333	Fort Lauderdale	.84
334,349	West Palm Beach	.84
335-336,346	Tampa	.87
337	St. Petersburg	.77
338	Lakeland	.83
339,341	Fort Myers	.81
342	Sarasota	.85
GEORGIA		
300-303,399	Atlanta	.89
304	Statesboro	.67
305	Gainesville	.74
306	Athens	.74
307	Dalton	.70
308-309	Augusta	.76
310-312	Macon	.78
313-314	Savannah	.79
315	Waycross	.71
316	Valdosta	.71
317	Albany	.75
318-319	Columbus	.80

Location Factors

STATE	CITY	Residential
HAWAII		
967	Hilo	1.22
968	Honolulu	1.23
STATES & POSS.		
969	Guam	1.62
IDAHO		
832	Pocatello	.88
833	Twin Falls	.72
834	Idaho Falls	.72
835	Lewiston	.97
836-837	Boise	.89
838	Coeur d'Alene	.84
ILLINOIS		
600-603	North Suburban	1.09
604	Joliet	1.11
605	South Suburban	1.09
606-608	Chicago	1.15
609	Kankakee	1.00
610-611	Rockford	1.03
612	Rock Island	.96
613	La Salle	1.01
614	Galesburg	.98
615-616	Peoria	1.01
617	Bloomington	.96
618-619	Champaign	.98
620-622	East St. Louis	.97
623	Quincy	.97
624	Effingham	.98
625	Decatur	.97
626-627	Springfield	.98
628	Centralia	.96
629	Carbondale	.96
INDIANA		
460	Anderson	.91
461-462	Indianapolis	.96
463-464	Gary	1.00
465-466	South Bend	.91
467-468	Fort Wayne	.91
469	Kokomo	.92
470	Lawrenceburg	.86
471	New Albany	.85
472	Columbus	.92
473	Muncie	.91
474	Bloomington	.94
475	Washington	.90
476-477	Evansville	.90
478	Terre Haute	.90
479	Lafayette	.91
IOWA		
500-503,509	Des Moines	.92
504	Mason City	.77
505	Fort Dodge	.76
506-507	Waterloo	.80
508	Creston	.81
510-511	Sioux City	.86
512	Sibley	.73
513	Spencer	.75
514	Carroll	.75
515	Council Bluffs	.81
516	Shenandoah	.74
520	Dubuque	.84
521	Decorah	.77
522-524	Cedar Rapids	.93
525	Ottumwa	.83
526	Burlington	.87
527-528	Davenport	.98
KANSAS		
660-662	Kansas City	.95
664-666	Topeka	.79
667	Fort Scott	.85
668	Emporia	.72
669	Belleville	.74
670-672	Wichita	.81
673	Independence	.75
674	Salina	.73
675	Hutchinson	.68
676	Hays	.74
677	Colby	.76

STATE	CITY	Residential
KANSAS (CONT'D)		
678	Dodge City	.73
679	Liberal	.68
KENTUCKY		
400-402	Louisville	.93
403-405	Lexington	.84
406	Frankfort	.82
407-409	Corbin	.67
410	Covington	.93
411-412	Ashland	.94
413-414	Campton	.68
415-416	Pikeville	.77
417-418	Hazard	.67
420	Paducah	.89
421-422	Bowling Green	.89
423	Owensboro	.82
424	Henderson	.91
425-426	Somerset	.67
427	Elizabethtown	.86
LOUISIANA		
700-701	New Orleans	.85
703	Thibodaux	.80
704	Hammond	.78
705	Lafayette	.77
706	Lake Charles	.79
707-708	Baton Rouge	.78
710-711	Shreveport	.78
712	Monroe	.73
713-714	Alexandria	.73
MAINE		
039	Kittery	.80
040-041	Portland	.88
042	Lewiston	.88
043	Augusta	.83
044	Bangor	.86
045	Bath	.81
046	Machias	.82
047	Houlton	.86
048	Rockland	.80
049	Waterville	.75
MARYLAND		
206	Waldorf	.84
207-208	College Park	.84
209	Silver Spring	.85
210-212	Baltimore	.89
214	Annapolis	.85
215	Cumberland	.86
216	Easton	.68
217	Hagerstown	.85
218	Salisbury	.75
219	Elkton	.81
MASSACHUSETTS		
010-011	Springfield	1.05
012	Pittsfield	1.02
013	Greenfield	1.01
014	Fitchburg	1.08
015-016	Worcester	1.11
017	Framingham	1.12
018	Lowell	1.13
019	Lawrence	1.12
020-022, 024	Boston	1.19
023	Brockton	1.11
025	Buzzards Bay	1.10
026	Hyannis	1.09
027	New Bedford	1.11
MICHIGAN		
480,483	Royal Oak	1.03
481	Ann Arbor	1.03
482	Detroit	1.10
484-485	Flint	.97
486	Saginaw	.94
487	Bay City	.95
488-489	Lansing	.96
490	Battle Creek	.93
491	Kalamazoo	.92
492	Jackson	.94
493,495	Grand Rapids	.83

Location Factors

STATE	CITY	Residential
MICHIGAN (CONT'D)		
494	Muskegon	.89
496	Traverse City	.81
497	Gaylord	.84
498-499	Iron Mountain	.90
MINNESOTA		
550-551	Saint Paul	1.13
553-555	Minneapolis	1.18
556-558	Duluth	1.10
559	Rochester	1.04
560	Mankato	1.00
561	Windom	.84
562	Willmar	.85
563	St. Cloud	1.07
564	Brainerd	.96
565	Detroit Lakes	.97
566	Bemidji	.94
567	Thief River Falls	.92
MISSISSIPPI		
386	Clarksdale	.61
387	Greenville	.68
388	Tupelo	.64
389	Greenwood	.65
390-392	Jackson	.73
393	Meridian	.66
394	Laurel	.62
395	Biloxi	.75
396	McComb	.74
397	Columbus	.64
MISSOURI		
630-631	St. Louis	1.01
633	Bowling Green	.90
634	Hannibal	.87
635	Kirksville	.80
636	Flat River	.93
637	Cape Girardeau	.86
638	Sikeston	.83
639	Poplar Bluff	.83
640-641	Kansas City	1.02
644-645	St. Joseph	.93
646	Chillicothe	.85
647	Harrisonville	.94
648	Joplin	.82
650-651	Jefferson City	.88
652	Columbia	.87
653	Sedalia	.85
654-655	Rolla	.88
656-658	Springfield	.84
MONTANA		
590-591	Billings	.88
592	Wolf Point	.84
593	Miles City	.86
594	Great Falls	.89
595	Havre	.81
596	Helena	.88
597	Butte	.83
598	Missoula	.83
599	Kalispell	.82
NEBRASKA		
680-681	Omaha	.89
683-685	Lincoln	.78
686	Columbus	.69
687	Norfolk	.77
688	Grand Island	.77
689	Hastings	.76
690	Mccook	.69
691	North Platte	.75
692	Valentine	.66
693	Alliance	.65
NEVADA		
889-891	Las Vegas	1.00
893	Ely	.90
894-895	Reno	.97
897	Carson City	.97
898	Elko	.97

STATE	CITY	Residential
NEW HAMPSHIRE		
030	Nashua	.90
031	Manchester	.90
032-033	Concord	.86
034	Keene	.73
035	Littleton	.81
036	Charleston	.70
037	Claremont	.72
038	Portsmouth	.84
NEW JERSEY		
070-071	Newark	1.12
072	Elizabeth	1.15
073	Jersey City	1.12
074-075	Paterson	1.12
076	Hackensack	1.11
077	Long Branch	1.12
078	Dover	1.12
079	Summit	1.12
080,083	Vineland	1.09
081	Camden	1.10
082,084	Atlantic City	1.13
085-086	Trenton	1.11
087	Point Pleasant	1.10
088-089	New Brunswick	1.12
NEW MEXICO		
870-872	Albuquerque	.85
873	Gallup	.85
874	Farmington	.85
875	Santa Fe	.85
877	Las Vegas	.85
878	Socorro	.85
879	Truth/Consequences	.84
880	Las Cruces	.83
881	Clovis	.85
882	Roswell	.85
883	Carrizozo	.85
884	Tucumcari	.86
NEW YORK		
100-102	New York	1.36
103	Staten Island	1.27
104	Bronx	1.29
105	Mount Vernon	1.15
106	White Plains	1.19
107	Yonkers	1.21
108	New Rochelle	1.21
109	Suffern	1.13
110	Queens	1.27
111	Long Island City	1.30
112	Brooklyn	1.32
113	Flushing	1.29
114	Jamaica	1.29
115,117,118	Hicksville	1.19
116	Far Rockaway	1.28
119	Riverhead	1.20
120-122	Albany	.95
123	Schenectady	.96
124	Kingston	1.02
125-126	Poughkeepsie	1.06
127	Monticello	1.03
128	Glens Falls	.88
129	Plattsburgh	.92
130-132	Syracuse	.96
133-135	Utica	.93
136	Watertown	.91
137-139	Binghamton	.92
140-142	Buffalo	1.06
143	Niagara Falls	1.01
144-146	Rochester	.98
147	Jamestown	.88
148-149	Elmira	.87
NORTH CAROLINA		
270,272-274	Greensboro	.73
271	Winston-Salem	.73
275-276	Raleigh	.74
277	Durham	.73
278	Rocky Mount	.64
279	Elizabeth City	.61
280	Gastonia	.74
281-282	Charlotte	.74
283	Fayetteville	.71

STATE	CITY	Residential
NORTH CAROLINA (CONT'D)		
284	Wilmington	.72
285	Kinston	.62
286	Hickory	.62
287-288	Asheville	.72
289	Murphy	.66
NORTH DAKOTA		
580-581	Fargo	.80
582	Grand Forks	.77
583	Devils Lake	.80
584	Jamestown	.75
585	Bismarck	.80
586	Dickinson	.77
587	Minot	.80
588	Williston	.77
OHIO		
430-432	Columbus	.96
433	Marion	.93
434-436	Toledo	1.01
437-438	Zanesville	.91
439	Steubenville	.96
440	Lorain	1.01
441	Cleveland	1.02
442-443	Akron	.99
444-445	Youngstown	.96
446-447	Canton	.94
448-449	Mansfield	.95
450	Hamilton	.96
451-452	Cincinnati	.96
453-454	Dayton	.92
455	Springfield	.94
456	Chillicothe	.96
457	Athens	.90
458	Lima	.92
OKLAHOMA		
730-731	Oklahoma City	.80
734	Ardmore	.78
735	Lawton	.81
736	Clinton	.77
737	Enid	.77
738	Woodward	.76
739	Guymon	.67
740-741	Tulsa	.79
743	Miami	.82
744	Muskogee	.72
745	Mcalester	.74
746	Ponca City	.77
747	Durant	.76
748	Shawnee	.76
749	Poteau	.77
OREGON		
970-972	Portland	1.02
973	Salem	1.01
974	Eugene	1.01
975	Medford	.99
976	Klamath Falls	1.00
977	Bend	1.02
978	Pendleton	.99
979	Vale	.99
PENNSYLVANIA		
150-152	Pittsburgh	.99
153	Washington	.94
154	Uniontown	.91
155	Bedford	.89
156	Greensburg	.95
157	Indiana	.91
158	Dubois	.90
159	Johnstown	.90
160	Butler	.93
161	New Castle	.93
162	Kittanning	.94
163	Oil City	.89
164-165	Erie	.97
166	Altoona	.89
167	Bradford	.89
168	State College	.92
169	Wellsboro	.88
170-171	Harrisburg	.94

STATE	CITY	Residential
PENNSYLVANIA (CONT'D)		
172	Chambersburg	.89
173-174	York	.88
175-176	Lancaster	.91
177	Williamsport	.85
178	Sunbury	.89
179	Pottsville	.89
180	Lehigh Valley	.98
181	Allentown	1.01
182	Hazleton	.89
183	Stroudsburg	.92
184-185	Scranton	.95
186-187	Wilkes-Barre	.91
188	Montrose	.89
189	Doylestown	1.04
190-191	Philadelphia	1.14
193	Westchester	1.07
194	Norristown	1.03
195-196	Reading	.96
PUERTO RICO		
009	San Juan	.84
RHODE ISLAND		
028	Newport	1.08
029	Providence	1.08
SOUTH CAROLINA		
290-292	Columbia	.73
293	Spartanburg	.71
294	Charleston	.72
295	Florence	.66
296	Greenville	.70
297	Rock Hill	.65
298	Aiken	.83
299	Beaufort	.67
SOUTH DAKOTA		
570-571	Sioux Falls	.77
572	Watertown	.73
573	Mitchell	.75
574	Aberdeen	.76
575	Pierre	.76
576	Mobridge	.73
577	Rapid City	.76
TENNESSEE		
370-372	Nashville	.85
373-374	Chattanooga	.77
375,380-381	Memphis	.85
376	Johnson City	.72
377-379	Knoxville	.75
382	Mckenzie	.70
383	Jackson	.71
384	Columbia	.72
385	Cookeville	.68
TEXAS		
750	Mckinney	.75
751	Waxahackie	.76
752-753	Dallas	.83
754	Greenville	.69
755	Texarkana	.74
756	Longview	.68
757	Tyler	.74
758	Palestine	.67
759	Lufkin	.72
760-761	Fort Worth	.83
762	Denton	.77
763	Wichita Falls	.79
764	Eastland	.72
765	Temple	.75
766-767	Waco	.77
768	Brownwood	.68
769	San Angelo	.72
770-772	Houston	.85
773	Huntsville	.69
774	Wharton	.70
775	Galveston	.83
776-777	Beaumont	.82
778	Bryan	.74
779	Victoria	.74
780	Laredo	.73

Location Factors

STATE	CITY	Residential
TEXAS (CONT'D)		
781-782	San Antonio	.80
783-784	Corpus Christi	.77
785	Mc Allen	.75
786-787	Austin	.79
788	Del Rio	.66
789	Giddings	.70
790-791	Amarillo	.78
792	Childress	.76
793-794	Lubbock	.76
795-796	Abilene	.75
797	Midland	.76
798-799,885	El Paso	.75
UTAH		
840-841	Salt Lake City	.83
842,844	Ogden	.81
843	Logan	.82
845	Price	.72
846-847	Provo	.83
VERMONT		
050	White River Jct.	.73
051	Bellows Falls	.75
052	Bennington	.74
053	Brattleboro	.75
054	Burlington	.80
056	Montpelier	.82
057	Rutland	.81
058	St. Johnsbury	.75
059	Guildhall	.74
VIRGINIA		
220-221	Fairfax	.85
222	Arlington	.87
223	Alexandria	.90
224-225	Fredericksburg	.75
226	Winchester	.71
227	Culpeper	.77
228	Harrisonburg	.67
229	Charlottesville	.72
230-232	Richmond	.81
233-235	Norfolk	.81
236	Newport News	.79
237	Portsmouth	.77
238	Petersburg	.78
239	Farmville	.68
240-241	Roanoke	.72
242	Bristol	.67
243	Pulaski	.66
244	Staunton	.68
245	Lynchburg	.69
246	Grundy	.67
WASHINGTON		
980-981,987	Seattle	1.01
982	Everett	1.03
983-984	Tacoma	1.00
985	Olympia	1.00
986	Vancouver	.97
988	Wenatchee	.90
989	Yakima	.95
990-992	Spokane	1.00
993	Richland	.97
994	Clarkston	.96
WEST VIRGINIA		
247-248	Bluefield	.88
249	Lewisburg	.89
250-253	Charleston	.97
254	Martinsburg	.84
255-257	Huntington	.96
258-259	Beckley	.90
260	Wheeling	.92
261	Parkersburg	.91
262	Buckhannon	.91
263-264	Clarksburg	.90
265	Morgantown	.91
266	Gassaway	.92
267	Romney	.86
268	Petersburg	.88

STATE	CITY	Residential
WISCONSIN		
530,532	Milwaukee	1.05
531	Kenosha	1.03
534	Racine	1.01
535	Beloit	.99
537	Madison	.99
538	Lancaster	.97
539	Portage	.96
540	New Richmond	.98
541-543	Green Bay	1.01
544	Wausau	.95
545	Rhinelander	.95
546	La Crosse	.94
547	Eau Claire	.98
548	Superior	.97
549	Oshkosh	.95
WYOMING		
820	Cheyenne	.75
821	Yellowstone Nat. Pk.	.70
822	Wheatland	.71
823	Rawlins	.69
824	Worland	.69
825	Riverton	.70
826	Casper	.76
827	Newcastle	.68
828	Sheridan	.73
829-831	Rock Springs	.73
ALBERTA		
	Calgary	1.06
	Edmonton	1.05
	Fort McMurray	1.03
	Lethbridge	1.04
	Lloydminster	1.03
	Medicine Hat	1.04
	Red Deer	1.04
BRITISH COLUMBIA		
	Kamloops	1.01
	Prince George	1.01
	Vancouver	1.08
	Victoria	1.01
MANITOBA		
	Brandon	1.00
	Portage la Prairie	1.00
	Winnipeg	1.01
NEW BRUNSWICK		
	Bathurst	.91
	Dalhousie	.91
	Fredericton	.99
	Moncton	.91
	Newcastle	.91
	Saint John	.99
NEWFOUNDLAND		
	Corner Brook	.92
	St. John's	.93
NORTHWEST TERRITORIES		
	Yellowknife	.98
NOVA SCOTIA		
	Dartmouth	.93
	Halifax	.93
	New Glasgow	.93
	Sydney	.91
	Yarmouth	.93
ONTARIO		
	Barrie	1.10
	Brantford	1.12
	Cornwall	1.11
	Hamilton	1.12
	Kingston	1.11
	Kitchener	1.06
	London	1.10
	North Bay	1.08
	Oshawa	1.10
	Ottawa	1.12
	Owen Sound	1.09

LOCATION FACTORS

STATE	CITY	Residential
ONTARIO (CONT'D)		
	Peterborough	1.09
	Sarnia	1.12
	Sudbury	1.03
	Thunder Bay	1.08
	Toronto	1.15
	Windsor	1.09
PRINCE EDWARD ISLAND		
	Charlottetown	.88
	Summerside	.88
QUEBEC		
	Cap-de-la-Madeleine	1.11
	Charlesbourg	1.11
	Chicoutimi	1.11
	Gatineau	1.10
	Laval	1.11
	Montreal	1.11
	Quebec	1.13
	Sherbrooke	1.10
	Trois Rivieres	1.11
SASKATCHEWAN		
	Moose Jaw	.91
	Prince Albert	.90
	Regina	.92
	Saskatoon	.91
YUKON		
	Whitehorse	.90

Abbreviations

A	Area
ASTM	American Society for Testing and Materials
B.F.	Board feet
Carp.	Carpenter
C.F.	Cubic feet
CWJ	Composite wood joist
C.Y.	Cubic yard
Ea.	Each
Equip.	Equipment
Exp.	Exposure
Ext.	Exterior
F	Fahrenheit
Ft.	Foot, feet
Gal.	Gallon
Hr.	Hour
in.	Inch, inches
Inst.	Installation
Int.	Interior
Lb.	Pound
L.F.	Linear feet
LVL	Laminated veneer lumber
Mat.	Material
Max.	Maximum
MBF	Thousand board feet
MBM	Thousand feet board measure
MSF	Thousand square feet
Min.	Minimum
O.C.	On center
O&P	Overhead and profit
OWJ	Open web wood joist
Oz.	Ounce
Pr.	Pair
Quan.	Quantity
S.F.	Square foot
Sq.	Square, 100 square feet
S.Y.	Square yard
V.L.F.	Vertical linear feet
'	Foot, feet
"	Inch, inches
°	Degrees

Notes

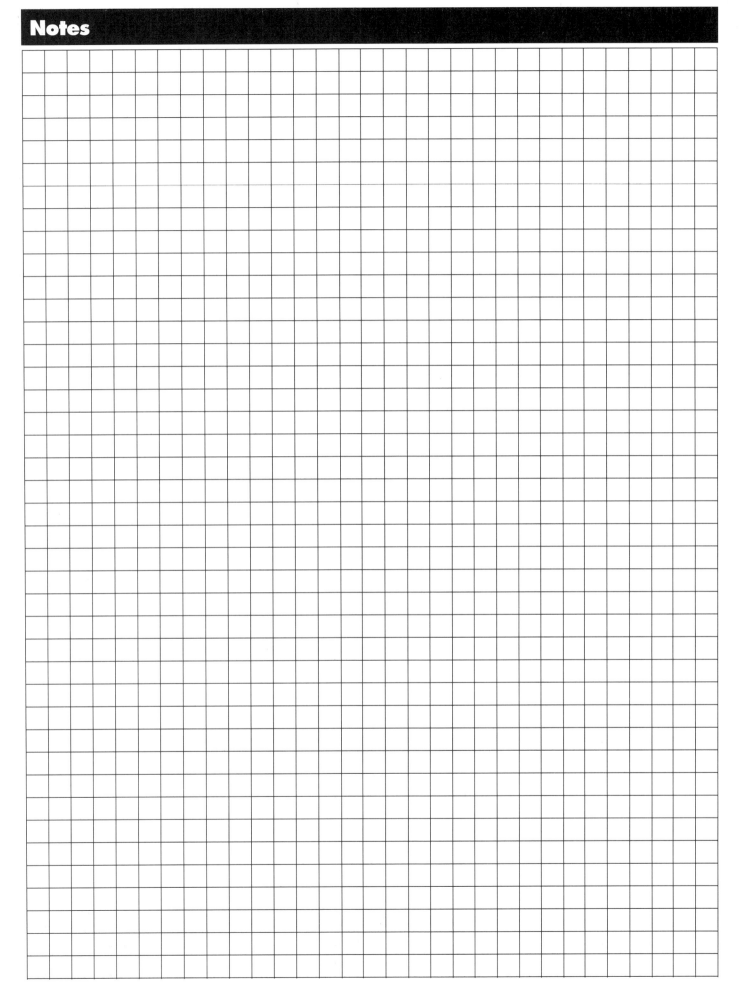

Notes

Reed Construction Data, Inc.

Reed Construction Data, Inc., a leading worldwide provider of total construction information solutions, is comprised of three main product groups designed specifically to help construction professionals advance their businesses with timely, accurate and actionable project, product, and cost data. Reed Construction Data is a division of Reed Business Information, a member of the Reed Elsevier plc group of companies.

The *Project, Product, and Cost & Estimating* divisions offer a variety of innovative products and services designed for the full spectrum of design, construction, and manufacturing professionals. Through it's *International* companies, Reed Construction Data's reputation for quality construction market data is growing worldwide.

Cost Information
RSMeans, the undisputed market leader and authority on construction costs, publishes current cost and estimating information in annual cost books and on the CostWorks CD-ROM. RSMeans furnishes the construction industry with a rich library of complementary reference books and a series of professional seminars that are designed to sharpen professional skills and maximize the effective use of cost estimating and management tools. RSMeans also provides construction cost consulting for Owners, Manufacturers, Designers, and Contractors.

Project Data
Reed Construction Data provides complete, accurate and relevant project information through all stages of construction. Customers are supplied industry data through leads, project reports, contact lists, plans and specifications surveys, market penetration analyses and sales evaluation reports. Any of these products can pinpoint a county, look at a state, or cover the country. Data is delivered via paper, e-mail, CD-ROM or the Internet.

Building Product Information
The First Source suite of products is the only integrated building product information system offered to the commercial construction industry for comparing and specifying building products. These print and online resources include *First Source,* CSI's SPEC-DATA™, CSI's MANU-SPEC™, First Source CAD, and Manufacturer Catalogs. Written by industry professionals and organized using CSI's MasterFormat™, construction professionals use this information to make better design decisions.

FirstSourceONL.com combines Reed Construction Data's project, product and cost data with news and information from Reed Business Information's *Building Design & Construction* and *Consulting-Specifying Engineer,* this industry-focused site offers easy and unlimited access to vital information for all construction professionals.

International
BIMSA/Mexico provides construction project news, product information, cost-data, seminars and consulting services to construction professionals in Mexico. Its subsidiary, PRISMA, provides job costing software.

Byggfakta Scandinavia AB, founded in 1936, is the parent company for the leaders of customized construction market data for Denmark, Estonia, Finland, Norway and Sweden. Each company fully covers the local construction market and provides information across several platforms including subscription, ad-hoc basis, electronically and on paper.

Reed Construction Data Canada serves the Canadian construction market with reliable and comprehensive project and product information services that cover all facets of construction. Core services include: *BuildSource, BuildSpec, BuildSelect,* product selection and specification tools available in print and on the Internet; Building Reports, a national construction project lead service; CanaData, statistical and forecasting information; *Daily Commercial News,* a construction newspaper reporting on news and projects in Ontario; and *Journal of Commerce,* reporting news in British Columbia and Alberta.

Cordell Building Information Services, with its complete range of project and cost and estimating services, is Australia's specialist in the construction information industry. Cordell provides in-depth and historical information on all aspects of construction projects and estimation, including several customized reports, construction and sales leads, and detailed cost information among others.

For more information, please visit our Web site at www.reedconstructiondata.com.

Reed Construction Data, Inc., Corporate Office
30 Technology Parkway South
Norcross, GA 30092-2912
(800) 322-6996
(800) 895-8661 (fax)
info@reedbusiness.com
www.reedconstructiondata.com

 Reed Construction Data

Contractor's Pricing Guides

For more information
visit Means Web Site
at www.rsmeans.com

Contractor's Pricing Guide:
Residential Detailed Costs 2005

Every aspect of residential construction, from overhead costs to residential lighting and wiring, is in here. All the detail you need to accurately estimate the costs of your work with or without markups–labor-hours, typical crews and equipment are included as well. When you need a detailed estimate, this publication has all the costs to help you come up with a complete, on the money, price you can rely on to win profitable work.

$39.95 per copy
Over 350 pages, with charts and tables, 8-1/2 x 11
Catalog No. 60335   ISBN 0-87629-762-9

Contractor's Pricing Guide:
Residential Square Foot Costs 2005

Now available in one concise volume, all you need to know to plan and budget the cost of new homes. If you are looking for a quick reference, the model home section contains costs for over 250 different sizes and types of residences, with hundreds of easily applied modifications. If you need even more detail, the Assemblies Section lets you build your own costs or modify the model costs further. Hundreds of graphics are provided, along with forms and procedures to help you get it right.

$39.95 per copy
Over 300 pages, illustrated, 8-1/2 x 11
Catalog No. 60325   ISBN 0-87629-763-7

Green Building: Project Planning & Cost Estimating

By RSMeans and Contributing Authors

Written by a team of leading experts in sustainable design, this is a complete guide to planning and estimating green building projects, a growing trend in building design and construction. It explains:

- How to select and specify green products
- How the project team works differently on a green versus a traditional building project
- What criteria your building needs to meet to get a LEED, Energy Star, or other recognized rating for green buildings

Features an extensive Green Building Cost Data section, which details the available products, how they are specified, and how much they cost.

$89.95 per copy
350 pages, illustrated, hardcover
Catalog No. 67338   ISBN 0-87629-659-2

Contractor's Pricing Guide:
Residential Repair & Remodeling Costs 2005

This book provides total unit price costs for every aspect of the most common repair & remodeling projects. Organized in the order of construction by component and activity, it includes demolition and installation, cleaning, painting, and more.

With simplified estimating methods; clear, concise descriptions; and technical specifications for each component, the book is a valuable tool for contractors who want to speed up their estimating time, while making sure their costs are on target.

$39.95 per copy
Over 300 pages, illustrated, 8-1/2 x 11
Catalog No. 60345   ISBN 0-87629-761-0

Plumbing Estimating Methods
New 3rd Edition

By Joseph Galeno and Sheldon Greene

This newly updated, practical guide walks you through a complete plumbing estimate, from basic materials and installation methods through to change order analysis. It contains everything needed for plumbing estimating and covers:

- Residential, commercial, industrial, and medical systems
- Updated, expanded information using *Means Plumbing Cost Data*
- Sample takeoff and estimate forms
- Detailed illustrations of systems & components

$59.95 per copy
Over 330 pages
Catalog No. 67283B   ISBN 0-87629-704-1

Residential & Light Commercial Construction Standards, 2nd Edition

By RSMeans and Contributing Authors

New, updated second edition of this unique collection of industry standards that define quality construction. For contractors, subcontractors, owners, developers, architects, engineers, attorneys, and insurance personnel, this book provides authoritative requirements and recommendations compiled from the nation's leading professional associations, industry publications, and building code organizations.

$59.95 per copy
600 pages, illustrated, softcover
Catalog No. 67322A   ISBN 0-87629-658-4

For more information
visit Means Web Site
at www.rsmeans.com

Annual Cost Guides

Means Building Construction Cost Data 2005

Offers you unchallenged unit price reliability in an easy-to-use arrangement. Whether used for complete, finished estimates or for periodic checks, it supplies more cost facts better and faster than any comparable source. Over 23,000 unit prices for 2005. The City Cost Indexes now cover over 930 areas, for indexing to any project location in North America.

$115.95 per copy
Over 700 pages, softcover
Catalog No. 60015   ISBN 0-87629-750-5

Means Open Shop Building Construction Cost Data 2005

The open-shop version of the *Means Building Construction Cost Data*. More than 22,000 reliable unit cost entries based on open shop trade labor rates. Eliminates time-consuming searches for these prices. The first book with open shop labor rates and crews. Labor information is itemized by labor-hours, crew, hourly/daily output, equipment, overhead and profit. For contractors, owners and facility managers.

$115.95 per copy
Over 700 pages, softcover
Catalog No. 60155   ISBN 0-87629-772-6

Means Plumbing Cost Data 2005

Comprehensive unit prices and assemblies for plumbing, irrigation systems, commercial and residential fire protection, point-of-use water heaters, and the latest approved materials. This publication and its companion, *Means Mechanical Cost Data*, provide full-range cost estimating coverage for all the mechanical trades.

$115.95 per copy
Over 550 pages, softcover
Catalog No. 60215   ISBN 0-87629-756-4

Means Residential Cost Data 2005

Speeds you through residential construction pricing with more than 100 illustrated complete house square-foot costs. Alternate assemblies cost selections are located on adjoining pages, so that you can develop tailor-made estimates in minutes. Complete data for detailed unit cost estimates is also provided.

$99.95 per copy
Over 600 pages, softcover
Catalog No. 60175   ISBN 0-87629-758-0

Means Electrical Cost Data 2005

Pricing information for every part of electrical cost planning: unit and systems costs with design tables; engineering guides and illustrated estimating procedures; complete labor-hour, materials, and equipment costs for better scheduling and procurement. With the latest products and construction methods used in electrical work. More than 15,000 unit and systems costs, clear specifications and drawings.

$115.95 per copy
Over 450 pages, softcover
Catalog No. 60035   ISBN 0-87629-754-8

Means Repair & Remodeling Cost Data 2005

Commercial/Residential

You can use this valuable tool to estimate commercial and residential renovation and remodeling. By using the specialized costs in this manual, you'll find it's not necessary to force fit prices for new construction into remodeling cost planning. Provides comprehensive unit costs, building systems costs, extensive labor data and estimating assistance for every kind of building improvement.

$99.95 per copy
Over 650 pages, softcover
Catalog No. 60045   ISBN 0-87629-760-2

Means Site Work & Landscape Cost Data 2005

Hard-to-find costs are presented in an easy-to-use format for every type of site work and landscape construction. Costs are organized, described, and laid out for earthwork, utilities, roads and bridges, as well as grading, planting, lawns, trees, irrigation systems, and site improvements.

$115.95 per copy
Over 600 pages, softcover
Catalog No. 60285   ISBN 0-87629-764-5

Means Light Commercial Cost Data 2005

Specifically addresses the light commercial market, which is an increasingly specialized niche in the industry. Aids you, the owner/designer/contractor, in preparing all types of estimates, from budgets to detailed bids. Includes new advances in methods and materials. Assemblies section allows you to evaluate alternatives in early stages of design/planning.

$99.95 per copy
Over 650 pages, softcover
Catalog No. 60185   ISBN 0-87629-765-3

Books for Builders

For more information
visit Means Web Site
at www.rsmeans.com

Builder's Essentials:
Plan Reading & Material Takeoff
By Wayne J. DelPico

A complete course in reading and interpreting building plans—and performing quantity takeoffs to professional standards.

This book shows and explains, in clear language and with over 160 illustrations, typical working drawings encountered by contractors in residential and light commercial construction. The author describes not only how all common features are represented, but how to translate that information into a material list. Organized by CSI division, each chapter uses plans, details and tables, and a summary checklist.

$35.95 per copy
Over 420 pages, illustrated, softcover
Catalog No. 67307   ISBN 0-87629-348-8

Builder's Essentials:
Best Business Practices for Builders & Remodelers:
An Easy-to-Use Checklist System
By Thomas N. Frisby

A comprehensive guide covering all aspects of running a construction business, with more than 40 user–friendly checklists. This book provides expert guidance on: increasing your revenue and keeping more of your profit; planning for long-term growth; keeping good employees and managing subcontractors.

$29.95 per copy
Over 220 pages, softcover
Catalog No. 67329   ISBN 0-87629-619-3

Means Estimating Handbook,
2nd Edition

Updated new Second Edition answers virtually any estimating technical question – all organized by CSI MasterFormat. This comprehensive reference covers the full spectrum of technical data required to estimate construction costs. The book includes information on sizing, productivity, equipment requirements, code-mandated specifications, design standards, and engineering factors.

This reference—widely used in the industry for tasks ranging from routine estimates to special cost analysis projects—has been completely updated with new and expanded technical information and an enhanced format for ease-of-use.

$99.95 per copy
Over 900 pages, hardcover
Catalog No. 67276A   ISBN 0-87629-699-1

Builder's Essentials:
Framing & Rough Carpentry,
2nd Edition

A complete, illustrated do-it-yourself course on framing and rough carpentry. The book covers walls, floors, stairs, windows, doors, and roofs, as well as nailing patterns and procedures. Additional sections are devoted to equipment and material handling, standards, codes, and safety requirements.

The "framer-friendly" approach includes easy-to-follow, step-by-step instructions. This practical guide will benefit both the carpenter's apprentice and the experienced carpenter, and sets a uniform standard for framing crews.
Also available in Spanish!

$24.95 per copy
Over 125 pages, illustrated, softcover
Catalog No. 67298A   ISBN 0-87629-617-7
Spanish 67298AS ISBN 0-87629-654-1

Interior Home Improvement
Costs, New 9th Edition

Estimates for the most popular remodeling and repair projects-from small, do-it-yourself jobs-to major renovations and new construction.

Includes:
• Kitchens & Baths
• New Living Space from Your Attic, Basement or Garage
• New Floors, Paint & Wallpaper
• Tearing Out or Building New Walls
• Closets, Stairs & Fireplaces
• New Energy-Saving Improvements, Home Theatres, and Much More!

$24.95 per copy
Over 250 pages, illustrated, softcover
Catalog No. 67308E   ISBN 0-87629-743-2

Exterior Home Improvement
Costs, New 9th Edition

Estimates for the most popular remodeling and repair projects-from small, do-it-yourself jobs, to major renovations and new construction.

Includes:
• Curb Appeal Projects-Landscaping, Patios, Porches, Driveways and Walkways
• New Windows and Doors
• Decks, Greenhouses, and Sunrooms
• Room Additions and Garages
• Roofing, Siding, and Painting
• "Green" Improvements to Save Energy & Water

$19.95 per copy
Over 250 pages, illustrated, softcover
Catalog No. 67309E   ISBN 0-87629-742-4

Books for Builders

For more information
visit Means Web Site
at www.rsmeans.com

Designing & Building with the IBC, Second Edition

By Rolf Jensen & Associates, Inc.

This new second edition (formerly titled, *From Model Codes to the IBC: A Transitional Guide*) is an essential reference on the latest building codes for architects and engineers, building officials and AHJs, contractors and homebuilders, and manufacturers and building owners. It provides complete coverage of the new *International Building Code®*(IBC) 2003, with expert advice and user-friendly, side-by-side comparison to the IBC 2000 and model building codes.

$99.95 per copy
Over 1,000 pages, softcover
Catalog No. 67328B   ISBN 0-87629-616-9

Historic Preservation
Project Planning & Estimating

by Swanke Hayden Connell Architects

Managing Historic Restoration, Rehabilitation, and Preservation Building Projects and Determining and Controlling Their Costs

The authors explain:
- How to determine whether a structure qualifies as historic
- Where to obtain funding and other assistance
- How to evaluate and repair more than 75 historic building materials
- How to properly research, document, and manage the project to meet code, agency, and other special requirements
- How to approach the upgrade of major building systems

$99.95 per copy
Over 675 pages, hardcover
Catalog No. 67323   ISBN 0-87629-573-1

Builder's Essentials: Advanced Framing Methods

By Scot Simpson

A highly illustrated, "framer-friendly" approach to advanced framing elements. Provides expert, but easy-to-interpret, instruction for laying out and framing complex walls, roofs, and stairs, and special requirements for earthquake and hurricane protection. Also helps bring framers up to date on the latest building code changes, and provides tips on the lead framer's role and responsibilities, how to prepare for a job, and how to get the crew started.

$24.95 per copy
250 pages, illustrated, softcover
Catalog No. 67330   ISBN 0-87629-618-5

Unit Price Estimating Methods
New 3rd Edition

This new edition includes up-to-date cost data and estimating examples, updated to reflect changes to the CSI numbering system and new features of Means cost data. It describes the most productive, universally accepted ways to estimate, and uses checklists and forms to illustrate shortcuts and timesavers. A model estimate demonstrates procedures. A new chapter explores computer estimating alternatives.

$59.95 per copy
Over 350 pages, illustrated, hardcover
Catalog No. 67303A   ISBN 0-87629-662-2

Means Spanish/English Construction Dictionary

By RSMeans, The International Conference of Building Officials (ICBO), and Rolf Jensen & Associates (RJA)

Designed to facilitate communication among Spanish- and English-speaking construction personnel—improving performance and job-site safety. Features the most common words and phrases used in the construction industry, with easy-to-follow pronunciations. Includes extensive building systems and tools illustrations.

$22.95 per copy
250 pages, illustrated, softcover
Catalog No. 67327   ISBN 0-87629-578-2

Means Illustrated Construction Dictionary, 3rd Edition

Long regarded as the Industry's finest, the *Means Illustrated Construction Dictionary* is now even better. With the addition of over 1,000 new terms and hundreds of new illustrations, it is the clear choice for the most comprehensive and current information.

The companion CD-ROM that comes with this new edition adds many extra features: larger graphics, expanded definitions, and links to both CSI MasterFormat numbers and product information.

Contains 19,000 construction words, terms, phrases, symbols, weights, measures, and equivalents. 1,000 new entries; 1,200 helpful illustrations, easy-to-use format with thumbtabs.

$99.95 per copy
Over 790 pages, illustrated, hardcover
Catalog No. 67292A   ISBN 0-87629-538-3

MeansData™

CONSTRUCTION COSTS FOR SOFTWARE APPLICATIONS
Your construction estimating software is only as good as your cost data.

A proven construction cost database is a mandatory part of any estimating package. The following list of software providers can offer you MeansData™ as an added feature for their estimating systems. See the table below for what types of products and services they offer (match their numbers). Visit online at **www.rsmeans.com/demosource/** for more information and free demos. Or call their numbers listed below.

1. **3D International**
713-871-7000
venegas@3di.com

2. **4Clicks-Solutions, LLC**
719-574-7721
mbrown@4clicks-solutions.com

3. **Aepco, Inc.**
301-670-4642
blueworks@aepco.com

4. **Applied Flow Technology**
800-589-4943
info@aft.com

5. **ArenaSoft Estimating**
888-370-8806
info@arenasoft.com

6. **ARES Corporation**
925-299-6700
sales@arescorporation.com

7. **BSD - Building Systems Design, Inc.**
888-273-7638
bsd@bsdsoftlink.com

8. **CMS - Computerized Micro Solutions**
800-255-7407
cms@proest.com

9. **Corecon Technologies, Inc.**
714-895-7222
sales@corecon.com

10. **CorVet Systems**
301-622-9069
sales@corvetsys.com

11. **Discover Software**
727-559-0161
sales@discoversoftware.com

12. **Estimating Systems, Inc.**
800-967-8572
esipulsar@adelphia.net

13. **MAESTRO Estimator Schwaab Technology Solutions, Inc.**
281-578-3039
Stefan@schwaabtech.com

14. **Magellan K-12**
936-447-1744
sam.wilson@magellan-K12.com

15. **MC2 - Management Computer**
800-225-5622
vkeys@mc2-ice.com

16. **Maximus Asset Solutions**
800-659-9001
assetsolutions@maximus.com

17. **Prime Time**
610-964-8200
dick@primetime.com

18. **Prism Computer Corp.**
800-774-7622
famis@prismcc.com

19. **Quest Solutions, Inc.**
800-452-2342
info@questsolutions.com

20. **RIB Software (Americas), Inc.**
800-945-7093
san@rib-software.com

21. **Shaw Beneco Enterprises, Inc.**
877-719-4748
inquire@beneco.com

22. **Timberline Software Corp.**
800-628-6583
product.info@timberline.com

23. **TMA SYSTEMS, LLC**
800-862-1130
sales@tmasys.com

24. **US Cost, Inc.**
800-372-4003
sales@uscost.com

25. **Vanderweil Facility Advisors**
617-451-5100
info@VFA.com

26. **Vertigraph, Inc.**
800-989-4243
info-request@vertigraph.com

27. **WinEstimator, Inc.**
800-950-2374
sales@winest.com

TYPE	1	2	3	4	5	6	7	8	9	10	11	12	13	14	15	16	17	18	19	20	21	22	23	24	25	26	27
BID					•		•	•		•	•				•		•		•	•	•					•	•
Estimating		•			•	•	•	•	•	•	•	•	•		•		•		•	•	•	•		•		•	•
DOC/JOC/SABER		•			•		•			•		•				•				•	•						•
IDIQ		•								•		•			•		•			•							•
Asset Mgmt.											•		•		•		•							•			•
Facility Mgmt.	•		•								•		•	•	•	•	•	•	•				•		•		
Project Mgmt.	•	•				•			•	•	•		•	•			•		•	•	•						
TAKE-OFF					•				•	•	•	•				•	•		•	•		•		•		•	•
EARTHWORK								•			•					•	•		•	•						•	
Pipe Flow				•							•						•										
HVAC/Plumbing					•			•		•	•						•										
Roofing					•			•		•	•						•										
Design	•				•						•						•			•	•						•
Other Offers/Links:																											
Accounting/HR		•			•						•								•	•	•						
Scheduling					•						•		•							•		•		•		•	•
CAD											•			•								•					•
PDA																						•		•			
Lt. Versions		•						•			•						•					•		•			•
Consulting	•	•			•			•		•	•		•		•	•	•		•			•		•	•		•
Training		•			•		•	•	•	•	•	•	•		•	•	•	•		•	•	•	•	•		•	•

Reseller applications now being accepted. Call Carol Polio Ext. 5107.

FOR MORE INFORMATION
CALL 1-800-448-8182, EXT. 5107 OR FAX 1-800-632-6732

**For more information
visit Means Web Site
at www.rsmeans.com**

New Titles

Builder's Essentials:
Estimating Building Costs for the Residential & Light Commercial Contractor
By Wayne J. DelPico

Step-by-step estimating methods for residential and light commercial contractors.
Includes a detailed look at every construction specialty—explaining all the
components, takeoff units, and labor needed for well-organized estimates:

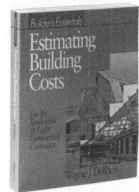

Correctly interpreting plans and specifications.
Developing accurate and complete labor and material costs.
Understanding direct and indirect overhead costs.
Using historical cost data to generate new project budgets.

Plus hard-to-find, professional guidance on what to consider so you can
allocate the right amount for profit and contingencies.

$29.95 per copy
Over 400 pages, illustrated, softcover
Catalog No. 67343   ISBN 0-87629-741-6

Building & Renovating Schools

This all-inclusive guide covers every step of the school construction process—from
initial planning, needs assessment, and design, right through moving into the new
facility. A must-have resource for anyone concerned with new school construction or
renovation, including architects and engineers, contractors and project managers,
facility managers, school administrators and school board members, building
committees, community leaders, and anyone else who wants to ensure that the
project meets the schools' needs in a cost-effective, timely manner. With square foot
cost models for elementary, middle, and high school facilities and real-life case studies
of recently completed school projects.

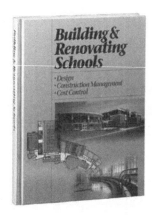

The contributors to this book – architects, construction project managers,
contractors, and estimators who specialize in school construction – provide start-to-
finish, expert guidance on the process.

$99.95 per copy
Over 425 pages
Catalog No. 67342   ISBN 0-87629-740-8

Means ADA Compliance Pricing Guide, New Second Edition
By Adaptive Environments and RSMeans

Completely updated and revised to the new 2004 Americans with Disabilities Act Accessible
Guidelines, this book features more than 70 of the most commonly needed modifications for
ADA compliance—their design requirements, suggestions, and final cost. Projects range
from installing ramps and walkways, widening doorways and entryways, and installing and
refitting elevators, to relocating light switches and signage, and remodeling bathrooms and
kitchens. Also provided are:

Detailed cost estimates for budgeting modification projects, including estimates for
each of 260 alternates.
An assembly estimate for every project, with detailed cost breakdown including materials,
labor hours, and contractor's overhead.
3,000 Additional ADA compliance-related unit cost line items.

Costs that are easily adjusted to over 900 cities and towns.

Over 350 pages
$79.99 per copy
Catalog No. 67310A   ISBN 0-87629-739-4